British Exploitation of German Science and Technology, 1943–1949

At the end of the Second World War, Germany lay at the mercy of its occupiers, all of whom launched programmes of scientific and technological exploitation. Each occupying nation sought to bolster their own armouries and industries with the spoils of war, and Britain was no exception. Shrouded in secrecy yet directed at the top levels of government and driven by ingenuity from across the civil service and armed forces, Britain made exploitation a key priority. By examining factories and laboratories, confiscating prototypes and blueprints, and interrogating and even recruiting German experts, Britain sought to utilise the innovations of the last war to prepare for the next.

This ground-breaking book tells the full story of British exploitation for the first time, sheds new light on the legacies of the Second World War, and contributes to histories of intelligence, science, warfare and power in the midst of the twentieth century.

Charlie Hall is Associate Lecturer in History at the University of Kent.

Routledge Studies in Second World War History

The Second World War remains today the most seismic political event of the past hundred years, an unimaginable upheaval that impacted upon every country on Earth and is fully ingrained in the consciousness of the world's citizens. Traditional narratives of the conflict are entrenched to such a degree that new research takes on an ever important role in helping us make sense of World War II. Aiming to bring to light the results of new archival research and exploring notions of memory, propaganda, genocide, empire and culture, Routledge Studies in Second World War History sheds new light on the events and legacy of global war.

Recent titles in this series

German-occupied Europe in the Second World War
Edited by Raffael Scheck, Fabien Théofilakis, and Julia Torrie

British Exploitation of German Science and Technology, 1943–1949
Charlie Hall

Unknown Conflicts of the Second World War
Forgotten Fronts
Chris Murray

A New Nationalist Europe Under Hitler
Concepts of Europe and Transnational Networks in the National Socialist
Sphere of Influence, 1933–1945
Edited by Johannes Dafinger and Dieter Pohl

The Swedish Jews and the Holocaust
Pontus Rudberg

https://www.routledge.com/Routledge-Studies-in-Second-World-War-History/book-series/WWII

British Exploitation of German Science and Technology, 1943–1949

Charlie Hall

LONDON AND NEW YORK

First published 2019
by Routledge
2 Park Square, Milton Park, Abingdon, Oxon OX14 4RN

and by Routledge
52 Vanderbilt Avenue, New York, NY 10017

First issued in paperback 2020

Routledge is an imprint of the Taylor & Francis Group, an informa business

British Library Cataloguing-in-Publication Data
A catalogue record for this book is available from the British Library

Library of Congress Cataloging-in-Publication Data
Names: Hall, Charlie, 1991- author.
Title: British exploitation of German science and technology, 1943–1949.
Description: Abingdon, Oxon; New York, NY: Routledge, 2019. |
Series: Routledge studies in Second World War history | Revision of author's thesis (doctoral)–University of Kent, 2017. | Includes bibliographical references and index.
Identifiers: LCCN 2018046758 | ISBN 9780815358381 (hardback) |
ISBN 9781351122559 (ebook)
Subjects: LCSH: World War, 1939–1945–Science. | Technology transfer–Great Britain–History–20th century. | Technology transfer–Germany–History–20th century. | Military research–Germany–History–20th century.
Classification: LCC D810.S2 H35 2019 | DDC 940.54/8–dc23
LC record available at https://lccn.loc.gov/2018046758

ISBN 13: 978-0-367-66219-6 (pbk)
ISBN 13: 978-0-8153-5838-1 (hbk)

Typeset in Times New Roman
by Deanta Global Publishing Services, Chennai, India

For my parents,

Vicki and Ian

Contents

Acknowledgements

The history, it is what it is.

H. Kane, Volgograd, 2018

This book has been some six years in the making, though the finished product is a very different animal to the project on which I first started working all that time ago. In the course of research and writing, I have been fortunate enough to be surrounded by a highly supportive network of friends, family and colleagues, to all of whom I owe a debt of gratitude. This book emerged from research I began as a doctoral student and so I would first like to thank my PhD supervisors, Professor Ulf Schmidt and Dr Stefan Goebel. Both offered endless advice, guidance and encouragement throughout the process and their influence on this work and on my ethos as a scholar more generally is enormous. The former's faith in me from a very early stage is the main reason that I chose to continue my studies beyond undergraduate level. I would also like to thank my two PhD examiners, Professor Brian Balmer and Dr Juliette Pattinson, partly for participating in what was a truly enjoyable viva, but also for helping me to develop my career thereafter. Their comments and suggestions helped me transform this work from thesis to book.

More generally, the School of History at the University of Kent has been my intellectual home for the past nine years and there is nowhere that I would rather have spent that time. The staff and students here have made it the best possible environment in which to research and write the book, as well as to learn and teach more broadly, and I owe them all my thanks. The funding provided by the School allowed me to complete my PhD and helped bring this project to maturity. In addition, my current team of close colleagues – Professor David Stirrup, Professor Jacqueline Fear-Segal, Professor Coll Thrush, Dr Kate Rennard and Dr Jack Davy – have allowed me to work on this book alongside my other duties, with only gentle jibes about where my priorities lie. My thanks also go to staff at The National Archives at Kew, the Imperial War Museum, the Churchill Archives Centre in Cambridge, and the British Library, whose assistance with all manner of queries has proven invaluable time and time again.

On a personal note, I am enormously grateful to my fellow *columbæ offeret* – Dr William Butler and Dr Mario Draper – who are both friends and colleagues, and who have acted as a sounding-board throughout the writing-up process, as well as knowing exactly when to change the subject. Words cannot express quite how much I owe to my parents, Vicki and Ian, to whom this book is dedicated. They nurtured my love of reading, writing, and history from a very young age and I hope that the publication of this book can act as testament to my gratitude for all that they have done, and continue to do, for me. It is very much their achievement as well as my own. My final and most significant thanks go to my wife, Annamarie, who has put up with the British exploitation of German science and technology for as long as I have, and with greater patience. There is no doubt in my mind that her support, encouragement and companionship ultimately made this book possible, and for that I will remain forever grateful.

Abbreviations

In text

All abbreviations, acronyms and codenames are as they appear in the original source material.

For definitions of some of the more significant terms listed here, please see the appended glossary.

30AU	No.30 Assault Unit	UK
ADI	Assistant Directorate of Intelligence	UK
Alsos	War Department Scientific Intelligence Mission	US/ALLIED
ASLIB	Association of Special Libraries and Information Bureaux	UK
BAOR	British Army of the Rhine	UK
BBRM	British Bombing Research Mission	UK
BBSU	British Bombing Survey Unit	UK
BIOS	British Intelligence Objectives Sub-Committee	UK
CAFT	Consolidated Advance Field Teams	ALLIED
CATOR	Combined Air Transport Operations Room	ALLIED
CCG (BE)	Control Commission for Germany (British Element)	UK
CCS	Combined Chiefs of Staff	ALLIED
CDEE	Chemical Defence Experimental Establishment (Porton Down)	UK
CIC	Combined Intelligence Committee	ALLIED
CIOS	Combined Intelligence Objectives Sub-Committee	ALLIED
CIPC	Combined Intelligence Priorities Committee	ALLIED
COGA	Control Office for Germany and Austria	UK
DCOS	Deputy Chiefs of Staff (Committee)	UK
DSIR	Department of Scientific and Industrial Research	UK
EAB	Economic Advisory Board	UK
EDU	Enemy Documents Unit	UK
EIPS	Economic and Industrial Planning Staff	UK
EPCOM	Enemy Publications Committee	UK
EPES	Enemy Personnel Exploitation Service	UK
ERDS	Enemy Research & Development Sub-Committee	UK
FIAT	Field Information Agency, Technical	ALLIED
FIAT (Br)	Field Information Agency, Technical (British Element)	UK
FIAT (US)	Field Information Agency, Technical (US Element)	US
FO	Foreign Office	UK
G-2	Military Intelligence	ALLIED

GED	German Economic Department	UK
G(T) & CW	General (T-Forces) & Chemical Warfare	UK
HMSO	His Majesty's Stationery Office	UK
I.E.T. Groups	Investigation of Enemy Technique Groups	*proposed*
IDCGS	Inter-Departmental Committee on German Scientists	UK
JEIA	Joint Export Import Agency	UK/US
JIC	Joint Intelligence (Sub-)Committee	UK
JIC-CCG	Joint Intelligence Committee, Control Council for Germany	UK
JIOA	Joint Intelligence Objectives Agency	US
LFA	*Luftfahrtforschungsanstalt* (Aeronautical Research Institute)	GERMAN
MAP	Ministry of Aircraft Production	UK
MEW	Ministry of Economic Warfare	UK
MFA&A	Monuments, Fine Art & Archives	ALLIED
NID	Naval Intelligence Division	UK
NKVD	People's Commissariat for Internal Affairs (security service)	SOVIET
OSRD	Office of Scientific Research and Development	US
RAE	Royal Aircraft Establishment (Farnborough)	UK
RAF	Royal Air Force	UK
RAT	Reparations Assessment Teams	UK
RDR Div.	Reparations, Deliveries and Restitution Division	UK
RM	Royal Marines	UK
RN	Royal Navy	UK
RNVR	Royal Naval Volunteer Reserves	UK
SCAEF	Supreme Commander Allied Expeditionary Force	ALLIED
SHAEF	Supreme Headquarters Allied Expeditionary Force	ALLIED
S.H. Parties	Sealing & Holding Parties	*proposed*
SIAS	Scientific Intelligence Advisory Service	UK
SPD	*Sozialdemokratische Partei Deutschlands* (Social Democratic Party)	GERMAN
SPOG	Special Projectile Operations Group (*Backfire*)	UK/US
STIB	Scientific and Technical Intelligence Branch	UK
STRB	Scientific and Technical Research Board	UK
T-Force	Target Force	ALLIED
TPA	Technical and Personnel Administration	UK
TIIC	Technical Industrial Intelligence Committee	US
USAAF	United States Army Air Force	US
USFET	United States Forces European Theatre	US
USSBS	United States Strategic Bombing Survey	US

Endnotes (primary sources)

All abbreviations are as they appear in the relevant archives.

IWM	Imperial War Museum
TNA	The National Archives, Kew
AB	Atomic Energy Authority (and predecessors) papers
ADM	Admiralty papers
AIR	Air Ministry papers
AVIA	Ministry of Aviation papers

BERCOMB	Berlin Commission (British) *[signal shorthand]*
BT	Board of Trade papers
CONCOMB	Control Commission for Germany (British Element) *[signal shorthand]*
CONFOLK	Control Office for Germany and Austria *[signal shorthand]*
CAB	Cabinet Office papers
DEFE	Ministry of Defence (and predecessors) papers
FO	Foreign Office papers
HW	Government Communications Headquarters (GCHQ) papers
LAB	Ministry of Labour papers
PREM	Office of the Prime Minister papers
RB	Research Branch
T	Treasury papers
TROOPERS	War Office *[signal shorthand]*
WO	War Office papers

Introduction

On 2 October 1945, almost five months after the Second World War in Europe had come to an end, a V-2 rocket was launched again from northern Germany. These so-called 'vengeance weapons' had first been unleashed just over a year earlier, in a desperate attempt by the Nazi regime to change the course of the war. They brought destruction and terror to residents of the targeted cities – London primarily, but also Paris, Antwerp and others – and in Britain, they prompted fears of a second Blitz. Ultimately, however, the V-2s were too little, too late for the Third Reich, and their limited quantities and issues of inaccuracy ensured they had only a very minimal bearing on the continuation of the conflict. That said, while the impact they had on their targets was fairly minimal, the impact they had on the future of warfare was highly significant – the V-2 was the first successful long-range rocket, and its arrival heralded the dawn of the ballistic missile age, thus decisively shaping international relations and military strategy for years to come, right up to the present day. Moreover, the V-2 was also the first man-made object to go into space when a rocket, launched by the Nazis, reached an altitude of 174 kilometres above sea level. In short, while the V-2 can be fairly described as ineffectual in its wartime deployment, the promise it held for the future was vast and tantalising.

It is, therefore, perhaps unsurprising that after the war ended, the victorious Allies sought to continue the work of the Third Reich's rocket scientists and bring the technology of the V-2 into their own armouries. It was for this reason that, in June 1945, General Dwight D. Eisenhower, Supreme Commander of the Allied Expeditionary Force in Europe, instructed Major-General Alexander Cameron, formerly of the Royal Engineers and British Anti-Aircraft Command, to launch a full investigation into the V-2. Under the codename Operation Backfire, this took place throughout the summer and early autumn of 1945 and culminated in three test-firings in October. All of these took place in the Lower Saxony coastal town of Cuxhaven and were run by the same German troops who had overseen the rocket attacks on London, albeit now under close British supervision. The first launch, on 2 October, took place on a clear and still afternoon and the rocket performed just as expected; rising in a straight line, levelling out and landing in the North Sea around two kilometres short and one kilometre to the left of its target. Britain, through the utilisation of German technology and expertise, had now entered the ballistic missile era.

Backfire was not the only operation of its type which took place at this time, nor was rocketry the only field in which the British (and the other Allies) were interested. For instance, 125 kilometres to the south-east of Cuxhaven, at Raubkammer, near Munster, experts from the British Chemical Defence Experimental Establishment at Porton Down spent three months conducting a wide range of trials and experiments with the newly discovered German nerve agents, primarily Tabun and Sarin. Elsewhere, Operation Surgeon, run by the Ministry of Aircraft Production and the Ministry of Supply, took over several for-mer Luftwaffe installations and set out to explore German progress in all manner of aeronautical topics. These large investigative projects which sprung up in the summer of 1945 were of huge significance, but only represent a small portion of British interest in German science and technology in the post-war period. British officials had entered Paris within a week of its liberation to explore formerly German-occupied laboratories in the French capital. Royal Navy experts were among the first Allied forces into Kiel and moved quickly to examine the subma-rine design and construction facilities there. After Germany surrendered, British investigators poured into the country and visited every laboratory, research site and factory of even passing interest, in their quest to learn all they could about German science and technology.

The process did not stop there. In January 1946, the first group of 23 German scientists recruited by Britain after the war arrived in Barrow-in-Furness in Cumbria, to work on submarine technology at the Vickers-Armstrongs shipyard. Many more followed over the next two or three years, travelling to Britain to work on rockets, aircraft, chemical warfare, and on a huge range of civil-industrial topics as well. When the British and French launched Concorde, the world's first supersonic commercial airliner, in 1976, few knew that much of the aerodynamics work involved in its design had been completed by German experts who came to Britain as part of a government scheme in the immediate post-war period. Other German specialists did not land such significant roles in Britain after the war, but did contribute to the British military or economy in other ways, often through written reports or interrogations which took place during periods of internment at special camps in Germany and Britain after the war. Some of these individu-als were not even detained because they were considered of value to Britain, but rather because it was considered important to keep them out of the employ of the Soviet Union. In this way, German science and technology became a source of much competition in the early Cold War period.

All of these examples form part of the wider British programme of exploitation of German science and technology, which had its roots in wartime scientific intel-ligence and which left legacies that ran throughout the latter half of the twentieth century. It took place at a time of great transition, not just in Britain, but across Europe and the world. The Second World War decisively reshaped the geopoliti-cal order, not least in giving rise to the two great superpowers of the United States and the Soviet Union, which would spend much of the ensuing fifty years staring each other down on the world stage. It also prompted a drastic reconfiguration of the European balance of power, while simultaneously diminishing the importance of that balance in determining global events for the first time in centuries. As a case in point, the formerly significant Germany was reduced – albeit temporarily – to the status of a weak and dependent nation, entirely at the mercy of its Allied conquerors. However, in the long term, it was perhaps Britain which suffered most acutely in the post-1945 adjustment of international power politics. The conventional narrative runs that Britain entered the war in 1939 as an old-fashioned Great Power, with a huge empire and access to enormous resources, both human and material, across the globe; that it spent around a year bravely 'standing alone' against Nazi tyranny, until the Soviet Union and the United States joined the fight; and that it emerged at the conclusion, limping and badly wounded, facing economic bankruptcy, decolonisation and increasing international irrelevance. Britain took on the role of the noble warrior who, having given everything to preserve freedom and democracy, now stepped aside to let others take up the mantle.

There is much wrong with this heavily mythologised conception of Britain's place in the world once the war had ended, but certain key elements are both true and essential for understanding the path which Britain elected to take after 1945. Critically, Britain ended the war in a much weaker position than when it had entered it and, dwarfed by its two wartime allies – the USA and the USSR – it was forced to find new ways to exert influence on global events. The post-war occupation of Germany provided one such opportunity. In this, Britain was on equal footing with the two superpowers; each, along with France, being responsible for the administration and control of a portion of German territory and the population therein. The policies Britain chose to pursue in this role as occupier would have a direct bearing on its future and this fact was not lost on those in positions of decision-making authority. Faced with a number of competing demands – economic shortage, diplomatic commitments, domestic political concerns – and a variety of potential international threats – German resurgence, Soviet hostility, American isolationism – British policymakers were forced to negotiate a safe route through these troubled waters in the governance of their occupation zone in Germany.

It was within this context that the programme of exploiting German science and technology had to operate. Since the beginning of the twentieth century, scientific and technological developments had played an increasingly significant role in the waging of war. The First World War had seen the first major use of tanks, aircraft, and chemical warfare, alongside considerable advances in communications, logistics and military medicine. Strategic bombing had been utilised throughout the 1930s, especially during the Spanish Civil War, and then formed a critical aspect of the Second World War. That latter conflict also saw the first deployment of ballistic missiles, jet aircraft, and the atomic bomb, even if these technologies did not have a substantial impact on the course of the war. Overall, by 1945, it seemed entirely probable that future conflicts would be decided by the utilisation of new and increasingly innovative developments in weaponry. This took on an increasingly urgent edge as the polarised conditions of the Cold War became ever more apparent.

During the Second World War, all belligerent nations invested heavily in military science and technology in an effort to develop an edge over their opponents.

In this, Germany was no exception, but nor was it any more successful than Britain or the United States, even if key developments such as the V-2 or nerve agents suggest otherwise. German military science during the war was very active but also deeply flawed – both the expulsion of Jewish experts during the 1930s and the desire to develop revolutionary 'wonder weapons' rather than simply improve existing technology conspired to limit the extent to which German laboratories and research establishments could really contribute to the war effort. Nonetheless, in the minds of Allied strategists, German military technology was far superior to their own and the perceived benefits to be gained by acquiring this equipment, and the expertise behind it, were extremely tempting. Thus, exploitation was born. The supposedly more advanced nation had been defeated, and its victorious occupiers could now claim the scientific and technological spoils of war; their own armouries and industries could make huge leaps forward by standing on the shoulders of their newly vanquished foe.

Britain did not hesitate to participate in the process of exploitation and was, in many ways, the first of the Allied powers to begin planning for it on a genuinely substantial scale. By the time the war in Europe ended in May 1945, Britain had put in place a major administrative framework for exploitation, complete with specially created committees and agencies, and with a dedicated logistical capacity to ensure that the programme could proceed with maximum efficiency as soon as territory was liberated from Nazi occupation. This expanded and gathered speed as the borders of the Third Reich itself were breached and became a significant part of the occupation machinery once hostilities ceased. With full British government sanction, German laboratories and factories were inspected and meticulously pillaged, machinery and prototypes were confiscated, and documents and blueprints were shipped back to Britain in their thousands. In addition, expert German personnel were detained, interrogated and, in many cases, recruited to work for the British state or for private companies. No area of expertise was left untouched, from the most highly sensitive military projects to the most mundane commercial production techniques; all was considered fair game under the terms of the British exploitation initiative. In time, the scheme was wound down, largely due to broader geopolitical considerations, but not before Britain (and the other Allies) had extracted all that they reasonably could from the ruins of defeated Germany.

For such a major undertaking, the British exploitation of German science and technology has only a very small footprint in the British popular imagination. The releases of relevant files in the National Archives at Kew are usually accompanied by brief flurries of newspaper articles, bearing headlines such as 'How Britain put Nazis' top men to work', but aside from these fleeting moments of publicity, it remains largely absent from conventional discourse.[1] In the process of writing this book, discussions I had with friends, colleagues and students about British exploitation were often met with surprise and curiosity and only infrequently with prior knowledge. In a 1997 piece for *History Today* on British reparations policy, the author describes British attempts to commandeer Nazi assets as 'intriguing and little-known'.[2] Precisely because of this absence of common knowledge about exploitation, the few histories on the subject which have emerged more recently, primarily of the broad, journalistic type, have sometimes tapped into a notion of 'forgotten heroes', when describing the men involved in the scheme.[3]

Other accounts have focused on the controversy of an Allied nation employing Nazis so soon after the war (particularly Project Paperclip in the United States) and either implying or outright asserting that this practice was part of a wide-ranging government cover-up. Many such works, desperate to stand out on bookshop shelves, scream their outrage from the front cover, with titles such as *The Paperclip Conspiracy*, *Blowback*, *Secret Agenda*, and *The Nazis Next Door*.[4] While these histories are usually based on some engagement with original source material, the conclusions drawn are often driven more by the desire to shock rather than by the need to formulate a strong argument. However, in the absence of more scholarly rigorous works on this topic, they have become the standard texts and, unfortunately, their sensationalist approach is often to the detriment of what is really interesting about the story. Exploitation was not a secret conspiracy, nor was it carried out by small bands of enterprising heroes; it was in fact a widespread government programme, acknowledged and endorsed at the highest level of British politics, with huge resources, and one which was integral to the beginning of the Cold War and which had far-reaching ramifications for science, strategy and diplomacy.

Among those who do know a little about post-war exploitation, there is another common misconception. This is that Britain was either not involved or that its attempts were so dwarfed by those of the United States that they are barely worth considering. Many people to whom I spoke during the course of my research made assertions along these lines and took some convincing to see it otherwise. In their defence, this appraisal is also reinforced by the existing literature. In 1990, John Gimbel, an established historian of the American occupation of Germany, published a book entitled *Science, Technology, and Reparations*, which examined the US exploitation scheme in depth, dealing primarily with material exploitation, that is, the confiscation of equipment, machinery and documents.[5] Six years later, following a conference which discussed Gimbel's work, an edited volume emerged, entitled *Technology Transfer out of Germany after 1945*, which scrutinised various aspects of the programme in greater depth.[6] In both cases, the focus was overwhelmingly on American exploitation, with the initiatives of other countries only mentioned as context. Prior to Gimbel, there had only been a handful of books on exploitation, mostly published in the 1960s and 1970s, which were based on anecdotal evidence and practically no documentary sources, largely because these files had not yet been released to the public. These accounts feature some fascinating stories, but the vast majority are almost impossible to verify and the works themselves have since faded into relative obscurity.[7] Both these and the more recent popular histories described above tend to have more interest in personnel exploitation and usually discuss the American programme in full while largely ignoring the parallel efforts of other nations. This imbalance is further reinforced in popular culture, where former Nazi scientists working for the Americans have become a recurring archetype, from Dr Strangelove in the 1964 film of the same name to Arnim Zola in *Captain America: The Winter Soldier* (2014), whereas their equivalents working for the British are essentially non-existent.[8]

In recent years, however, there has been an increase in scholarship on exploitation, both material and personnel, and this has included a call to look beyond the American scheme and understand the phenomenon in an international framework. In Michael J. Neufeld's 2012 article, 'The Nazi Aerospace Exodus', he laments that 'we know far too little about British and French policies' with regard to exploitation and urges scholars to unearth much more about these elements.[9] In part, this book aims to respond to this call to action and ultimately redress the US-centric imbalance, a process which is starting to happen, to some extent, throughout Cold War studies.[10] In so doing, it will contribute towards a very small, but growing, pool of literature which offers a more rounded picture of post-war utilisation of German science and technology.[11] Not only does this book focus exclusively on the British programme, which adds a valuable and hitherto neglected perspective to the discourse on exploitation, it also moves away from the obsession with controversy which has coloured so many previous works. Instead, rooted in extensive engagement with a broad range of archival sources, it shows that narratives of exploitation can be compelling and remarkable, without having to artificially inflate the sensationalist aspect. Moreover, it also offers a fully rounded account of British exploitation, covering both material and personnel exploitation, and is based in a suitably broad timeframe. The story begins with the scheme's wartime precursors, operating as early as 1943, and concludes with the official termination of the programme at the end of the occupation period in 1949, though it does also look beyond this to examine further-reaching legacies.

Despite this fairly brief time period in which the policy of exploitation was pursued by the British authorities, it is not sufficient to study it purely in isolation. Instead, it is necessary to see British exploitation of German science and technology after the Second World War as part of a much longer history, which stretches both back into the earlier part of the twentieth century and ahead into the Cold War. It is, for example, indicative of the growth of the military-industrial complex which President Eisenhower warned of in his final presidential address in 1961, but which had truly begun to emerge, especially in Britain, around the time of the First World War.[12] In turn, access to the vast scientific and technological resources of an enemy's military-industrial complex made the prospect of forceful exploitation all the more desirable. This urge was embodied by the actions of Nazi forces when they seized control of laboratories, factories and facilities in European countries which they occupied during the Second World War, such as France, Norway, and Czechoslovakia.[13]

These trends continued and expanded after 1945, becoming an increasingly ubiquitous part of the Cold War landscape. Defence expenditure in Britain in 1946 still stood at one-fifth of the gross national product, and more than half of government-funded research and development, and something like a quarter of the national total, was funded out of defence budgets.[14] The impact of exploitation was felt here too – in Brian Crim's recent work on Project Paperclip, he explores how German scientists became part of the 'national security state' in the post-war United States.[15] More generally, science had become an integral part of defence planning, as senior military figures tried to envisage what a future war would look like and pre-empt the kind of weapons which would be needed to win it. The Royal Navy, for instance, struggled to define a role for itself in an era when long-range land-based aircraft, rocketry, and the atomic bomb threatened to make naval warfare increasingly obsolete.[16] As a result, comprehending the contents of an enemy's arsenal became more and more essential and scientific intelligence as a commitment of espionage services around the world grew in prominence.[17] Here again, exploitation had a role to play – indeed, a technique entitled 'Foreign Military Exploitation' became a part of the Central Intelligence Agency's playbook throughout the Cold War era. This tactic entailed gathering enemy military technology by any means necessary – scavenging crash sites, purchasing from third parties, capturing on the battlefield – and then having it inspected by relevant experts to learn what they could about a hostile, or potentially hostile, nation's military capacity.[18]

Exploitation also exists within another context, beyond the remit of intelligence-gathering or military strategy; it is also a distinctly transnational phenomenon. Neufeld defines transnational as 'cross-border movements of people, knowledge or artefacts carried out by nation states and non-governmental actors outside the formal system of international relations between states'.[19] Measured against this definition, exploitation certainly fits the bill. As Jan Rüger has argued, technology transfer is perhaps one of the most compelling examples of transnational exchange, whether it comes in the form of material, personnel or ideas. He also notes that Anglo-German interactions are among the most heavily studied of any in the field, covering topics as diverse as education, music, art, medicine, publishing, religion, industry, finance, gymnastics and horse racing.[20] As a result, exploitation of science and technology can be seen as part of this rich heritage of transnational exchange between Britain and Germany over the past two centuries. Some take this further – for example, Volker Berghahn has asserted that post-war scientific and technological exploitation presents no more than a 'case of limited historical significance' within the broader trends of global technology transfer.[21] This book takes a rather different perspective and argues that the exploitation of German science and technology was an essentially unique case, albeit one which had both precedents and legacies, and should be treated as such.

In a similar vein, exploitation should not simply be seen as merely part of Allied reparations claims, though the relationship between the two should be examined (as it is here, in Chapter Ten). In this sense, it is also important to understand exploitation within the Allied occupation of Germany as a whole. This occupation was, in many ways, unprecedented, involving the wholesale but temporary destruction of a formerly independent nation's sovereignty, with territory divided into zones to be controlled by four different powers, each of which had a relatively free hand as to how to govern their respective zones, but which were also expected to co-ordinate certain aspects of their policy on a quadripartite basis. For Britain, the occupation presented a number of dilemmas as it sought to achieve broad goals such as denazification and post-war restitution, the preservation of European peace, and defence against Soviet aggression and encroachment, while also wrestling with a number of practical constraints, such as economic shortage, diplomatic uncertainty, and a more general loss of power and prestige. The situation was complex and problematic and one which had serious ramifications for international relations and global power structures for decades to come. It is heartening, therefore, that the post-war occupation of Germany, including the British involvement, has become a subject of particularly active interest among scholars in recent years.[22]

In order to fully understand Britain's programme of scientific and technological exploitation in Germany, there are two distinct but connected elements of British occupation policy which need to be appreciated. The first is the predominance of pragmatism. As highlighted above, the number of practical constraints which Britain faced in this period meant that it could only pursue policies which offered a direct, visible and fairly immediate benefit. Grand schemes which had been discussed favourably during the war, such as the reduction of Germany to a primitive pastoral state or the complete expurgation of Nazism from every sphere of public and private life, fell by the wayside, in favour of an approach, redolent of British colonial rule, based on a drive to secure British interests at the lowest possible cost and a willingness to work with existing native power structures to ensure stability.[23] Initially, this outlook favoured exploitation as a scheme which offered short-term economic and strategic gain, but later the programme fell afoul of similar calculations, as it was feared that extensive exploitation might weaken Germany to the point where it became susceptible to hostile nationalist or communist ideologies.

This links neatly with the second critical element – Britain's reappraisal of its relationship with the Soviet Union. Perhaps the biggest shift which Britain underwent during the early occupation period was the realisation that the Soviet Union's agenda in Europe was generally incompatible, and often directly at odds, with its own.[24] The transformation of the Soviets from ally to enemy in British eyes happened quite abruptly, and can seem surprising, but it should be remembered that, aside from temporary alliances of necessity during the two world wars, Britain and Russia had traditionally been rivals, if not true adversaries, on the world stage and their differences had only become more pronounced since the 1917 Russian Revolution and ensuing civil war. At the end of the Second World War, Germany's role changed from that of common enemy, with a unifying effect on the Allies, to that of major bone of contention, reviving the division between the powers of East and West. Negotiating this geopolitical rearrangement became a serious preoccupation for all the occupying forces and Britain was no exception. As this book will consistently demonstrate, British fear, suspicion and dislike of the Soviet Union became the dominant factor in dictating exploitation policy and ensured the survival and even the expansion of the programme on numerous occasions, despite it often running counter to other British objectives in occupied Germany.

Ultimately, the British exploitation of German science and technology after the Second World War was both a unique, standalone programme which is worthy of study in its own right, and an integral part of broader topics, such as the development of the military-industrial complex and scientific intelligence, the Allied occupation of Germany, and the start of the Cold War. This book approaches the subject with a view to these dual interpretations, presenting both a fluent narrative of the programme's origins, preparation, execution, and demise, while also considering it within its various relevant contexts. The first of these contexts forms the basis of Chapter One, which looks at the role which both scientific research and development, and scientific intelligence, played in the conduct of the Second World War. It then moves on to discuss two organisations, which had their origins in scientific espionage but which became early pioneers of exploitation, laying the groundwork for the later exploitation initiative in terms of both stated objectives and *modus operandi*. As such, this chapter demonstrates the fertile soil in which the seeds of exploitation were sown, and therefore goes some way to explaining why the programme developed so rapidly in its early period.

Chapters Two and Three form two consecutive parts of a chronological narrative which describes exploitation during its most active phase. Chapter Two begins with the discussions that took place in Whitehall and Washington DC during 1943 and 1944, which led to the formal decision to pursue an exploitation policy and to the creation of the committees and agencies necessary to carry it out. The activities of these agencies during the war, first in parts of western Europe liberated from Nazi control, such as France and the Low Countries, and then in Germany itself once the borders of the Reich were breached in the winter of 1944–45, forms the rest of Chapter Two. Chapter Three picks up shortly after the war's end, at the point when the Supreme Headquarters Allied Expeditionary Force was disbanded and the joint Anglo-American exploitation scheme was split into two distinct and unilateral elements. The rest of the chapter then explores the way in which Britain conducted exploitation, dealing with the administrative and logistical framework first, then the experiences of the investigators who represented the programme on the ground in Germany and finally turning to the potential for competition and co-operation which existed between those tasked with exploitation and those responsible for other important duties under the auspices of the occupation.

While understanding the process by which exploitation was carried out is important, it is also necessary to discuss the scientific spoils of war which Britain obtained as a result. Chapter Four begins this discussion by exploring the material spoils – that is, the physical equipment, objects and documents – which were seized by Britain during the course of the programme. This includes a general summary of the various forms which these material spoils took, as well as two case studies of areas of major interest for the British authorities in Germany. The first of these is rocketry and aeronautics and the second is chemical and biological warfare – both represent highly sensitive military technologies as well as fields in which German research was often more advanced than that of the Allies. Chapter Five then moves on to explore the short-term element of personnel exploitation, covering the early detention centres which were used in Germany and in Britain, as well as the broader interrogation and analysis techniques which were utilised therein. Chapter Six, in turn, charts the evolution of personnel exploitation to involve the recruitment and long-term employment of German scientists and technicians in Britain, initially in the defence sphere and later in civil industry too. It also provides an account of the experiences of these expert *émigrés* and the way in which they were received by their new colleagues and neighbours.

Chapters Seven and Eight move away from the fairly straightforward chronological narrative of the preceding sections to discuss the international dimension of exploitation. In Chapter Seven, the British programme is considered in reference to its parallels operated by the USA, France and other nations across Europe and the world. This shows examples both of collaboration and, perhaps more often, of competition, as the Allies and others struggled against one another to obtain the greatest advantage from German science and technology. It also offers a chance to look beyond the immediate occupation period and to see how the legacies of exploitation had a long reach, with the transnational exchange of German expertise continuing well into the 1960s and 1970s. Chapter Eight provides an opportunity to explore one of the core contentions of this book in greater depth. Here, the growing antipathy and deteriorating relations between Britain and the Soviet Union are documented with regard to exploitation and it becomes increasingly evident that the policy was overwhelmingly influenced by the terms of this new and volatile geopolitical paradigm.

Chapters Nine and Ten also serve to situate exploitation within a broader context, but from a slightly different angle. Chapter Nine examines how exploitation coexisted with the other policies enacted by Britain as part of its occupation of Germany, particularly in relation to denazification, demilitarisation, and the control of German science. This helps to ground exploitation as just one of a number of concomitant initiatives that the British had to balance in the administration of their zone of control. Chapter Ten broadens the terms still further and considers exploitation in terms of its relationship with the wider quest for postwar reparations, as well as unpacking the legal and moral framework within which it operated. Furthermore, for all the descriptions of exploitation as a topsecret programme, this chapter also outlines the place it occupied in the public domain, exploring how it was reported in the contemporary press, discussed in Parliament and, on occasion, actively promoted by the government departments responsible. As such, this chapter reminds us that the conventional conceptions of exploitation as clandestine, unjust and even illegitimate deserve to be revisited and challenged. Finally, in the conclusion, the first of two brief epilogues considers the way in which exploitation was wound down and finally terminated at the end of the occupation period, on the eve of the creation of West and East Germany as independent nations, while the second uncovers some of the impacts and legacies of exploitation, which stretch beyond the stated timeframe of this study.

As the first comprehensive account of British exploitation of German science and technology, this book will tell a previously overlooked side to the story, thus reasserting Britain's role in a discourse which has traditionally been dominated by narratives of American supremacy. It will also provide a fresh take on the origins of the Cold War arms race, which very much began in the scramble for the scientific and technological spoils of defeated Germany. Moreover, this book will show that exploitation was central to the transitional period which Britain underwent during the 1940s, as it moved from fighting the Germans to fighting the Soviets, developed new military and diplomatic strategies on account of the advent of new technologies, and was essentially demoted from a first-rate to a second-rate power on the world stage. Studying exploitation therefore provides us with a lens through which to analyse these changes as it both helped to shape, and was in turn shaped by, these larger transformations. Ultimately, it will tell a previously untold story about the post-war nexus of science, technology, strategy, intelligence and power, and will also shed new light on Britain's place in the world at one of the most important junctures of the twentieth century.

Notes

1 Stewart Payne, 'How Britain Put Nazis' Top Men to Work', *Daily Telegraph*, 30 August 2007, 16.
2 Ian Locke, 'Post-War Germany – Britain's Lost Opportunity?', *History Today*, 47:8 (1997), 11.
3 Sean Longden, *T-Force: The Forgotten Heroes of 1945* (London: Constable, 2009); Nicholas Rankin, *Ian Fleming's Commandos: The Story of 30 Assault Unit in WWII* (London: Faber & Faber, 2011).
4 Tom Bower, *The Paperclip Conspiracy: The Battle for the Spoils and the Secrets of Nazi Germany* (London: Paladin, 1988); Christopher Simpson, *Blowback: America's Recruitment of Nazis and its Impact on the Cold War* (London: Weidenfeld & Nicolson, 1988); Linda Hunt, *Secret Agenda: The United States Government, Nazi Scientists and Project Paperclip, 1945–90* (New York: St Martin's Press, 1991); Eric Lichtblau, *The Nazis Next Door: How America Became a Safe Haven for Hitler's Men* (New York: Houghton Mifflin Harcourt, 2014).
5 John Gimbel, *Science, Technology, and Reparations: Exploitation and Plunder in Postwar Germany* (Stanford, CA: Stanford University Press, 1990).
6 Matthias Judt and Burghard Ciesla (eds.), *Technology Transfer out of Germany after 1945* (Amsterdam: Harwood, 1996).
7 Michel Bar-Zohar, *The Hunt for the German Scientists* (London: Arthur Barker, 1967); James McGovern, *Crossbow and Overcast* (London: Arrow, 1968); Clarence G. Lasby, *Project Paperclip: German Scientists and the Cold War* (New York: Atheneum, 1971).
8 *Dr Strangelove or: How I Learned to Stop Worrying and Love the Bomb*, dir. Stanley Kubrick (USA: Columbia Pictures, 1964); *Captain America: The Winter Soldier*, dir. Anthony Russo, Joe Russo (USA: Disney, 2014).
9 Michael J. Neufeld, 'The Nazi Aerospace Exodus: Towards a Global, Transnational History', *History and Technology*, 28 (2012), 61.
10 Grace Huxford, *The Korean War in Britain: Citizenship, Selfhood and Forgetting* (Manchester: Manchester University Press, 2018), 182.
11 Douglas O'Reagan, *Taking Nazi Technology: Allied Scientific Espionage and Exploitation of German Technology after the Second World War* (Baltimore, MD: Johns Hopkins University Press, forthcoming).
12 Katherine Epstein, *Torpedo: Inventing the Military-Industrial Complex in the United States and Great Britain* (Cambridge, MA: Harvard University Press, 2014), 1–3; G.C. Peden, *Arms, Economics and British Strategy: From Dreadnoughts to Hydrogen Bombs* (Cambridge: Cambridge University Press, 2007).
13 Peter Liberman, *Does Conquest Pay? The Exploitation of Occupied Industrial Societies* (Princeton, NJ: Princeton University Press, 1996), 53; Peter Liberman, 'The Spoils of Conquest', *International Security*, 18 (1993), 139–41.

14 Robert Bud and Philip Gummett, 'Introduction: Don't You Know There's a War On?', in Bud and Gummett (eds.), *Cold War, Hot Science. Applied Research in Britain's Defence Laboratories, 1945–1990* (London: Science Museum, 2002), 1, 7.

15 Brian E. Crim, *Our Germans: Project Paperclip and the National Security State* (Baltimore, MD: Johns Hopkins University Press, 2017).

16 Tim Benbow, 'The Royal Navy and Sea Power in British Strategy, 1945–1955', *Historical Research*, 91 (2018), 379.

17 Christopher Andrew, 'Intelligence in the Cold War', in Melvyn P. Leffler and Odd Arne Westad (eds.), *The Cambridge History of the Cold War*, vol. II (Cambridge: Cambridge University Press, 2010), 417–37; Huw Dylan, *Defence Intelligence and the Cold War: Britain's Joint Intelligence Bureau, 1945–1964* (Oxford: Oxford University Press, 2014), 4; Daniel W.B. Lomas, *Intelligence, Security and the Attlee Governments, 1945–51* (Manchester: Manchester University Press, 2016).

18 James E. David, 'Scavenging for Intelligence: The U.S. Government's Secret Search for Foreign Objects during the Cold War', *National Security Archive Briefing Books*, 616 (2018) [accessed online 4 September 2018, https://nsarchive.gwu.edu/briefing-book/intelligence/2018-01-31/scavenging-intelligence-us-governments-secret-search-foreign].

19 Neufeld, 'Nazi Aerospace Exodus', 60.

20 Jan Rüger, 'OXO: Or, the Challenges of Transnational History', *European History Quarterly*, 40 (2010), 659–61, n.33.

21 Volker Berghahn, 'Technology, Reparations, and the Export of Industrial Culture: Problems of the German-American Relationship, 1900–1950', in Judt and Ciesla, *Technology Transfer*, 4.

22 Camilo Erlichman and Christopher Knowles (eds.), *Transforming Occupation in the Western Zones of Germany: Politics, Everyday Life and Social Interactions, 1945–55* (London: Bloomsbury, 2018); Christopher Knowles, *Winning the Peace: The British in Occupied Germany, 1945–1948* (London: Bloomsbury, 2017); Susan L. Carruthers, *The Good Occupation: American Soldiers and the Hazards of Peace* (Cambridge, MA: Harvard University Press, 2016); Lee Kruger, *Logistics Matters and the US Army in Occupied Germany, 1945–1949* (Basingstoke: Palgrave, 2016); Francis Graham-Dixon, *The Allied Occupation of Germany: Denazification, the Refugee Crisis and the Path to Reconstruction* (London: I.B. Tauris, 2013); Jessica Reinisch, *The Perils of Peace: The Public Health Crisis in Occupied Germany* (Oxford: Oxford University Press, 2013); Stefan-Ludwig Hoffman, 'Germany is No More: Defeat, Occupation, and the Postwar Order', in Helmut Walser Smith, *The Oxford Handbook of Modern German History* (Oxford: Oxford University Press, 2011), 593–614; Richard Bessel, *Germany 1945: From War to Peace* (London: Pocket Books, 2010); Atina Grossmann, *Jews, Germans and Allies: Close Encounters in Occupied Germany* (Princeton, NJ: Princeton University Press, 2007).

23 Knowles, *Winning the Peace*, 36.

24 Anne Deighton, *The Impossible Peace: Britain, the Division of Germany, and the Origins of the Cold War* (Oxford: Clarendon, 1993), 5.

1   The scientific war

In order to fully understand the British post-war exploitation programme, it is necessary to trace its roots back to the Second World War. Specific conditions were evidently necessary for this innovative policy of comprehensive scientific and technological utilisation to come about. The post-war period was shaped decisively by the nature of the war which preceded it and this is abundantly clear in the case of exploitation – various elements of the way the Second World War was fought, especially the rising significance of scientific research and development, as well as changing perceptions of military capacity in comparison to one's enemies, and indeed allies, played an influential role in the development of this new initiative, which straddled the boundaries between intelligence gathering, scientific and technical development, and international relations. The war also provided an opportunity for the tactics later employed by agents of exploitation to be tried and tested – the experimental conditions of the contemporary battlefield allowed a strategy to emerge and be refined in preparation for much broader deployment as the war came to an end. Overall, this chapter describes the foundations upon which the major programme of British post-war scientific and technical exploitation was built by exploring how science and technology shaped the war in general and military intelligence, in particular, and by examining two forerunner operations – the Naval Intelligence Division's 30 Assault Unit and the Manhattan Project's Alsos Mission – the successes and failures of which directly influenced the form which the main exploitation initiative later assumed.

Science and strategy

It would be an exaggeration to suggest that the course of the Second World War was decisively altered by advances in military science and technology; indeed, it is arguable that the war was won with the same weapons with which it was begun – tanks, aircraft, artillery, and submarines.[1] Yet, while no state was able to achieve a radical transformation of military technology before 1945, the war certainly proved a fertile breeding ground for new developments – for example, radar, jet engines and the atomic bomb – and also saw scientific considerations exerting greater influence over military strategy.[2] This was not a process which began in 1939, however, as it had already occurred to a substantial extent in the

First World War. This earlier global conflict also drove considerable technological innovation and saw the advent of new weapons and tactics, among them tanks, gas warfare and aerial bombing, which modified the way the war was fought. By 1918, Britain had become 'a gigantic military-academic-industrial complex, co-opting and managing much of the nation's scientific workforce'.[3] Victory came to be seen to depend upon a 'process of continual experimentation'.[4] The technological advances which arose as a result had ramifications which stretched beyond the outcome of the immediate battles and campaigns. In the spirit of 'swords into ploughshares', such technology could be demilitarised and absorbed into civilian industry or into public service. In addition, technology brought the frontline much closer to home, whether in the form of attacks on civilian targets or the more extensive dissemination of information by a technologically advanced press – it marked the true advent of total war. New weapons demanded not only new military tactics and strategy but also new politics, diplomacy and even morality.[5]

As a result, the exploitation initiative cannot be understood without first exploring the roots of the modern military's preoccupation with science and technology and the wider influence that this has had. While the entanglement of science and strategy has a long heritage, it reached new levels of synthesis during the Second World War and thus became a permanent fixture of conflict in the modern age. In his foreword to Irvin Stewart's 1948 history of the American Office of Scientific Research and Development (OSRD), its wartime head, Vannevar Bush, wrote that:

> This is probably the most significant military fact of our decade: that upon the correct evolution of the instrumentalities of war, the strategy and tactics of war must now be conditioned.[6]

This new strategic thinking began taking root in the first years of the war and was recognised quickly within British military officialdom. As early as January 1940, a memorandum from the Air Ministry to the Joint Intelligence Sub-Committee noted 'that the direct application of the results of scientific research to warfare has increased and is increasing needs no demonstration.'[7]

One particular example where this was true was on the aerial battlefield, which had expanded exponentially in the interwar years, in terms of both scale and importance. Success in this theatre now relied heavily on the technology of mass production to ensure that an air force would not be overwhelmed by its enemy's numerical superiority, while the bombing of civilian targets, building upon the precedent set in the Spanish Civil War, quickly became a common feature of the ongoing conflict.[8] Popular pressure to defend against these devastating raids was understandably immense and this drove a process of technical one-upmanship between the warring powers, with a particular focus on detection and advance warning, specifically on radar. This was arguably one of the most important scientific developments of the Second World War and one that would continue to be modified and improved for the duration of the conflict and beyond.[9] Another element of the air war was the push to make aircraft faster and thus far more effective in aerial combat as well as safer from attack from the ground. This drove the development of the jet engine, which was largely experimental right up until the end of the war, with only small sections of any national air force occupied by jet aircraft. As such, like radar, jet research progressed beyond 1945 and soon the jet engine came to dominate not only military but also civilian aviation.[10]

These are just two examples of the myriad advances in science and technology which took place during the Second World War and the impact of which was felt far beyond 1945. There are, of course, countless others, many of which will be discussed, either directly or tangentially, over the course of this book. However, no discussion of military technology in this period can be complete without mention of the atomic bomb, the development of which has been described as 'an organisational, engineering, and intellectual undertaking that had no precedent'.[11] This new weapon not only brought about a swift end to the war in the Pacific (though there is some debate as to how necessary or justified its use truly was) but also, by creating the high-stakes paranoia of the Cold War, helped shape international relations for the next fifty years.[12] More than anything, it showed that the possible results of applied research in modern warfare were potentially limitless and ensured that from then on science and warfare would be indefinitely and inextricably entwined.

The impact of science and technology on modern warfare cannot be judged on the merits of individual developments alone, no matter how significant or far-reaching they have proven to be. Instead, it is important to examine the way in which science factored into the waging of war as a whole, thus creating the necessary preconditions for the Cold War arms race and the exploitation initiative. At the end of the First World War, despite the significant contributions which science had made, the influence of scientists over military planning and policy-making was rolled back and deemed to have been only a temporary necessity.[13] However, during the 1930s, Britain witnessed the rise of the scientific expert in public life, as well as the steady expansion of military research establishments.[14] As another global war loomed, scientists drew closer once again to the corridors of power but were still kept at arm's length during discussions of fundamental national security.[15] Once war broke out, even this barrier was obviated and scientists quickly became an integral part of the war effort, including involvement at the very highest levels of decision-making. This inclusion did not happen overnight, however, and it was not until September 1940, after Winston Churchill had replaced Neville Chamberlain as Prime Minister, that the Scientific Advisory Committee to the War Cabinet was first convened.[16] It seems that, as in the First World War, 'the need to mobilise science was more apparent to scientists than it was to administrators'.[17]

Despite this relatively slow uptake of expertise, the foundations for Britain's expansive scientific war effort were in place even before the outbreak of war. In the interwar years, Britain almost certainly devoted more resources to war-like research, development and invention than any other world power and, up until 1941, British scientists and engineers had made perhaps the single largest contribution to the development of weapons of war.[18] By 1939, it had the largest arms industry in the world, which has led David Edgerton to describe Britain as a liberal-militarist 'warfare state', which displayed 'an obsession with masses of machines, specifically machines designed to destroy enemies both physically and economically'. In short, he argues that the British government and military, especially the navy and air force, were uniquely structured so as to fight 'a war of science and invention; the next war, not the last'.[19] Certainly, by the end of the war, the 'very scale of effort and complexity of organisation' in military science had been revolutionised; by the late 1940s, 'more than half of government-funded research and development, and something like a quarter of the national total, was funded out of defence budgets'.[20] In addition, the government research corps expanded to unprecedented proportions during the war (and then diminished very little after 1945).[21] However, these facts were not always acknowledged by the British public (and many politicians) at the time and, instead, a pervasive myth of backwardness and inadequacy was propagated, meaning that pressure to innovate, modernise and 'catch up' remained constant, yet divorced from reality.

Often this useful falsehood was promulgated by scientific experts at the very highest echelons of British government, with the hope that it would help secure greater investment for research and development. One such expert was Frederick Lindemann, Viscount Cherwell – nicknamed 'The Prof', he has often been described as Churchill's closest advisor during the war and he used the ear of the Prime Minister (who was himself deeply sympathetic towards scientific innovation) to ensure new research played a key role in the British war effort.[22] Another influential wartime scientist was Henry Tizard, President and Rector of Imperial College London and a leading figure in the development of radar, who was largely responsible for fostering improved Anglo-American collaboration on military science, even before the USA entered the war.[23] In 1940, he led the eponymous Tizard Mission to America, which overcame mutual fears of lax security and chauvinistic beliefs in the superiority of their respective items of equipment and facilitated the exchange of virtually all of their most carefully guarded military technical secrets.[24] One such item, the cavity magnetron (which facilitated airborne radar), was described by American experts in 1946 as 'the most valuable cargo ever brought to our shores'.[25] This endeavour even presaged the exploitation programme in some ways, especially in the use of prototypes, blueprints and instructions to facilitate technology transfer (though in this instance it was mutually consensual). Prominent individuals such as Cherwell and Tizard, though these are only two examples from a large and esteemed cohort of influential wartime British scientists, were invaluable in ensuring that Britain, and its allies, did not fall behind their enemies in terms of the technical capabilities of their arsenals.

This situation in Britain cannot be viewed in isolation. It was motivated and encouraged, at least in part, by the enthusiastic adoption of research and development in the German war effort.[26] The Nazi war machine had to rely heavily on science from the very beginning in order to expand and modernise their armed forces from the level prescribed by the Treaty of Versailles to a level adequate for waging aggressive war across Europe and beyond.[27] In Ian Kershaw's biography of Hitler, he notes that Nazi Germany was initially so successful because it combined the imperialism of the nineteenth century with the technological potential of the twentieth.[28] This fusion of battlefield technology with military planning and military-economic preparation has also been described as a 'strategic synthesis' that resulted in a devastatingly effective unit.[29] After the war, British authorities commented that Germany was the only belligerent nation which carried out this 'prostitution of science' to such an extreme.[30] Civilian industrial concerns were closely involved with supporting the regime and fighting the war, especially in the form of large cartels such as IG Farben, which were able to work towards German victory while also making a healthy profit themselves. In addition, the scale of the Final Solution meant that it could only be perpetrated with the use of modern scientific and industrial techniques. In return, the enormous reserves of slave labour offered up by the concentration camp system made available a considerable workforce to a range of mass production projects, from rubber manufacture to the construction of missiles.[31] That said, it should also be noted that ideological concerns were often prioritised over, and could therefore hamper, scientific progress in Nazi Germany, as in the dismissal and persecution of a number of esteemed Jewish scientists, many of whom subsequently fled to Britain and the USA. The fact that the Allied nations mobilised their scientific, technical and engineering expertise without subverting the basic principles that allowed for the effective use of this talent may well have given them the edge in the scientific war.[32]

Away from the cold realities of the Third Reich's utilisation of science, Hitler and the Nazis also cultivated a remarkable belief in secret so-called 'wonder weapons'. These were the product of applied research too, and of far more interest to the British establishment. They were threatened as early as September 1939, when Hitler gave a speech to a Nazi rally in Danzig where he boldly stated: 'The moment might very quickly come for us to use a weapon with which we could not be attacked.'[33] It is undeniable that such a statement was mostly posturing, but it nonetheless created a great panic in the offices of British government, fuelled mostly by the dread of 'death rays' and other rumoured fantastical weapons.[34] When these failed to materialise, rational thought quickly returned to the civil service in Whitehall and the necessarily generous pinch of salt was administered to Hitler's comments.

However, Hitler was not simply giving voice to his wildest fantasies. The Third Reich was a regime which embraced science and technology to achieve its ends and, in reality, Germany did have an impressive pool of brilliant minds, including some of the world's top atomic physicists, and many of the most remarkable technologies of the Second World War had their origins in the Third Reich.[35] When Hitler gave his speech in Danzig, he was probably thinking more of technology on which research was substantially advanced – recoilless guns or rocketry, for instance – rather than absurd weapons which more accurately belonged to the realm of science fiction.[36] Later in the war, these new developments were sometimes easier to mass-produce than older technologies, on account of particular material shortages, which Germany suffered on a huge scale.[37] As a result, among the German public, the legend of the 'wonder weapons' remained potent, lasting in the popular consciousness until the very end of the war, when, as Allied armies crossed their borders, it was the German people's only faint hope for a reversal of fortunes.[38] The impact of the 'wonder weapon' threat did not evaporate immediately in Britain either. It seemed likely that Hitler's comments were not wholly groundless and it became clear to military scientists and policymakers alike that Britain would be incredibly vulnerable to attack by a weapon about which they knew nothing. It now became of utmost importance that nothing drastically new was added to the German arsenal without the British knowing about it – not only would this help them to defend against potential secret weapons but it would also allow them to add the technology to their own armouries. From now on, the race for scientific developments would have to be run in conjunction with the race for scientific intelligence and thus a new facet of modern warfare was born.

Spying on science

In F.H. Hinsley's expansive study of British intelligence during the Second World War, he notes that, along with order-of-battle information and operational intentions, one task of absolute priority for the intelligence agencies was 'to ensure that the enemy should not spring a surprise through some secret weapons or new type of aircraft or armament'.[39] More broadly, historians of intelligence have noted that rapid technological change, and the increased risk of surprise attack which this entails, essentially necessitates the contemporaneous growth of intelligence communities.[40] This gathering of details on new and future weaponry goes by the name of 'scientific and technical intelligence' and this can take many forms. Several of the techniques developed during the Second World War, a few of which are briefly detailed here, would later be adopted, often in a slightly modified form, by the exploitation initiative. This was a critical chapter in the story of scientific intelligence as it fed directly into the early arms race of the Cold War – arguably the period of the greatest revolution in military technology in history.[41]

However, in wartime Britain, scientific and technical intelligence was merely a nascent branch of a nebulous military intelligence network, the co-ordination of which repeatedly proved to be a substantial challenge for senior figures throughout the establishment. As with Cherwell and Tizard's contributions to the marriage of science with warfare in Britain, scientific and technical intelligence too relied partly on the brilliance and determination of individuals to show its true worth. Perhaps the most prominent example therein is Professor R.V. Jones. Reginald Victor Jones, with his recently earned doctorate in Natural Philosophy, was the first civilian scientist to be attached to a military intelligence agency in Britain, when in September 1939 he joined the Secret Intelligence Service (SIS). From this position, he established himself as the head of scientific intelligence at the Air Ministry, in the form of the Assistant Directorate of Intelligence (Science), for the duration of the war – a role which, thanks to the RAF being arguably the most technologically competent wing of the British armed forces, placed him at the forefront of British military scientific intelligence.[42]

His appointment was perhaps characteristic of the brand of total war which the Second World War necessitated (as had the First World War before it), where many civilian experts had to be drafted in to support the military in a number of roles – a tradition which would later be integral to the exploitation initiative too. The military lacked a significant scientific establishment of its own so was forced to rely on the resources of private research institutions and universities to compensate for this potential weakness. Jones, in his highly acclaimed memoir, *Most Secret War*, acknowledges that his initial appointment, made a mere matter of weeks before the outbreak of war, came about because the existing intelligence services admitted that they could not provide adequate information on German scientific developments.[43] It is interesting to note that a framework for scientific intelligence in Britain predated a similar structure for fusing science and strategy (the Scientific Advisory Committee) by almost a full year. Nevertheless, in the early part of the war, British gathering of scientific intelligence remained a largely piecemeal process and an effective, co-ordinated policy continued to be elusive.

A large part of the problem was a perennial reluctance among the relevant agencies and ministries to adopt a proactive strategy, instead preferring to allow Jones and his colleagues to struggle on, gleaning information from wherever they could, in a decidedly haphazard manner. There was a surprising reliance on 'open-source intelligence', such as studying German scientific journals from before, and sometimes during, the war for clues as to German progress in various fields.[44] Even at the Air Ministry, where there was often the greatest scope for technical innovation, intelligence-gathering methods were mostly passive and opportunistic. Downed Luftwaffe aircraft were examined as a means of keeping tabs on development in that field – this served remarkably well in the earlier part of the war, as the German air force's overconfidence in its own superiority led it to adopting a policy of modifying older, often pre-1939, models, as opposed to embracing new innovative designs. This made it very easy for the British scientific intelligence experts to build on existing knowledge and simply keep abreast of these relatively small alterations. Understandably, this did lead to shocks, such as the Luftwaffe's deployment of the Focke-Wulf FW 190 in 1941, which was both faster and carried heavier armour and armaments than its contemporaries and was itself not fully appreciated until one force-landed in Wales in June 1942. This reactive approach was also hindered by the sheer variety of different aircraft in use by the Luftwaffe, many of which were only ever so slightly different from each other; the product of competition among aeronautical firms in Germany, all vying for the attention and favour of senior Nazis.[45]

Another source which wartime British scientific intelligence utilised was information passed on by officials in neutral countries or resistance operatives in Nazi-occupied territories – a method almost as passive as waiting for planes to crash. In many cases, even when intelligence which hinted at a major threat from new technology in the German war effort was received, it was paid no heed. In the case of long-range rocketry, because progress was slow in this field in Britain at the time the information was assessed, it was assumed that German developments must be similarly limited.[46] This method did have its successes though, perhaps most notably the Oslo Report, which consisted of details of current and future German weapons projects compiled by an anti-Nazi German physicist in November 1939, mailed to the British embassy in Oslo and subsequently passed on to the SIS. It was picked up by R.V. Jones who vouched for its veracity and accuracy and it did indeed prove invaluable, especially in developing effective countermeasures to be used against the Luftwaffe during the Battle of Britain – and Jones even admitted revisiting it in the few dull moments he experienced during the war, 'to see what should be coming along next'.[47]

Despite occasional victories like the Oslo Report, the approach of waiting for intelligence to fall into their laps proved an infuriatingly slow one for the British intelligence services. Moreover, it was felt that 'the evidence thus offered can rarely be complete and the deductions may be faulty.'[48] Continuing this passive process was motivated by a belief that it would be unwise, if not impossible, to infiltrate agents into Germany to obtain this information first-hand. If an agent was to be effective in collecting details about complex scientific issues, they would need to be very extensively briefed; if such a knowledgeable operative was then captured whilst on a mission in Germany, any information they might disclose could seriously jeopardise some high-priority military projects in Britain – this risk was judged to be too high.[49] In the face of these numerous shortcomings in the British scientific intelligence system, which Henry Tizard considered to be greatly inferior to the German equivalent, the successes of Jones and his peers seem even more impressive.[50] One significant triumph was the discovery, and subsequent jamming, of the Luftwaffe *Knickebein* radio beam system, used for directing night-bombing raids, in 1940. Jones considered this his greatest wartime victory, partly because of its significant contribution to Britain's air security, but also because it finally 'put scientific intelligence on the map' and made it 'an essential component of defence'.[51]

This growing fascination with the contents and capabilities of an enemy's arsenal, and particularly the realisation of German superiority in certain fields, provided much of the early drive for exploitation. It should also be noted that even while the war against Germany was ongoing, Britain also began to take an interest in the technological armouries of the Soviet Union – an ally for now, but believed by many to be the greatest future threat to European peace and British interests.[52] The experiences of scientific intelligence in the Second World War did not just provide the motive for exploitation, they also provided some of the means. For instance, one particular operation pioneered a more proactive approach to scientific intelligence-gathering and took place on that most important scientific battlefield of the early years of the war – radar. In early 1941, confirmation reached Jones at the Air Ministry that Germany was employing a new radar array, mentioned in the Oslo Report and known as *Würzburg*, which could assess the height of aircraft, essential to deploying an effective defensive response, either by fighter or by anti-aircraft batteries.[53] Aerial photographs were taken but little more could be ascertained about this important technological development without close examination. This need gave rise to the scientific intelligence mission known as Operation Biting or, in more common parlance, the Bruneval Raid.

On the night of 27 February 1942, 120 specially trained British Combined Operations commandos were dropped by parachute near the small town of Bruneval on the northern French coast. They moved to their target – a villa in the area, which housed a radar installation – which they successfully attacked, allowing them to seize *Würzburg* radar equipment and take prisoners, before evacuating by sea from a nearby beach.[54] This daring raid had numerous repercussions – it was trumpeted in the British press in order to boost morale, which was at low ebb thanks to military losses in North Africa and the Far East;[55] it ensured that parachute assault would become a major part of Britain's military capacity from then on; and it gave scientific intelligence experts a remarkable opportunity to unravel one of the most important mysteries of German air defence.

Perhaps of the greatest relevance to future exploitation were the techniques used to carry out the raid. It was the first modern example of a behind-enemy-lines incursion for which scientific and technical intelligence was the prime objective.[56] Although no civilian expert could take part on account of the strictly specialist military nature of the mission, an Air Force radar mechanic, Flight Sergeant C.W.H. Cox, a peacetime cinema projectionist who had never been in a plane or on a ship before, volunteered and, along with a team of Royal Engineers, played an essential role in dismantling the radar equipment at Bruneval and ensuring the elements of greatest value were returned to Britain for examination.[57] This inclusion of a non-combatant technician in a commando unit would later come to characterise the exploitation initiative, especially in its actions before the end of the war, which would share much DNA with operations like the Bruneval Raid. Another interesting similarity between Operation Biting and future exploitation endeavours was the capture of trained German technicians as prisoners and their subsequent interrogation by intelligence services back in Britain. In the case of Bruneval, the radar operator who was detained turned out to know very little about the equipment he worked with, though he was quite forthcoming with what he did know. This questioning, paired with a thorough investigation of the device itself, elicited plenty of useful information about the *Würzburg* radar system and, like so many other scientific discoveries made by British Intelligence about the German war machine, they were shocked to discover how much more advanced German capabilities were.[58] This reaction would continue to be all too common throughout much of the early exploitation process.

This may have been Britain's first foray into proactive intelligence-gathering, but it had been part of Germany's military strategy since the outbreak of war. A central element of Blitzkrieg tactics had been the involvement of the *Abwehr*'s intelligence commando units, the brainchild of Admiral Wilhelm Canaris, head of Hitler's foreign intelligence service at the outbreak of war. They travelled with, or sometimes ahead of, the first wave of ground troops and had a remit to seize preordained targets and anything else of intelligence value which they came upon.[59] Just as the Bruneval Raid commandos were representative of the new technique of airborne assault, these *Abwehrkommando* teams could only exist thanks to the advent of rapid motorised ground warfare. Made up of handpicked operatives, many of whom had been born outside the Reich so they could bring valuable knowledge of foreign cultures and languages into their activities, the commandos were extensively trained, with a particular focus on speed and mobility – for instance, most were experienced cross-country motorcyclists.[60] Although their background was military, they were often equipped with civilian clothes to allow them to advance while attracting only minimal attention, and they were prepared for unarmed combat and the use of foreign radio sets to communicate with their superiors or mislead the enemy.[61]

When Canaris first envisaged his *Abwehrkommando*, he saw their primary role as being a preparatory one, laying down groundwork, through espionage and sabotage, for the advance of conventional ground troops. This was first employed in the invasion of Poland in 1939, where they captured numerous targets of industrial as well as military significance, including coal mines, factories and a rail junction.[62] For naval operations, they were often tasked with securing ports and harbours, as well as ciphers and top-secret documents before the defenders could destroy them.[63] They were also responsible for obtaining traditional intelligence, particularly 'political, economic and military information', which could have benefits when implementing occupation of foreign territories and opposing any resistance therein.[64] They also had some similar responsibilities to R.V. Jones' fledgling scientific intelligence department at the Air Ministry, in the gathering of information on military technology in the enemy countries which Germany invaded.[65] Unsurprisingly, it did not take too long for word of these commando units, the like of which had never before been seen in modern warfare, to reach Britain. The first details came from a British accountant, Trevor James Glanville, who was working for the Special Operations Executive (SOE, a British sabotage organisation) in Yugoslavia when it fell to the Nazis in 1941 and was subsequently taken prisoner, but who eventually returned to Britain with tales of these special German commandos. This, along with the success of Bruneval and other such raids, was enough to convince the British military intelligence establishment of the need for, and feasibility of, a special commando unit of their own, to be consciously modelled on the *Abwehrkommando* example.[66]

Forerunners to exploitation

For a British intelligence commando unit to come into being, it once again took the work of a particularly talented individual, who by merit of education and experience, was in the right place at the right time. In this instance, that individual was Ian Fleming who, for the duration of the war, held the position of Assistant to the Director of Naval Intelligence, but who would later receive much greater acclaim for creating the most famous spy in fiction, James Bond. During his time at the Naval Intelligence Division (NID), he had proven over and again that he was a master strategist and he continuously displayed a remarkable level of operational creativity. It was he who began to take real note of the operations of the German intelligence commando units, especially those led by the infamous Otto Skorzeny during the German invasion of Crete, where they avoided much of the main fighting, instead striving to secure British military headquarters and the sensitive documents stored within. As a result, plans for a British equivalent began to take shape in Fleming's mind.[67] On 20 March 1942, only three weeks after the Bruneval Raid, Fleming submitted a memo to the Joint Intelligence Sub-Committee (JIC) outlining the methods and successes of the German commandos and suggesting that:

> We would do well to consider organising such a Commando within the NID, for use when we reassume the offensive on the Continent, in Norway or elsewhere. The unit would be modelled on the same lines as its German counterpart and would be placed under the command of CCO [Chief of Combined Operations], perhaps a month before a specific objective is attacked.[68]

The JIC met shortly after to discuss Fleming's proposals and decided that an 'Intelligence Assault Unit', whose responsibility would be to gather 'enemy material and documents of immediate operational value and other archives, documents and equipment of importance', would be very useful.[69] Their approval ensured that this new unit was considered for use during the preliminary planning of Operation Sledgehammer – a proposed invasion of continental Europe which never materialised, but would later be successfully resurrected as Operation Overlord. With approval given, Fleming was free to begin assembling, training and preparing his commandos for imminent deployment. He called them his 'Red Indians', on account of their fast and light movement and aggressive raiding tactics, but they were given the official naval designation of 30 Commando, later changed to 30 Assault Unit, or 30AU.[70]

Their story has been widely and extensively told, perhaps best in David Nutting's *Attain by Surprise*, which gathers many fascinating wartime recollections from members of the Unit, but their role is worth exploring here, if only briefly, as its impact on future exploitation endeavours was indelible.[71] The composition of the Unit was split between two 'Wings' – one from the Royal Navy (RN) and one from the Royal Marines (RM). The RN component was made up primarily of 'specialist officers in the various branches of the intelligence and research departments of the Admiralty' while the RM element provided the bulk of the manpower, 'taking [the specialists] to their targets, protecting them and assisting them with their work at the targets and then escorting their withdrawal'. Fleming justified the slightly unorthodox dual composition thus: 'the functions of each Wing are therefore equally essential to the success of the Unit as a whole – i.e. without the RN Wing there would be no purpose in the Unit; without the RM Wing the RN Wing would not long survive in the field.'[72]

The selection of Marines, already members of an elite division of the military, to be part of 30 Assault Unit was a rigorous process and few met all the numerous criteria – at its inception, 30AU numbered fewer than 40 men. Their officers had to be able to speak another language besides English – German, French, Dutch, Flemish, possibly Norwegian – and to have a general knowledge of the countries in which they might be operating; there should be a good number of trained parachutists in case a naval insertion was not possible; some should have specialised technical or mechanical knowledge or a familiarity with the relevant documents and material; and all should be capable fighting men 'in order to be able to meet any eventualities'.[73] Even with all the specifications met, Ian Fleming knew that the success of these intelligence assault units would hinge on their training – even if they were able to reach key targets before the enemy could conceal or destroy them, it would be in vain if they were unable to identify items of value or to utilise them. Here once again, unconventionality was the order of the day. A veteran chief inspector at Scotland Yard was engaged to instruct the men in the theory and practice of blowing safes, picking locks and breaking and entering. They were then subject to a series of demonstrations in the use of gelignite, plastic explosive, booby traps, mines and small arms weapons. Away from these practical sessions, with which the Marines would have been far more familiar, were the sessions devoted to the recognition and capture of the so-called 'treasure trove' of modern war – ciphers, codebooks, intelligence reports, secret orders, new weapons, radar sets, and so on.[74]

With this training in mind, it is unsurprising that 30 Assault Unit were considered to be 'armed and expert authorised looters', even by those directly involved in their formation or deployment.[75] They were viewed with caution or hostility by many of the more narrow-minded members of the naval establishment, or were seen as foolish, even before they had been given a chance to prove themselves in action. Rear-Admiral Jan Aylen, who served with 30AU in Germany and elsewhere, recalled that the Royal Navy's Deputy Engineer-in-Chief chided him for joining that 'hare-brained skylark on the Continent', but was soon requesting personal feedback from the operations, once they began to show their worth.[76]

30 Assault Unit's first action almost proved its critics right, though through no fault of its own. Fleming had insisted on involving a small cohort of his commandos in the ill-fated raid on Dieppe on 19 August 1942. They were tasked with entering the *Kriegsmarine* headquarters in the French port and seizing codebooks and ciphers; instead, their landing craft was struck by a shell before they reached shore and they were forced to swim back out to the ships anchored some distance from the coast.[77] It was an ignominious beginning but it taught Fleming and NID some valuable lessons which could be applied when 30AU next went into action, only three months later, in Operation Torch, the Allied invasion of French North Africa. Again, due largely to factors beyond their control, it was a slow and uncertain start and they did not achieve their primary objective, of capturing the Vichy Admiralty building intact and securing the ciphers within.[78] However, in continued exploits in North Africa, they did acquire an unbroken Enigma machine and accompanying codebooks, which allowed the Allies to intercept and decode German radio communication in the area for the next six weeks.[79]

Undeterred by their difficulties and encouraged by their successes, their momentum picked up and, despite a couple of returns to Britain to be debriefed and re-briefed, and changes in leadership and composition, they began to truly justify their existence in operations in Sicily and Italy in 1943, where they obtained 'a substantial quantity of documents and equipment of operational value'.[80] Using information they gleaned from Italian industrial concerns, which were fulfilling military contracts for the German navy, they were able to furnish the NID with specifications for new designs of torpedoes, sea mines and depth charges.[81] The shift from operational and order-of-battle intelligence to include scientific intelligence was well underway and by 1944, 30 Assault Unit, increased to a strength of 50, were concerned with the entire range of the enemy's armoury.[82]

Unsurprisingly, this newly expanded remit necessitated much more careful planning than had been utilised before. Although some of the Unit's greatest finds had been made opportunistically and on the fly, by officers who were trained to know what to look for, as all of continental Europe was set to be the next theatre, including the research establishments and arms factories of Germany itself, pre-approved target lists and orders of priority were going to be essential in order to derive maximum benefit from their activities. This was effected by the NID asking various other divisions of the Navy to submit requests for intelligence on, or examples of, technology in which they were especially interested. They were duly swamped by a deluge of responses – the Director of Anti-Submarine Warfare wanted information on sonar and hydrophones, the Gunnery Division wanted to know more about automatic guns and all calibres of ammunition, the Director of Torpedoes and Mining requested details of external markings on these devices and their launching mechanisms, and the Signals Division were keen to learn more about infrared and ultraviolet technologies in detection. This is just a small cross-section of the great quantity of requests which were filed, and which became known as the 'Black List'.[83]

As a result of the sheer volume of demands placed upon 30 Assault Unit and their administrative support in Room 39 at the Admiralty, it soon became apparent that the list would need to be prioritised. The top priority, A.1, was only afforded to intelligence of 'immediate operational importance in the prosecution of the war against Germany'. This included items such as codebooks, ciphers, and anything pertaining to the Enigma machines; the importance of which was 'sufficient to justify the mounting of special operations and the incurring of heavy casualties on the part of 30 Assault Unit'.[84] It was a remarkably frank account which measured the value of intelligence in the terms of commandos' lives. Multiple 'Black Lists', subdivided in terms of priority, would later become a central feature of the broader exploitation effort, the guiding documents for the hundreds of investigation teams which swarmed across Europe in the aftermath of D-Day. In many respects, they were based on 30AU's template, especially in terms of their flexibility, the significance of which Fleming and the NID were quick to acknowledge. As a result of 'aerial bombing or evacuation' on the continent, it was decided that 'all indications of probable sources of materiel and intelligence required should be reviewed continually in the light of aerial reconnaissance photographs, the interrogation of prisoners of war and enemy civilians, and captured enemy documents.'[85]

The degree of influence which 30 Assault Unit had on exploitation agencies that came later is often hidden because these subsequent efforts very quickly eclipsed and overtook 30AU. It is important to note that 30 Assault Unit did serve very successfully in France, the Low Countries and Germany as the advancing line of Allied liberation made its way across Europe, but their remit narrowed and their greatest contributions were already behind them.[86] In part, their small number – only around 150 at the time of D-Day – meant that they were not able to handle 'so many operations over so large an area' and they worked 'sometimes under fire, during all the hours of daylight as well as keeping turns on two-hour watches at night'.[87] This reduced role was not, however, simply the result of a shortage of manpower, or of other larger organisations overshadowing them, but was also down to the culmination of a gradual falling out of favour which had its origins in the initial scepticism shown towards the Unit since its formation.

30 Assault Unit was viewed fairly critically even by those who worked closely with them. Lieutenant-Commander Robert Harling, one of Ian Fleming's most trusted assistants, described them as 'merry, courageous, amoral, loyal, lying toughs, disinclined to take no for an answer from foe or *Fräulein*' – something of a mixed bag of praise and critique – which in turn led Fleming himself to dub them '30 Indecent Assault Unit'.[88] They could not always even justify their unorthodox methods with results, as negligence often meant that many of their prizes were lost *en route* back to Britain.[89] Their officers were accused of being 'high-handed' and 'unscrupulous' in their efforts to secure vehicles and supplies for their men, and it was even suggested that the 'Assault' in their name be changed to 'Intelligence' as they apparently disliked 'being told to do a bit of assaulting'.[90]

Once Overlord and the main campaign in continental Europe got underway, it was considered that the already controversial tactics of 30 Assault Unit, widely regarded as a 'private army', did not fit in with the massive co-ordinated organisation structure of the Supreme Headquarters of the Allied Expeditionary Force (SHAEF). They were created to serve the need for daring smash-and-grab intelligence raids, not the meticulous and gradual accumulation of all of Germany's scientific and technical knowledge.[91] The agencies which took their place mistrusted 30AU, haranguing them constantly to make sure that they shared all the information they accrued and making it abundantly clear that they did not want to fight alongside them.[92] Nonetheless, even during this phase of relative decline, Fleming's commandos enjoyed a major victory in helping to secure the immensely valuable *Walterwerke* submarine plant in Kiel on 5 May 1945 and had their swan song in capturing the entire German naval archive at Tambach Castle near Coburg, Bavaria, shortly after the war had ended. There was even a lingering possibility that they might be transferred to the Far East where their experience in seizing 'targets of opportunity' could prove very useful, but the end of the war in the Pacific in August 1945 put paid to that idea.[93]

30 Assault Unit was not the only precursor to exploitation active during the war. While Fleming's men were participating in Operation Torch in North Africa, another operation was coming together in the hot, dry desert of New Mexico where, at a secret laboratory complex at Los Alamos, American and British scientists were working to develop an atomic bomb. Known as the Manhattan Project, and largely subsuming the similarly purposed 'Tube Alloys' programme in Britain, this research was yet another facet of the ongoing wartime arms race between the Allies and Axis powers. The secrecy afforded to all work on the atomic bomb was unequalled and naturally all those involved feared that parallel

German efforts, similarly hidden from view, would exceed their own; certainly, German physicists who had fled from the Nazi regime and come to work in the Allied countries were convinced that 'German science was the best in the world, and that if a bomb could be built, the Germans could – and would – build it'.[94] Ultimately, the Allies were terrified that they would fall victim to atomic warfare before they were in a position to unleash it themselves.[95]

To try and avoid this fate, or at least to better understand similar work going on in hostile countries, the head of the Manhattan Project, Brigadier-General Leslie R. Groves, ordered the creation of a War Department Scientific Intelligence Mission, better known by its codename, Alsos (derived from the classical Greek word for a 'sacred grove'). Headed militarily by Colonel Boris T. Pash, a US Army career soldier of Russian descent, and scientifically by Samuel A. Goudsmit, a Dutch-American physicist, Alsos was inter-Allied in make-up but the bulk of its staff, and the ultimate command, lay with the Americans. This was not because the Americans were necessarily better-equipped to handle the demands of the task (in fact, British scientific intelligence was rather more advanced than that of their allies across the Atlantic), but rather that the US created a team first and it seemed better diplomacy for British operatives to ask permission to join that, than to create their own rival unit. It was in this way that the British, who were perhaps more accustomed to a senior partner role, were effectively demoted to juniors.[96]

Alsos began operation in the Mediterranean and was later deployed across Western Europe and into Germany. In his memoirs, Brigadier-General Groves has stated that the mission's primary purpose was:

> to obtain intelligence of atomic developments in Italy and Germany. Nevertheless it was logical to expect that, in the course of its work, the mission would also come upon data about other enemy projects; accordingly, it was directed to exploit to the fullest sources in a number of fields of technical interest.[97]

This mention of 'other enemy projects' extended to any other scientific and technical research taking place in Germany, or under German supervision, which had a potential military application. It included both chemical and biological warfare, as well as ordnance and aircraft technology. Atomic energy and weaponry would remain their primary concern throughout their period of operation, but the additional intelligence they gathered, often just by having a keen eye for items of scientific interest, proved to be immensely valuable.

While in the upper echelons of Allied command preparation was being made for a vast operation to examine *all* aspects of German scientific and technical endeavour, as shall be explored in the next chapter, Alsos was able to get into the field much sooner. It was a more streamlined operation, with fewer men and a unilateral command structure, but more than that, it was designed to supplement existing intelligence organisations, not duplicate them, and could therefore count on shared resources and co-operation at the front, and could minimise administrative hassle or superfluous personnel.[98] In addition, so great was the fear of a

German atomic attack to which the Allies would be completely vulnerable, that the objectives of Alsos always had highest priority. It is no surprise that it has retrospectively been described as the first large-scale scientific intelligence mission in history,[99] but at the time it was thought to be so unprecedented as to be an experiment of sorts.[100]

The Alsos operatives, many of whom were trained scientists, were also aided in their mission by their own inherent knowledge of the state of atomic research in Germany, and elsewhere, in the years leading up to the war. They were furnished with long lists of 'targets' to investigate, many of which were, in fact, individual specialists, but these lists could quickly be cut down as the agents themselves knew which scientists were important and which were not. This was a level of insight which the military chiefs could not comprehend, but Goudsmit has asserted that 'any reputable scientist working in the same field would have known the same thing'.[101] The involvement of these experienced and knowledgeable experts also furnished the project with the perfect interrogators to elicit maximum information from the German scientists who they detained. This was particularly relevant because Alsos agents were not only permitted to pursue and investigate civilian research but actively encouraged to do so.[102] The realisation had already dawned on those with a vested interest in the fruits of German scientific effort during the war that the best results may be gleaned from non-military institutions – this would later become a common aspect of the whole exploitation initiative.

The fears of a German atomic bomb were far from baseless. Nazi Germany was home to some of the world's most renowned nuclear physicists, including Otto Hahn, who had isolated pure uranium-235 in December 1938, and Werner Heisenberg, who had recognised its potential if weaponised.[103] The Manhattan Project's intelligence officials had discounted Japan, which they felt lacked the necessary scientific and technical prowess to develop an atomic bomb, but strongly believed that Nazi Germany, with its extensive resources and eager support for research into new ways of killing, not to mention its unrivalled cadre of physicists, could add such a weapon to its arsenal sooner rather than later. Add to this the constant misinformation spewed forth by Joseph Goebbels' Ministry of Public Enlightenment and Propaganda which spoke of secret super-weapons, and it appears unsurprising that Alsos continuously expected to find evidence of a German atomic bomb project nearing completion.

In reality, the German project was still very much in its harmless infancy, but the Alsos investigators only found this out gradually as they moved across Europe in the immediate wake of the advancing Allied forces. Colonel Pash laid claim to being among the first column of Allied troops to enter Paris on 25 August 1944, his small jeep nestling among much larger vehicles and surrounded by crowds of the liberated French public, being showered with adulation. The high-spirited local populace were not allowed to become a distraction or hindrance and Pash and his colleague Major Horace Calvert soon tracked down eminent French physicist Frédéric Joliot-Curie in his laboratory at the Collège de France. He was flown immediately back to London for interrogation and more investigators, including Goudsmit and British 'Tube Alloys' specialist, Michael Perrin, came over to Paris to run through all of Joliot-Curie's papers and equipment with a fine-toothed comb. All they really managed to learn was that the German physicists had some solid ideas about nuclear fission but were not far along in the process of developing it into a weapon.[104]

These conclusions were based primarily on Joliot-Curie's poorly informed suspicions and there was also a sense that the Nazis were unlikely to have conducted their most important military research in Paris. The search would have to continue. The next investigation of importance took place in Strasbourg, on the border between France and Germany, where the University had housed much recent endeavour in the field of nuclear physics. Alsos operatives entered the city and University on 25 November 1944 and pored over the wealth of the material available there. They found out much about other areas of science and technology – medicine, aircraft and naval matters, for example – and were able to detain and interrogate seven senior German physicists and chemists. The products of this further confirmed their suspicions that the German atomic bomb project was still mired in the early experimental stages.[105]

The final real prize of Alsos's atomic investigations lay only sixty miles or so across the German border, at the small town of Hechingen, just south of Stuttgart. Aerial reconnaissance and word-of-mouth from scientists in neutral countries suggested that Werner Heisenberg, by now the most desirable figure of the German atomic establishment, and many of his similarly esteemed colleagues, were based at the Kaiser Wilhelm Institute for Physics at Hechingen (relocated from Berlin after severe bombing) and were continuing their research there. By now, the Americans and the British were confident that Nazi Germany did not have the capability to launch an atomic strike, but they pressed on nonetheless. This was because Hechingen was now one of many targets in south-western Germany included in Operation Harborage – an Anglo-American effort to secure equipment, documents and individuals from sites which would later fall under French occupation. Groves feared that anything which the French seized might very soon become accessible to the Soviets, and mistrust of the USSR was growing day by day.[106] Harborage therefore marked the beginning of a policy of 'denial', wherein scientific and technical targets were seized by the British or Americans, not necessarily for their own use but simply to prevent the French or Soviets laying claim to them, which would persist in exploitation throughout the occupation and set the tone for the Cold War. In this instance, the Americans were even sufficiently suspicious of their British partners (who had thus far given more than they had got in scientific intelligence) to refuse to share with them the key documents found at Hechingen and instead shipped them straight back to the United States.[107]

Perhaps the greatest prizes seized in Hechingen and its environs were ten of the most prominent of the German atomic scientists, including Heisenberg and Otto Hahn, who were then detained, incommunicado, at Farm Hall, in Godmanchester, near Cambridge, from the time of their capture in July until they were released in December 1945.[108] This detention was highly secretive, at least at first, and the scientists' quarters were all fitted with listening devices, unbeknownst to the men themselves.[109] This eavesdropping shed some interesting light on these scientists' world-view, particularly when news broke of the bombing of Hiroshima, but was hardly legally, let alone ethically, sound. Nonetheless, it would come to set the tone for the detainment of numerous scientific and technical personalities of interest throughout the post-war period.[110] Alsos did some more investigative work after Hechingen, largely concerned with completing records and denying personnel and materiel to the Soviets, but ceased to exist as an organisation on 15 October 1945, despite urgings by many, particularly in the USA, that it should become a permanent scientific intelligence agency.[111] Even in its fairly short period of operation, Alsos had visited over 70 targets, including 16 universities and four concentration camps, and had filed approximately 400 scientific reports.[112]

Though at first glance it may seem that the story of Alsos has little to contribute to the history of exploitation – it swiftly established that there was no threat from the German bomb project and was disbanded fairly promptly at the end of the war – its influence on organisations which came into being simultaneously or afterwards should not be understated. First, the very fact that it discovered the absence of a direct wartime threat so soon and yet pressed on is significant. Those involved with Alsos, whether on the ground or in a supervisory role, understood that there were benefits to unravelling German science and technology beyond mere direct tactical utility. As early as March 1944, this was displayed in a memo pertaining to Alsos's impending actions in Western Europe, which included 'planning of own strategy if similar weapons or tactics are available to our own forces' and 'direction of our own war research projects' among the agency's main mission aims, alongside more immediate military objectives.[113]

Later on, as we have seen, denial policy also became part of their operating remit. This in turn transformed and expanded into something different. As the Second World War ended and the Cold War began, many Western eyes stopped perceiving Germany as an enemy and instead focused suspicion on the Soviet Union. The work Alsos had carried out in investigating the research undertaken in Germany during the war, and by observing which German physicists the Soviets snapped up, now allowed them to make measured estimates of the progress of similar atomic research in Russia.[114] This is just one example of how the exploitation initiative provided a smooth transition from investigating a current enemy to sizing up a future one.[115]

In this way, and in many others, the actions of Alsos played a major role in shaping the structure of future exploitation organisations. Their methodology, as outlined by Goudsmit, was as follows:

> It was the task of the scientists to obtain and analyse all pertinent information having to do with German science. From such information they had to deduce just what places, institutions, buildings, and people in enemy territory were important for giving us the information we wanted. It then became the task of Colonel Pash and his men to see that we got to these people and these places before anyone else got there. They also had to supply us with all relevant intelligence collected by other groups in the American and British armed forces.[116]

This could describe the *modus operandi* of any one of the numerous exploitation agencies which came into being in the last year or so of the war, or immediately afterward. Two elements of this were particularly crucial – the first was the combination of civilian scientist investigators with conventional military operatives on the ground, not common even in an era of total war but absolutely necessary if exploitation was to be both genuinely beneficial and logistically feasible and the second was the interaction between the multiple organisations operating alongside one another in the field.[117] Both of these policies were sensibly adopted by the agencies which Alsos had preceded.

More than just an inheritance of ideas and methods, Alsos was also able to pass on some hugely valuable concrete intelligence to help the fledgling operations of their successors get off the ground. Perhaps the most significant example of this was the so-called 'Osenberg List'. Dr Werner Osenberg was the head of the *Wehrforschungsgemeinschaft* (the Nazi Military Research Association) when he was captured by Alsos operatives near Göttingen in early 1945. The information he furnished, complemented extensively by his list of 15,000 leading German scientists and technicians (found in scraps, which someone had supposedly attempted to flush down a toilet at Bonn University), allowed all future exploitation teams to identify key targets more quickly and to pursue them more accurately. Unsurprisingly, this would prove time and again to be utterly invaluable.[118]

To conclude, exploitation was very clearly a product of the conflict in which it originated. As this chapter has shown, the Second World War was characterised by the growing influence of science and technology on modern warfighting. The course of the war had proven that the appropriate application of research and development had the very real potential to bestow both tactical and strategic advantage on the state in question. In turn, this made understanding the armouries of potentially hostile nations absolutely essential and, while this was first realised during the Second World War, it was the fog of suspicion and secrecy fostered by the Cold War in which scientific intelligence really came of age.[119] There is no question that exploitation emerged from this atmosphere as both a fascination with the contents of an enemy's arsenal and as the source of a potential head-start in any future arms race.

30 Assault Unit and Alsos both played relatively small, but still crucial, parts in the developing story of scientific and technical exploitation. They were pioneering forces which represented the changing priorities of modern warfare and were well-suited to this new relationship between science and strategy, which had reached maturity during the war. It is clear that the exploitation initiative would not have existed in the form that it did, if at all, without the experience of 30AU and Alsos, and it would have faced much greater difficulty had it not been able to build on the problems faced, and solutions devised, by these two agencies. It is for this reason that the issues discussed in this chapter, though perhaps initially seeming peripheral, are relevant to the wider history of exploitation and their influence will be detected throughout this book. As the war entered its final destructive

stage, the operational techniques of these small, daring enterprises began to fall from favour and it became clear that the era of comprehensive exploitation on a grand scale was dawning.

Notes

1 Richard Overy, *Why the Allies Won* (London: Jonathan Cape, 1995), 242.
2 Ad Maas and Hans Hooijmaijers (eds.), *Scientific Research in World War II: What Scientists did in the War* (Abingdon: Routledge, 2009); Cathryn Carson, 'Knowledge Economies: Toward a New Technological Age', in Michael Geyer and Adam Tooze (eds.), *The Cambridge History of the Second World War*, vol. III (Cambridge: Cambridge University Press, 2015), 196–219.
3 Roy MacLeod, 'Scientists', in Jay Winter (ed.), *The Cambridge History of the First World War*, vol. II (Cambridge: Cambridge University Press, 2014), 451.
4 Roger Cooter, Mark Harrison and Steve Sturdy (eds.), *War, Medicine and Modernity* (Stroud: Sutton, 1998), 77; David Edgerton, *The Shock of the Old: Technology and Global History since 1900* (Oxford: Oxford University Press, 2011).
5 Jonathan D. Moreno, *Undue Risk: Secret State Experiments on Humans* (New York: Routledge, 2001), 16.
6 Vannevar Bush, foreword to Irvin Stewart, *Organising Scientific Research for War: The Administrative History of the Office of Scientific Research and Development* (Boston, MA: Little Brown, 1948).
7 TNA, CAB 21/1421, 'Co-ordination of Scientific Intelligence', 23 January 1940.
8 Richard Overy, *The Bombing War: Europe, 1939–1945* (London: Penguin, 2013); Dietmar Süss, *Death from the Skies: How the British and Germans Survived Bombing in World War Two* (Oxford: Oxford University Press, 2014).
9 R.V. Jones, *Most Secret War: British Scientific Intelligence, 1939–1945* (London: Coronet, 1978), 43.
10 Robert Bud and Philip Gummett (eds.), *Cold War, Hot Science: Applied Research in Britain's Defence Laboratories, 1945–1990* (London: Science Museum, 2002), 5.
11 Bruce Cameron Reed, *The History and Science of the Manhattan Project* (New York: Springer, 2014), 1.
12 Wilson D. Miscamble, *The Most Controversial Decision: Truman, the Atomic Bombs, and the Defeat of Japan* (Cambridge: Cambridge University Press, 2011); Campbell Craig and Sergey Radchenko, *The Atomic Bomb and the Origins of the Cold War* (New Haven, CT: Yale University Press, 2008); Andrew J. Rotter, *Hiroshima: The World's Bomb* (Oxford: Oxford University Press, 2008); J. Samuel Walker, *Utter Destruction: Truman and the Use of Atomic Bombs against Japan* (Chapel Hill, NC: University of North Carolina Press, 2004).
13 Nicholas J. Vig, *Science and Technology in British Politics* (London: Pergamon, 1968), 1.
14 Daniel Todman, *Britain's War: Into Battle, 1937–1941* (London: Penguin, 2017), 19; David Edgerton, *Warfare State: Britain, 1920–1970* (Cambridge: Cambridge University Press, 2006), 144.
15 David Zimmerman, *Top Secret Exchange: The Tizard Mission and the Scientific War* (Montreal: McGill-Queen's University Press, 1996), 4–5.
16 Philip Gummett, *Scientists in Whitehall* (Manchester: Manchester University Press, 1980), 30.
17 David Mellor, *The Role of Science and Industry: Australia in the War of 1939–1945* (Canberra: Australian War Memorial, 1958), 57.
18 David Edgerton, 'British Scientific Intellectuals and the Relations of Science, Technology, and War', in Paul Forman and Jose M. Sanchez-Ron (eds.), *National*

Military Establishments and the Advancement of Science and Technology (Dordrecht: Kluwer, 1996), 5.

19 David Edgerton, *Britain's War Machine: Weapons, Resources and Experts in the Second World War* (London: Allen Lane, 2011), esp. 7–8.

20 Bud and Gummett (eds.), *Cold War, Hot Science*, 1.

21 Edgerton, *Warfare State*, 167–8.

22 Adrian Fort, *Prof: The Life and Times of Frederick Lindemann* (London: Jonathan Cape, 2003); Frederick Smith, 2nd Earl of Birkenhead, *The Prof in Two Worlds: The Official Life of Professor F.A. Lindemann, Viscount Cherwell* (London: Collins, 1961).

23 Ronald W. Clark, *Tizard* (London: Methuen, 1965).

24 Zimmerman, *Top Secret Exchange*, 4.

25 James Phinney Baxter III, *Scientists against Time* (New York: Little Brown, 1946), 142.

26 Monika Renneberg and Mark Walker (eds.), *Science, Technology and National Socialism* (Cambridge: Cambridge University Press, 2003).

27 Mark Walker, *Nazi Science: Myth, Truth, and the German Atomic Bomb* (New York: Plenum, 1995).

28 Ian Kershaw, *Hitler, 1889–1936: Hubris* (London: Penguin, 1998), 449.

29 Adam Tooze, *The Wages of Destruction: The Making and Breaking of the Nazi Economy* (London: Penguin, 2008), 371.

30 Paul Julian Weindling, *Nazi Medicine and the Nuremberg Trials: From Medical War Crimes to Informed Consent* (Basingstoke: Palgrave Macmillan, 2004), 44.

31 Diarmuid Jeffreys, *Hell's Cartel: IG Farben and the Making of Hitler's War Machine* (London: Bloomsbury, 2008), 8.

32 Zimmerman, *Top Secret Exchange*, 4.

33 Speech cited in Ian V. Hogg, *German Secret Weapons of the Second World War: The Missiles, Rockets, Weapons and New Technology of the Third Reich* (London: Greenhill, 1999), 9.

34 On rumours see David Coast and Jo Fox, 'Rumour and Politics', *History Compass*, 13 (2015), 222–34; Brian Balmer, *Secrecy and Science: A Historical Sociology of Biological and Chemical Warfare* (Abingdon: Routledge, 2012), 70–1.

35 John Cornwell, *Hitler's Scientists: Science, War and the Devil's Pact* (London: Penguin, 2013).

36 Hogg, *German Secret Weapons*, 10.

37 Hermione Giffard, 'Engines of Desperation: Jet Engines, Production and New Weapons in the Third Reich', *Journal of Contemporary History*, 48 (2013), 822.

38 Ian Kershaw, *The End: Germany, 1944–1945* (London: Penguin, 2012), 73.

39 F.H. Hinsley, *British Intelligence in the Second World War*, vol. III, (London: HMSO, 1984), 329. See also: Michael S. Goodman, 'Jones' Paradigm: The How, Why, and Wherefore of Scientific Intelligence', *Intelligence and National Security*, 24 (2009), 236–56.

40 Richard J. Aldrich, 'British intelligence and the Anglo-American 'Special Relationship' during the Cold War', *Review of International Studies*, 24 (1998), 331.

41 Paul Maddrell, *Spying on Science: Western Intelligence in Divided Germany, 1945–61* (Oxford: Oxford University Press, 2006), 7; Brian Balmer, *Britain and Biological Warfare: Expert Advice and Science Policy, 1930–65* (Basingstoke: Palgrave, 2001), 25–7.

42 James Goodchild, 'R.V. Jones and the Birth of Scientific Intelligence', Ph.D. dissertation, University of Exeter (2013); James Goodchild, *A Most Enigmatic War: R.V. Jones and the Genesis of British Scientific Intelligence, 1939–1945* (Warwick: Helion, 2017).

43 Jones, *Most Secret War*, 25.

44 Pamela Richards, *Scientific Information in Wartime: The Allied-German Rivalry, 1939–45* (Westport, CT: Greenwood, 1994), 54–5.
45 Hinsley, *British Intelligence*, 329–30.
46 McGovern, *Crossbow and Overcast*, 11.
47 R.V. Jones, 'Scientific Intelligence', *RUSI Journal*, 92 (1947), 353.
48 TNA, CAB 21/1421, 23 January 1940.
49 Jeffrey T. Richelson, *Spying on the Bomb: American Nuclear Intelligence from Nazi Germany to Iran and North Korea* (New York: W.W. Norton, 2006), 31.
50 TNA, CAB 21/1421, 3 February 1940.
51 Jones, *Most Secret War*, 155.
52 Peter Hennessy, *The Secret State: Whitehall and the Cold War* (London: Penguin, 2003), 5.
53 Ibid., 254.
54 George Millar, *The Bruneval Raid: Stealing Hitler's Radar* (London: Cassell, 2002); Hastings, *Secret War*, 255–60.
55 For example see 'Parachutists in Action', *The Times*, 2 March 1942, 4–5.
56 Goodchild, 'R.V. Jones and the Birth of Scientific Intelligence', 262.
57 Jones, *Most Secret War*, 307–8.
58 Ibid., 315.
59 Heinrich Hohne, *Canaris*, trans. J. Maxwell Brownjohn (London: Secker & Warburg, 1979), 414.
60 A. Cecil Hampshire, *The Secret Navies* (London: William Kimber, 1978), 175.
61 TNA, WO 204/12455, 'Marine Einsatz Kommando (MEK) 80', 30 January 1945.
62 Hohne, *Canaris*, 354.
63 TNA, HW 8/104, 'History of 30 Commando – Notes on German Intelligence Assault Units', 1946.
64 TNA, WO 204/12911, 'Abwehrkommandos: Activities, Staffing, Accommodation etc.', February 1945.
65 Franz Kurowski, *The Brandenburger Commandos: Germany's Elite Warrior Spies in World War II* (Mechanicsburg, PA: Stackpole Books, 2005), 4.
66 Rankin, *Ian Fleming's Commandos*, 131.
67 John Pearson, *The Life of Ian Fleming* (London: Jonathan Cape, 1966), 161–62.
68 TNA, ADM 223/500, '30 Assault Unit and 30 Commando: Papers', 20 March 1942.
69 Hampshire, *Secret Navies*, 175.
70 Pearson, *Life of Ian Fleming*, 62. The name of this unit changed several times during its lifespan, and specific operational codenames added further variations. To avoid confusion, 30 Assault Unit, or 30AU, will be used throughout here.
71 David Nutting, *Attain by Surprise: The Story of 30 Assault Unit* (Chichester: David Colver, 1997).
72 TNA, DEFE 2/1107, '30 Assault Unit: mobilisation, control, disbandment, Honours and Awards', 23 September 1944.
73 TNA, ADM 202/308, '30 Assault Unit – Unit Diaries'.
74 Pearson, *Life of Ian Fleming*, 166.
75 TNA, ADM 223/500, 4 November 1942.
76 Nutting, *Attain by Surprise*, 202.
77 Rankin, *Ian Fleming's Commandos*, 11–15.
78 Bower, *Paperclip Conspiracy*, 86–87.
79 Longden, *T-Force*, 8.
80 TNA, HW 8/104, 'History of 30 Commando', 1946.
81 Longden, *T-Force*, 7.
82 Pearson, *Life of Ian Fleming*, 171.
83 Hampshire, *Secret Navies*, 177–78.
84 TNA, ADM 223/501, '30 Assault Unit: Targets', 1944.

85   TNA, ADM 223/349, 'No 30 Assault Unit: Target Lists for Operations in Germany', 15 December 1944.
86   Rankin, *Ian Fleming's Commandos*.
87   Lt-Cdr. Jim Glanville, cited in Nutting (ed), *Attain by Surprise*, 218.
88   Pearson, *Life of Ian Fleming*, 172.
89   Bower, *Paperclip Conspiracy*, 88.
90   TNA, DEFE 2/1107, 11 October 1944.
91   Longden, *T-Force*, 95.
92   TNA, WO 219/1668, March 1945.
93   TNA, DEFE 2/1107, 30 June 1945.
94   Alan D. Beyerchen, *Scientists under Hitler: Politics and the Physics Community in the Third Reich* (New Haven, CT: Yale University Press, 1977), 201.
95   Leo J. Mahoney, 'A History of the War Department Scientific Intelligence Mission (Alsos), 1943–1945', PhD dissertation, Kent State University (1981), 2.
96   Jones, *Most Secret War*, 601.
97   Leslie R. Groves, *Now It Can Be Told: The Story of the Manhattan Project* (London: Andre Deutsch, 1963), 191.
98   Ibid.
99   McGovern, *Crossbow and Overcast*, 96.
100 TNA, AVIA 10/70, 'Alsos mission: report', 4 March 1944.
101 Samuel A. Goudsmit, *Alsos* (New York: American Institute of Physics Press, 1996), 25.
102 Groves, *Now It Can Be Told*, 192.
103 Bower, *Paperclip Conspiracy*, 80.
104 Mahoney, 'History of the War Department Scientific Intelligence Mission', 186.
105 Groves, *Now It Can Be Told*, 221–2.
106 Ibid., 234.
107 Jones, *Most Secret War*, 602.
108 Walker, *Nazi Science*, 207ff.
109 Charles Frank (ed.), *Operation Epsilon: The Farm Hall Transcripts* (Bristol: Institute of Physics Publishing, 1993), 1.
110 See Chapters Five and Six.
111 Groves, *Now It Can Be Told*, 249.
112 John D. Hart, 'The Alsos Mission, 1943–1945: A Secret U.S. Scientific Intelligence Unit', *International Journal of Intelligence and Counter-Intelligence*, 18 (2005), 510.
113 TNA, AVIA 10/70, 4 March 1944.
114 Henry S. Lowenhaupt, 'On the Soviet Nuclear Scent', *Studies in Intelligence*, 11 (1967), 14.
115 Richelson, *Spying on the Bomb*, 62.
116 Goudsmit, *Alsos*, 18.
117 TNA, AB 1/110, 'Investigation of nuclear physics developments in Germany', 30 August 1945.
118 Gimbel, *Science, Technology, and Reparations*, 17.
119 Balmer, *Secrecy and Science*, 7–14.

2   The origins of exploitation

While the inchoate operations of 30 Assault Unit (30AU) and Alsos were unfolding rapidly and effectively in Western Europe, driven, initially at least, by pure military utility, an idea was dawning in the minds of officials in London and Washington alike that something on a much grander scale could be possible and perhaps necessary. The aim of this broader scheme would not just be to hasten the end of the war against Germany and bring about Japan's defeat, but also part of a larger strategy to ensure peace and security in Europe, particularly in the face of any future Soviet aggression. This new programme would be designed as an attempt to glean every last morsel of scientific or technical intelligence from Germany while it lay at its most vulnerable – invaded by foreign armies, society in disarray, ordinary people living in chaos and uncertainty and the Nazi political system stumbling towards its complete implosion. This chapter will chart the progress of the exploitation initiative in the final chaotic year of the war; a significant and formative prelude to the more comprehensive investigation of German science and technology which unfolded after the war's end.

As such, it will begin by exploring the original germ of the idea to exploit on a large scale and the gradual formulation of policy which arose from this early thinking and the changing military circumstances. What this policy entailed was the establishment of a complex but effective administrative framework to handle the sizeable task of full-scale exploitation of German science and technology. Once this framework was successfully formulated and once the Allies had a foothold in Western Europe in summer 1944, no time was wasted in despatching teams of expert investigators to the continent as it was liberated from Nazi occupation. These units raced forward in the wake of the advancing frontline troops and in parallel with the men of 30AU and Alsos and seized all the scientific and technological spoils on which they could lay their hands. In the last months of the war, these exploitation operatives followed the regular armies across the German border and, as the conflict reached its denouement, they began examining the myriad developments of the Third Reich's scientific and industrial war machine. This was to be the age-old notion of 'to the victor, the spoils' reconceived for the modern age and on an unprecedented scale – it was to be methodical, systematic and irresistible.

The exploitation idea

Considering its numerous successes, it is no surprise that the actions of Alsos quickly caught the attention of British officials, especially those who were tasked with planning for the post-war future. In September 1944, Colonel Geoffrey Vickers, Director-General of the Enemy Branch of the Ministry of Economic Warfare (MEW), wrote to Major-General Kenneth Strong, the British G-2 intelligence chief for General Dwight D. Eisenhower's Supreme Headquarters Allied Expeditionary Force (SHAEF), noting the missions which Alsos had been conducting. He incorrectly described it as an exclusively American force and one primarily concerned with chemical warfare (revealing the secrecy surrounding the atomic bomb at this time), but nonetheless felt that 'there might be considerable advantages to all concerned if there were a similar British team which could work in the same or adjacent fields and could arrange for an exchange of ideas and information with Alsos'. He added that such a team surely could not 'fail to be of the greatest value to those Sections of the proposed Military Government which are charged with responsibility for the administration of German industry and technical research'.[1]

This was not, however, Colonel Vickers' earliest involvement with the idea of exploitation, as he was also a member of the Joint Intelligence Sub-Committee (JIC), which existed as a subsidiary of the Chiefs of Staff committee.[2] The first indication that there was a perceived demand for intelligence on Germany after the war came about through the JIC, in a meeting in January 1944, when Vickers noted that the MEW would be keen to ascertain how much impact economic measures against Germany had made during the war and suggested that 'other departments would also have much of a corresponding nature which they wished to know'.[3] Indeed, on 29 March, the Enemy Research and Development Sub-Committee produced a paper innocuously entitled 'Post-Hostilities Equipment Policy', which outlined the central thrust of the British exploitation initiative, set out in a manner which would remain largely unchanged until the end of the war and beyond. The first element of this was the official recognition, long after R.V. Jones and his colleagues in scientific intelligence had reached the very same conclusion, that 'much German equipment is as good, or better, than ours'.[4] In May, Vickers remarked that the Admiralty had begun putting together a list of German equipment to be examined, while the JIC's chairman, Victor Cavendish-Bentinck, recommended that a representative from the Ministry of Supply be involved in any further deliberations, which suggests that technology was, by this point, very definitely under consideration.[5]

Experience from the end of the First World War also played a role in shaping future exploitation policy. In August 1944, Brigadier William van Cutsem, the former Deputy Director of Military Intelligence, who had worked closely with the Special Operations Executive during the war and was a member of a number of bodies concerned with the post-war future of Germany, filed a report on German war material, in which he recounted several tales of failure from the post-1918 period. In the case of a certain new machine gun, he recalled that 'every effort was made to obtain a specimen or at least full technical information, but without success'. Similarly, 'attempts to secure information regarding processes in the manufacture of chemical warfare products under the relevant article in the peace treaty failed dismally. The answers provided were dubbed by an Allied expert as merely "a child's guide to knowledge" and perfectly useless.' In terms of war material factories, van Cutsem noted that incomplete information had been gathered on them and, in addition, 'by the time control had started many firms had already gone over to the manufacture of peace time commodities under different names'. To avoid repeating these mistakes, van Cutsem recommended that information should be gathered both quickly and comprehensively (arguably this was easier in 1945 than it had been in 1918–19 due to the unconditional German surrender and the total occupation of enemy territory by the victorious Allies).[6]

The process of formulating exploitation policy was accelerated by the impending Allied invasion of Europe under Operation Overlord. Cavendish-Bentinck appointed a JIC Special Sub-Committee on Intelligence Priorities 'to draw up a list of the principal intelligence targets for the assault phase'.[7] This sub-committee met on 19 May and consisted of representatives from the War Office, MEW, Ministry of Supply and Air Ministry, as well as Ian Fleming of the Admiralty who, through his involvement with 30 Assault Unit, had already contributed indirectly to the preparations for exploitation. Noting that 'SHAEF were anxious to have a clear directive as to what was required from ... an inter-Service inter-Allied body', the sub-committee resolved to draw up lists of intelligence targets – the essential first step of the exploitation programme. In fact, the main outcome of the single meeting of this special sub-committee was, in an act of all too familiar bureaucratic self-perpetuation, to recommend the establishment of another committee to gather information 'either of great value to the Allies for operational purposes at present, or of such a nature as to constitute a dangerous potential threat in the future'.[8] This would become, with the involvement of the Americans, the Combined Intelligence Priorities Committee (CIPC), the first true exploitation agency.

Science and technology were now a major part of this planning, as shown in a MEW memo from September 1944. This described how 'a nation at war or planning for war stimulates, to a very high degree, research and technical developments in all its major activities. This speeding-up process produces in months what would normally take years under peacetime conditions.' The memo then continued by exploring the possibilities this would open up at the end of the war:

> After the capitulation of Germany we will have before us the results of this speeding-up process. Since this has been accomplished by the organised effort and best talent of Germany exerting all its efforts in this direction, it seems logical to assume that there are available many ideas, developments and techniques military and industrial that would benefit the Allies.[9]

It is important that these estimations of German superiority be qualified. The Allies did not believe that Germany was ahead of them in all fields, or else their eventual defeat in the war would seem rather too improbable, but the realisation was made that, in some aspects, earlier wartime Allied arrogance had been misplaced. Vannevar Bush, the influential head of the American Office of Scientific Research and Development (OSRD) throughout the war and beyond, put it most clearly when he explained that modern industrial societies advance unevenly and variously; at any given time, each will be ahead of its rivals in some, but not all, of the countless areas of endeavour.[10]

In some cases, the desire to exploit was driven forward particularly eagerly by perceived German progress in just one of these areas. For Britain, with its understandably pronounced fear of bombing, the most prominent was almost certainly the field of rocketry, in particular the revolutionary long-range offensive power of the German V-weapons.[11] Between 8 September 1944 and 29 March 1945, some 1,403 V-2 rockets were launched against Britain and 1,054 exploded on the mainland, with approximately half of these hitting their target of London. The total death toll inflicted by the V-2 rockets in Britain was 2,754, with a further 6,523 people seriously injured, as well as considerable damage to property.[12] Although the destruction wreaked by these new weapons, and by their flying-bomb predecessors, the V-1s, was significantly less than that inflicted by Luftwaffe raids during the height of the Blitz, the terror of a one-ton explosive warhead dropping to Earth faster than the speed of sound was unrivalled, and shook the courage of even the hardiest of Blitz survivors.[13]

As mentioned in the previous chapter, prior to the shock and awe inspired by these attacks, many senior military and political officials in Britain had remained dubious of Nazi Germany's ability to unleash an entirely new type of weapon, especially as the tide of war turned against them. One voice speaking out against this complacency was that of the Crossbow committee, which had been set up to handle all issues pertaining to rocket warfare in 1943 and which became increasingly prominent once the German V-weapon attacks began. Very rapidly, British views on rocketry began to change – by November, the Joint Committee on Research and Development Priorities declared that the Allies 'were now on the threshold of great changes in the sphere of ordnance. There were some who believed that the days of the heavy gun were numbered.'[14] The bold statements and predictions did not stop there. Allied technical experts believed that the V-weapons had not only changed the nature of contemporary warfare but also offered a terrifying spectacle of what a future war might be like.[15] It was correctly suspected that German experts had been working on a bomb to cross the Atlantic to attack US soil and this hinted at the potential of all manner of intercontinental missiles, which, it was estimated, could replace raids by manned bombers in ten years or less. In any case, Britain, with its direct experience of attack and advanced air defence system, was widely expected to become a 'vigorous competitor' in the post-war missile race.[16]

For the time being, however, the focus still remained very much on winning the war. This, as we have seen with the deployment of 30 Assault Unit and Alsos, was the primary motivator in establishing an exploitation initiative. This applied not just to the V-weapons and long-range rocketry, though this was an area of particular importance, but to all manner of German science and technology. The 'Post-Hostilities Equipment Policy' stated that, with any confiscated material, 'it is for our consideration whether it should be used either by ourselves or our allies, either in Europe or in the Far East.' In terms of the defeat of Germany, it would not be unfair to claim that this outcome was expected widely enough within the Allied establishment to justify planning extensively for it, even before Operation Overlord had been successfully mounted. The reality vindicated their predictions, with Allied progress across Europe, though often slow and sometimes beset by major difficulties, fairly inexorable all the way into Germany. As a result, the focus in the European theatre was merely hastening an outcome which most people in the Allied countries (and many in Nazi Germany) saw as inevitable. The Pacific theatre posed a different problem – some feared the conflict there would last a further three years after the defeat of Germany, while even the more conservative estimates did not think it would wrap up until well into 1946, and almost all forecasts assumed that a bloody battle for the Japanese Home Islands lay on the horizon.[17] Either way, it was acknowledged that specific technologies, such as carrier-borne aircraft and swimming tanks, would be very helpful in securing victory, and so any technical benefits gleaned from Germany might prove pivotal.[18]

However, bringing about a swift Allied victory in the war in both theatres was only one reason why exploitation was able to build momentum. No one involved expected the scheme to conclude as soon as both Germany and Japan had capitulated. In fact, many assumed, correctly, that exploitation proper would not begin until the former enemy countries had been defeated and were completely open to unchallenged Allied investigation.[19] One individual who acknowledged the potential which exploitation had, beyond mere military utility, was Deputy Chief of the Imperial General Staff, Lieutenant-General Sir Ronald Weeks.[20] In March 1944, he made the following statement to the Enemy Research and Development Sub-Committee:

> It is considered that the obtaining of German research records and as much information as possible of design and development projects in hand, is one of the most vitally important of our immediate post-war aims; not only would the confiscation of this information deprive Germany of many years of painstaking work, but it would also be of the greatest value to us. It may be that this is the only form of reparation which it will be possible to exact from Germany. Everything possible to ensure that it is exacted must be carefully planned now.[21]

What is particularly interesting about Weeks' statement is that he touches on two separate driving forces behind exploitation – one of which was the seeking of reparations, the age-old process of restitution exacted by victor over vanquished foe, and the other was the punishment of Germany through the removal of its valuable science and technology, in reality a precautionary measure against any future resurgence as much as a punitive one.

At the end of the Second World War, Allied statesmen and officials alike sought to avoid repeating the mistakes of the Treaty of Versailles in 1919, when unmanageably harsh financial reparations demands had crippled Germany and thus played so substantially into the hands of Hitler and the Nazis in their quest for power.[22] Stripping Germany of its scientific and technical resources, including skilled manpower, after 1945 and utilising them for the victors' own ends was one way to achieve this while still securing some recompense. In the words of Alec Cairncross, loot and slavery, the traditional forms of reparations, quickly returned to favour.[23] This approach also offered some security – before the polarised mentality of the Cold War had become truly entrenched, many feared that a German resurgence was still the biggest threat to peace in Europe and that comprehensive German disarmament should therefore feature very highly in Allied post-war priorities. This was influenced initially by historical precedent, as well as by deep-seated racial stereotyping of the German people as aggressive and militaristic and, later, by concerns that Germany could become a socialist satellite of the Soviet Union.[24] In the USA, meanwhile, the Morgenthau Plan (as advanced by Secretary of the Treasury Henry Morgenthau, Jr.) proposed that Germany be totally demilitarised and deindustrialised and reduced to a simple pastoral state – Morgenthau felt that, in light of the German treatment of others, 'they should in their turn be exploited'.[25] In this way, both reparations and retribution came to feature very heavily in early discussions around the exploitation initiative – it was a policy which would continue to be inextricably linked to divisive politics for its entire lifespan.[26]

As time went on, more perceptive British officials came to acknowledge that the real threat to European peace and stability was more likely to come from the Soviet Union than from Germany. The somewhat euphemistic term 'policing of Europe' began to crop up often in official memoranda and directives, referring mostly to a general defence against a possible Soviet hegemony over the continent.[27] As we have seen, even before the war ended, and especially once the future zonal divisions of Germany had been decided upon, British and American exploitation teams scurried to seize the best scientific and technical spoils from areas which would later fall under the impenetrable blanket of Soviet control. Science and technology, and the new weapons which they could elicit, would be of particular value to the Western Allies in order to counter what they saw as the Soviets' 'overwhelming superiority on land'.[28] In short, Britain and the US were under no pretences that the Soviets would pursue exploitation, ruthlessly and on a grand scale, and they knew they could not afford to be left behind while the Soviet Union, an ally for now but almost certainly a future rival or even foe, vastly increased its war potential.[29]

The final motivation for exploitation was somewhat more prosaic than defence of the peace or preservation of the values which Western democracies held dear. It was a purely financial aim, driven less by government officials and more by industry chiefs, who felt they had contributed, at great commercial cost, to the Allied victory and now wanted to seek reimbursement from the vanquished enemy. Initially, this pressure came mostly from the military industries, and by March 1944, the Army was already stating as one of its long-term aims of exploitation, 'the establishment of a well-founded and virile British armament industry'.[30] By September, the Ministry of Economic Warfare was planning to collect 'factual intelligence' on 'German industry, economic transport, food and agriculture, fuel, labour conditions, economic administration, prices and price control, and the employment of foreign workers in Germany'.[31] The influence which civilian industry was able to exert on what was essentially a military initiative was, to some extent, inevitable – it was the product of the complete mobilisation of total war, and no exploitation of any kind could conceivably go ahead without the uniquely specific knowledge of civilian experts.[32]

With so many motives for exploitation evident, it is no wonder that it soon became an essential part of planning for the assault on Europe and the end of the war. One of the most important manifestations of this was in the Draft Armistice, the document which it was planned to present to the German high command to bring about an end to the conflict. As early as October 1943, G.W. Turner of the Ministry of Supply wrote to Colonel C.W.G. Walker of the Post-Hostilities Planning Sub-Committee suggesting that the terms of this Draft Armistice include an instruction to the German government to 'prepare and provide at once a detailed statement of all research and development carried out by or on behalf of the German Government since the outbreak of war'. He felt they should also ensure 'the provision of information, either in the shape of statistics or otherwise, necessary to enable the controlling powers to exercise direction over economic affairs'.[33] The Draft Armistice was later replaced by a proposed Instrument of Surrender; Article 5 of which ordered that the German authorities should hold intact, and make accessible to Allied representatives, all arms and ammunition, aircraft, naval vessels, military establishments, travel and communication facilities and, most critically:

> all factories, plants, shops, research institutions, laboratories, testing stations, technical data, patents, plans, drawings and inventions, designed or intended to produce or to facilitate the production or use of the articles, materials and facilities ... to further the conduct of war.

It went on to instruct that the German authorities also furnish the labour force necessary to operate any of these facilities and to ensure that their records were 'maintained and kept up-to-date'.[34]

Of course, this rested very heavily on the uncertain premise that this directive could be successfully circulated throughout a bombed-out and dislocated Germany, and that those who received it were co-operative and obedient. Many exploitation officials had little faith in this being the case; in fact, a large number felt 'almost certain that the Germans will take every possible step to prevent the United Nations from learning their technical secrets'.[35] As a result, in a memo concerning the proposed actions of the Technical Sub-Division of G-2 Military Intelligence during the SHAEF occupation of Germany, it was considered that one of their roles would be to 'ferret out' any items which the secretive enemy had tried to 'bury' and that 'this should be done aggressively with the full military and economic strength of the Allies backing it up'. In accordance with the widespread fear of an aggressive German resurgence, the memo goes on to say that 'the enemy should not be allowed to retain any advantage, whether military or industrial, resulting from his preparations for, or activities during, hostilities.'[36]

Now that the reasoning behind exploitation had been solidly accounted for, and the official policy had both acknowledged this need and put measures into place to facilitate it, all that was left to do was to prepare for it in logistical terms. Of first consideration was the form which the agency, or agencies, responsible for exploitation on the ground would take. In April 1944, the Enemy Research and Development Sub-Committee, submitted a report to its parent group, the Joint Technical Warfare Committee, entitled 'Investigation of German Research and Development', which highlighted many of the key points of a broad exploitation plan. It started by showing remarkable prescience and recognising that there were many unknown variable factors involved when concocting such a scheme – these included: the parts of Germany to which British personnel would have access, the extent of inter-Allied co-operation, the extent of inter-departmental co-operation, and the way in which Germany would become accessible to investigators, whether by formal armistice, gradual military retreat or anarchic collapse. In short, the report ran, 'the final plan of action must be sufficiently elastic to adjust itself to these various possibilities'.[37]

Nonetheless, these uncertainties did not stop the report elucidating a very clear programme for exploitation which, though in its early stages, would later come to characterise the whole initiative, in a fairly unchanged format. Firstly, it proposed, Sealing and Holding (S.H.) Parties would proceed to specified technical targets 'when Germany becomes accessible in whole or in part and *immediately* military circumstances permit'. They would then, as their name suggested, secure the targets and defend them against counter-attack and sabotage for as long as necessary. These S.H. Parties would be almost exclusively military, though would have to include at least one technical officer for guidance, and would get first priority on travel, moving against their targets swiftly and as simultaneously as possible, 'to prevent the news of the action spreading and the enemy taking "evasive action" in unoccupied establishments'.[38] They would also be responsible for 'seeing not only that nothing in the factory is disturbed, but also that the key personnel do not abscond from the neighbourhood'.[39]

The S.H. Parties would not operate in isolation; instead, they were to work in co-ordination with Investigation of Enemy Technique (I.E.T.) Groups. These would proceed, with a military escort, directly to the facilities secured by the S.H. Parties and conduct thorough examinations there. Furthermore, in response to concerns that 'German security is good and there may be important establishments we know nothing about,' I.E.T. Groups would also proceed to and establish themselves in 'the administrative headquarters of the German R&D organisation'. From there, they could establish targets which had not been included in the preliminary plan and ensure that S.H. Parties were dispatched with haste to seal and hold these too. In order for this to work, it was considered that these I.E.T.

Groups 'would have to be composed of very highly competent scientists and technicians'.[40] Although the eventual form that the exploitation programme took was a little less neat and dichotomous than this proposal, many of the features became central to its success, most notably the complementary use of both conventional military troops and civilian scientific experts, and the allowance that some of the most important finds would be previously unidentified 'targets of opportunity'. It was clear that the officials responsible for planning and preparing for exploitation had done their job to the very best of their ability. Now all they could do was wait for the Allied armies to make sufficient headway on the continent to allow for the creation and dispatch of their proposed investigation teams.

Wartime deployment

Within a month of the D-Day landings in Normandy on 6 June 1944, it was decided that 'much valuable information might be lost during operations in occupied and enemy territory unless special measures were adopted to secure it'.[41] This concern, raised by SHAEF, fed directly into the creation of the proposed Sealing and Holding Parties, which were given the actual designation of 'Target Forces' and were more commonly referred to as T-Forces. The Director of Naval Intelligence proudly noted that 'the Admiralty had been a pioneer amongst the Allies' as SHAEF were 'thinking in terms of a force of some two divisions trained and manned on lines similar to 30 Assault Unit'.[42] Though sometimes for the wrong reasons, 30AU had caught the attention of a number of senior SHAEF planners and now provided a model for the training and technique of the T-Forces. These new units were comprised of intelligence specialists, prisoner-of-war interrogators, linguists, engineers, bomb disposal experts and a bulk of combat personnel and were attached to the 6th, 12th and 21st Army Groups.[43]

The prescribed role for the T-Forces, following on from the original designs for the S.H. Parties was to be primarily a military and logistical one. It was split into four key tasks, thus:

a   Moving in the immediate wake of the assaulting forces
b   Locating and securing intact the targets concerned
c   Preserving them from destruction, loot, robbery and, if necessary, counterattack until the completion of their examination by teams of experts or until the removal of the essential installations or documents
d   In enemy territory, providing armed escorts for the expert investigators[44]

For many regular soldiers, the troops of the T-Forces were a welcome sight, arriving to defend properties which were often some distance from the frontline, allowing the ordinary forces to move on to their next objective or to relax at the conclusion of a particularly tough advance. This is not to suggest that the work of T-Forces was easy or static. In reality, 'the highly mobile T-Forces ultimately had to cover the whole area of operations in the course of a few hectic weeks.' As such, they were subject to much of the same criticism as 30 Assault Unit, no doubt largely down to the fact that they were an unusual formation that did not conform to the same norms as the majority of regular fighting men, and were often referred to as a 'private army'.[45]

With the T-Forces surging forward and seizing all manner of targets, based on a haphazard set of priorities provided by a number of different agencies, and roughly co-ordinated by their headquarters division, it was now necessary to furnish the structure and manpower to facilitate proper investigations and to ensure that nothing of scientific and technical value was missed. On 12 June 1944, less than a week after D-Day, General Eisenhower cabled the Combined Chiefs of Staff, stating that:

> Need has arisen for an Anglo-American Inter-Service Organisation to deal with Anglo-American requirements for technical intelligence ... [to] include such material, personnel and information of military importance, either of great value to the Allies for operational purposes, or constituting a dangerous potential threat in the future as to justify urgent action on the part of the Allies in seizing them.[46]

This organisation, which would have responsibility for drawing up lists of targets, arranging the dispatch of troops (primarily T-Forces) to seal and hold them, and of investigators to assess them, and of the dissemination of the resulting intelligence to all concerned parties, was the Combined Intelligence Priorities Committee (CIPC).

The CIPC was made up of representatives from seven British and seven American departments – for the British these were the Foreign Office, the Ministries of Economic Warfare, Supply, and Aircraft Production, and the Intelligence sections of all three Armed Services; for the Americans: the State Department, the Foreign Economic Administration, the Office of Strategic Services, the OSRD and the three Forces' Intelligence divisions.[47] As this list of constituent agencies would suggest, the intelligence with which CIPC was concerned lay 'between normal technical intelligence obtained by established means during operations and intelligence of a clearly non-military nature. It thus includes political and economic items so long as they are of military importance.'[48]

In this role, CIPC was remarkably short-lived. On 22 August 1944, all of its members and most of its responsibilities were migrated to the Combined Intelligence Objectives Sub-Committee (CIOS), which met fortnightly from that date until the dissolution of SHAEF, after the cessation of European hostilities, on 13 July 1945. It was chaired by the American Brigadier-General Thomas J. Betts and had as its Deputy Chairman British Professor Reginald Patrick Linstead, an esteemed organic chemist then serving as deputy director of scientific research at the Ministry of Supply (and who would later go on to become Dean of the Royal College of Science and Rector of Imperial College, London).[49] Linstead's involvement was indicative of the influence of experts on policy, even at such a senior and secretive level, which had proved to be a highly successful characteristic of Britain's war effort up to this point.[50] It is important to note that CIOS was, of itself, not an investigating agency and did not have a permanent staff of exploitation personnel of its own. Rather, its role was primarily a supervisory one – it was responsible for aiding its member agencies, co-ordinating and facilitating their efforts and settling any disputes between them. In its own official retrospective report, published just two months after it ceased to exist, CIOS was described both as 'an instrument' and as the 'means whereby intelligence and information were pooled, and the burden of investigation shared'. As such, its Secretariat, crucial to serving these purposes, which began as just one British and one American officer, had expanded by the end of the war to a total of 25 officials and 58 enlisted and civilian personnel.[51]

The separation which existed between CIOS in London and the T-Forces in the field caused some issues. Despite the diverse make-up of the T-Force units, they were widely considered, perhaps a little unfairly, to be 'merely dumb soldiery' and therefore not properly equipped to assess the worth of any particular target, especially if it was seized by opportunity and had not featured on the initial designated lists.[52] In response to this, the idea of 'CIOS Forward Observers' was mooted – they would not be experts in any one particular area but would be located forward in the field 'to ascertain the definite existence of CIOS targets, their physical condition, to reassess priorities, and to follow up all lines of investigation'.[53] C.H. Noton, of the Ministry of Economic Warfare, was particularly eager to see these reconnaissance officers deployed, as he feared that T-Force in 21 Army Group was being overstretched in the 'near area' of operations and would thus have no men to spare to secure targets when important German cities, such as Cologne, fell to the Allies; a 'misdistribution [which] would be caused entirely by the absence of advisers' who could inform the T-Force commander 'which CIOS targets should continue to be guarded pending the despatch of an investigating team from London'.[54]

The eventual form which this forward exploitation reconnaissance took, in provision for the expected sudden expansion in the number of targets available which would come when Germany's resilient Rhine defences finally collapsed, was that of the Consolidated Advance Field Teams (CAFTs). These attached 'a limited number of qualified specialists' to the advancing spearheads, with the following remit:

> As a target was seized, the CAFT assessors quickly appraised it and advised the combat commanders whether its importance merited the assignment of guard troops ... Targets meriting further investigation were [then] exploited by specially qualified investigators dispatched by CIOS.[55]

It is important to remember that, although the process described here suggests a smooth collaborative effort, shared by the different agencies and teams to ensure that comprehensive exploitation was conducted, the wartime reality was a little different. One American aerospace expert who was directly involved later recalled the 'confusion, chaos and mutual distrust' which characterised the 'teams composed of scientists, engineers, soldiers and sometimes fools' as they 'dashed competitively about … impounding documents, drawings, laboratory equipment, [and even] whole laboratories'.[56] In this period, CIOS rather tellingly received the nickname CHAOS.[57]

All the endeavours of all the exploitation organisations were guided by lists of targets. The entire initiative would have been completely unable to function without them. Included as targets were such things as industrial firms, factories, laboratories and research facilities, scientific and technical institutes, military installations, universities, and individual scientists and technicians, all of which the T-Forces were tasked with securing.[58] The targets of top priority, those which were valuable 'for operational purposes, or constituting a dangerous potential threat in the future', featured on the so-called 'Black List'[59]. Devising this list, and making sure it was accompanied by sufficient information about the targets it contained, was arguably the greatest challenge faced by CIOS in this period.[60] Data for the Black List was 'provided by the Service Intelligence Directorates, MEW and anyone else, e.g. scientists of the Supply Ministries, who can help'.[61] In some cases, if intelligence about specific items was not available, the requests filed by the relevant departments would be formulated in broad terms in the hope that CIOS teams might come across pertinent material while investigating other targets.[62] As investigations progressed and more intelligence was duly gathered, the Black List was altered and updated to reflect this, especially as intensified Allied bombing and the encroaching Soviet advance on the Eastern Front meant that many important German scientific and technical facilities were relocated in great haste.[63]

To give some idea of scale, the Black List in circulation in August 1944 contained a total of 1,118 targets; of which 167 were considered 'Priority One', 271 were 'Priority Two' and 680 were 'Priority Three'. The total number were spread across all of northern and western Europe, with targets mentioned in Germany, France, Belgium, the Netherlands, Norway and Denmark. Over half of the total targets listed (658) were in Germany, while for Norway there were only nine. Furthermore, every target belonged to one of twenty-eight different categories, which ranged from obvious military concerns, like 'Directed or Controlled Missiles' and 'Chemical Warfare' to more general subjects, such as 'Metallurgy' and 'Physical and Optical Instruments and Devices'; there were also separate designations for more miscellaneous groupings, including 'Documents and Personnel' and 'Instruments and Equipment'. These categories, and the careful separation of targets not only by country but also by sub-national location, were essential to ensure the right experts were allocated to investigate facilities which their knowledge and experience best-equipped them to assess.[64]

Despite the three priority levels within the Black List, it still only contained targets of direct and immediate importance. A separate Grey List was created to deal with the multitudes of less pressing subjects and CIOS described it thus:

> The difference between Black List and Grey List targets is essentially one of military urgency. The interpretation of the word 'military' in the Grey List should be very wide and would include targets of a general economic,

industrial, commercial or political interest, even if their purely military value were only secondary.[65]

Practically speaking, targets of economic interest alone were not considered to be important enough 'to justify the employment of armed forces solely on their behalf' and no T-Force protection was accorded them.[66] In terms of science and technology, the rough distinction between Black and Grey Lists was that the former concentrated interest more on 'end-products, e.g. a new tank, torpedo, or jet engine plane, and the power and limitations of that instrument' while the latter 'generally focused on industrial techniques, methods, "know-how"'.[67] This interest in items of indirect utility was not purely academic, it was noted from 'past experience' that 'the control of Libraries, as well as Archives, will be of great importance in order to break up the German Military Machine.' These libraries and archives included those of major Nazi governmental bodies, such as the Propaganda Ministry and the German Labour Front, as well as Wehrmacht, SS, and SA headquarters.[68] Certainly, the broader remit of the Grey List meant that it had a much grander scale than the Black List – a CIOS report from 28 December 1944 predicted that the eventual number of Grey List targets might be as high as 10,000.[69]

An interesting case study of the distinction between the Grey and Black Lists, and the process involved in splitting targets between the two, is that of the German Patent Office. Victor Cavendish-Bentinck, chairman of the Joint Intelligence Sub-Committee (JIC), recommended to CIOS that the Patent Office be considered a Black List target. Linstead replied that he felt it was 'a very important long range target, but not one of urgent military importance'. He conceded however that its inclusion on the Black List would hinge on 'the likelihood of it containing secret military patents', as the British Patent Office did, and he agreed to look into it.[70] Just over two weeks later, Brigadier John G. Foster, the Chief of SHAEF's Legal Branch, received word from his Special Legal Unit that the German Patent Office did indeed handle secret military patents and, furthermore, that they were considered of such high value that, in Germany, 'disclosure of secret patents constitutes an act of treason'. As a result, the German Patent Office became a Black List target, albeit of the lowest priority.[71]

While the bulk of this planning was geared towards the exploitation of Germany, CIOS was keen to have investigators at work in the occupied countries, where plenty of research directed by the Nazis had taken place. Even if nothing of particular utility could be unearthed there, it was still hoped that it would give a clearer idea of what the assessors would find once the borders of the Reich were breached and thus allow a more accurate and detailed Black List to be devised. It was believed that France, Holland, Czechoslovakia, Norway, Denmark, Belgium and Luxembourg might all have sources of 'useful technical information related to the German war effort' to examine. Indeed, Dutch firms developed radios for the Luftwaffe and the Belgian electronics industry collaborated so wholeheartedly with the Nazi authorities that the German firm AEG began to fret about its Belgian competition.[72] British officials had no qualms about ruthlessly exploiting research conducted in countries which were ostensibly their allies and fellow members of the United Nations, whose governments were often in exile in Britain. Their proposed policy was that:

> The examination of establishments and records in these countries on their liberation should be carried out along the same general lines as those laid out for Germany. It is considered the Allies should take a firm stand in this on the grounds that all European research has been for the Germans and is therefore a legitimate prize of war.[73]

The term 'legitimate prize of war' was burdened with a considerable degree of significance in international law and would later form part of the major debate about the legality of Allied exploitation programmes.[74]

The first CIOS-sponsored mission on the European continent was to Paris on 28 August 1944, a mere three days after the French capital had been liberated. From the outset the mission was beset by difficulties – eight individuals who were supposed to take part did not even show up at Northolt Air Field for the flight to France, delaying it by over an hour. Of the 52 specialists who did attend, several had not been properly briefed and 'many individuals did not know why they were going to Paris'. Upon arrival, the transport provisions from Chartres Airport to Paris, and within the city itself, were described as 'inadequate'. When the investigators needed to contact their parent agencies in Britain, communication was difficult and it took over five hours before contact was established with them all. Despite these numerous shortcomings, CIOS remained optimistic, summarising that 'on the whole it appears that the mission was successful'. After all, this was to be the first of many such missions and not only had 'much valuable information … been obtained' but in addition, 'the lessons learned by everyone concerned in the Paris operation should go far to making subsequent operations more efficient'.[75]

The first report to emerge as a result of the Paris mission was on 'Radar and Guided Missiles', and particularly German research and development, in this field. To this end, the investigators had decided to 'confer with engineers of the French radio companies and to inspect the work they had done for the Germans to attempt to learn the information they and their companies had gleaned from the Germans in either direct or indirect association'. This proved to be a fairly successful method although the assessors did encounter difficulties with 'companies which preferred not to reveal their own research', as, despite bold statements about seizing the 'legitimate prizes of war' from all quarters, British and American investigators did not possess the authority to compel private French enterprises to share their work with them.[76]

The Paris operations were swiftly followed up by further CIOS investigations in Brussels, Eindhoven, Vlissingen and Strasbourg as soon as these target areas were occupied by Allied armies. Eindhoven was visited by a previously selected field team within 24 hours of its liberation.[77] In the case of Strasbourg, a large city which had multiple targets of interest, CIOS formed an experimental 'Intelligence

Assault Force' to carry out the most efficient exploitation, reflecting the flexibility of the scheme and the pressure the agencies were under to act effectively in a fairly small time window.[78] In addition, CIOS also made early perfunctory excursions into Germany; at Aachen, which was the first German city to come under Allied control on 21 October 1944, and at nearby Stolberg.[79] Unsurprisingly, one of the main concerns of the CIOS teams at this stage, while the V-weapons were still striking targets in Britain and elsewhere, with steadily increasing accuracy, was to stop these attacks and learn more about the advanced rocketry involved at the same time. In early November, CIOS investigator Lieutenant-Colonel Greatbatch travelled to Holland and, near Eindhoven, met with a Dutch Resistance fighter who had supposedly witnessed some V-2 launches in the area. His account suggested that the launches could be made non-vertically from ordinary roadways, but were very costly in German lives, estimating (enormously inaccurately) that 'on average, one man of the crew died from severe burns for each launching that took place'. He also passed on a rumour, which remained unverified and seems highly unlikely, that some of the V-2 warheads were filled with anthrax.[80] This is just one example of the degree of uncertainty and conjecture on which CIOS teams often had to rely during their work.

This was by no means the only difficulty which the exploitation initiative faced during its early phases of operation. Despite the best intentions of the co-ordinators in London, there was still considerable distance between the straightforward soldiers of the T-Forces and the scientific and technical experts sent in by CIOS. The CAFT system meant that 'large numbers of specialists ... were able virtually to remain at their desks and drawing boards until required for a specific task' which, though it was more efficient, meant that many arrived in the field 'expecting apparently not only a chain of hotels in which to live, but also unlimited supplies of transport, clerical facilities, interpreters and the like'.[81] Instead, the best they could hope for were 'rough and ready' investigators' messes and load-carrying vehicles, and even these were far from easy to come by. The CIOS assessors' complaints about everything from comfort to cleanliness did not often sit well with the fighting men of the T-Forces, many of whom had been directly involved in D-Day and the subsequent fierce combat across western Europe.[82]

This was by no means a one-sided story though. The civilian investigators were not always simply unused to 'war at the sharp end', but genuinely did suffer in ways which not only made them very uncomfortable but also prevented them from doing their job effectively. In December 1944, Arthur R. Stella, an investigator sent to mainland Europe by the Economic Advisory Branch (EAB), wrote to C.H. Noton – the Ministry of Economic Warfare's representative on CIOS – enclosing a report which he had entitled 'Comments on difficulties experienced as a civilian investigating targets in or near the battle zones'. He detailed how, as a civilian, he was given none of the appropriate battlefield clothing, none of the relevant medical inoculations, no financial aid and no means to purchase small comforts, such as cigarettes, biscuits or sweets. In addition, his lack of rank and full military apparel meant that he was often denied access to crucial targets of interest. This resulted in him having to borrow equipment and supplies from his military companions, having to 'depend on the kindness of other members of the team for small comforts' and he even admitted to assuming 'fictitious army ranks in certain cases'. Though it is perhaps hard to feel too much sympathy for Mr Stella, who cannot truly have expected too many comforts so close to the frontline, it is worth noting that his civilian status actually hindered the progress of exploitation. Stella concluded his report by recommending that no other EAB member 'be sent to the forward areas without being fully provided with all the requisites which, from the above, appear to be indispensable for the successful carrying out of duties in the conditions referred to'.[83]

Logistics also provided a major difficulty for the planners of exploitation. Moving such a considerable quantity of men from Britain to targets near the frontline, when transport was so direly needed by pretty much every other element of the armed forces was a particular challenge. This situation was grossly exacerbated by the actions of the agencies and investigators themselves, who all too regularly changed plans and travel details after the necessary Army Group approval had been obtained. Complaints were raised that these last-minute alterations were 'seriously jeopardising the co-operation being shown by ... Air Transport Companies', especially when the inclusion of CIOS investigators often resulted in the removal of lower-priority passengers on these transports. This, in turn, led to fears that it would become very difficult to 'retain the co-operation of the transport companies in genuinely urgent cases where additional investigators must be booked'.[84] Furthermore, another resource which was almost in as high demand as transport was interpreters. German speakers, ideally technically trained, were an essential part of the exploitation process, but they were distinctly scarce and, in many cases, improvisation was necessary, such as drafting in any T-Force transport personnel who could speak German.[85] However, as more and more targets needed investigation, CIOS had to start looking to the civilian population at home, where they hoped to find 'a number of older German-speaking personnel who would be unfit for any form of military service but who would be suitable for interpreting work under static conditions and could hence accompany the investigators'.[86]

One problem that the exploitation planning staff had anticipated, but which did not materialise to the expected extent for the men on the ground, was German efforts to impede Allied investigations. T-Force reported that 'in no place were [files] found intentionally disarranged or concealed', though there was some evidence of attempts to burn documents or transport them to secret locations.[87] For example, fourteen tons worth of boxed documents relating to the V-2 programme had been hidden in an abandoned mine on the northern edge of the Harz Mountains in April 1945, but the purpose of this concealment was not so much to prevent the Allies from acquiring the material as it was to give the German experts a bargaining chip when negotiating their own post-war futures.[88] Many such individuals could obviously see that the end of the war was coming and were shrewd and pragmatic enough to do what they could to inflate their value to the imminent victors. There were certainly other groups who posed a greater threat to the successful execution of the exploitation mission. A report on T-Force activity with the 2nd British Army commented that 'the principal damage was done by displaced persons in various states of inebriation, with our own troops running a bad second' and that these proved to be 'a far greater menace to the security of targets than any attempt on the part of the enemy to destroy their contents'. In fact, the report concluded, the careless destruction of material by these groups 'was to a large extent nullified by the whole-hearted co-operation of German directors and scientists and by the skill of interrogators'.[89]

In fact, the French often threw up more obstacles to CIOS operations in France than the Germans did to those taking place in Germany. Wing-Commander T. Jackson of the Overseas Aircraft Control wrote to Air Commodore C.M. Grierson, SHAEF's Assistant Chief of Air Staff, recounting the story of a CIOS party which arrived in France to investigate aircraft factories. Jackson took the captain heading this party to the French Air Ministry to get clearance, and once there:

> he caused a certain amount of alarm and despondency by suddenly announcing that MAP [Ministry of Aircraft Production] had instructed him not to confine himself to the collection of intelligence concerning the enemy, but also to study French production, and interrogate personnel in French factories regarding types of aircraft now being produced by the French for the French.[90]

The French were then understandably reluctant to grant any permission for such investigations and this was far from an isolated incident. Reports stated that many CIOS field teams had been stranded in Paris, unable to secure clearance to visit any of their targets within France. Moreover, it was believed that French manufacturers had received formal instructions to release no information to foreign military, naval or civilian personnel without specific authority from the French government. The only hope for a thawing of relations was considered to be 'if proper machinery and contacts are established and further that an element of reciprocity exists'.[91] The idea was that if the French were allowed to learn from the British and Americans, then perhaps they would be more willing to share their own secrets.

French reticence was not the only cause of strained relations between exploitation personnel and the people of recently liberated countries. CIOS operatives' often close contact with the local populace meant they were sometimes treated as reliable channels of information on Allied conduct of the war, especially when the targets under investigation were located particularly close to the frontline. In November 1944, three CIOS officers who had been investigating a V-weapon launching site near Watten, in the Pas-de-Calais, left the area and word got out to the local people that their departure was due to the fact that the US Army Air Force (USAAF) was set to bomb the site imminently. The officers were chastised for creating 'alarm and despondency' in the area and were told to 'refrain from making such depressing forecasts' in the future, thus highlighting the delicate nature of interactions between exploitation operatives and local residents.[92]

Despite not coming into existence until the end of August and having to operate while the war was still raging, CIOS achieved a considerable amount during 1944. It dispatched 197 investigators from all fourteen of its constituent departments across Europe, to examine 115 Black List items and many more targets of opportunity. No casualties were suffered at all during this period.[93] Of course, CIOS was aided throughout by the logistical capabilities of the T-Forces, without whom it was assumed 'a great amount of invaluable information would irretrievably have been lost'.[94] The product of this first wave of exploitation was that 211 reports were filed with the CIOS Secretariat; reports which were then made available to all concerned parties. In fact, dissemination of the reports had to be increased on account of greater interest 'arising from the circulation of the lists of titles'. In concluding the final report on CIOS activity in 1944, the British Deputy Chairman R.P. Linstead commented that continued exploitation was 'essential' if information was to be obtained on 'new weapons which may imperil the future security of the United Nations, and on new technical discoveries which may assist in our own developments'.[95]

Looking to the future, Linstead went on to say that 'the results of the hard work in planning and exploitation are now becoming apparent and the record to date promises well for the future investigations in Germany proper.'[96] All attention within CIOS, T-Forces and the numerous other bodies concerned with exploitation was now focused in this direction. Their investigations in the occupied countries and the few brief incursions across the crumbling borders of the Reich had been impressive in their own right, but had also most definitely whetted appetites for the spoils of Nazi science and technology which would be accessible as sizeable amounts of German territory fell into Allied hands, undoubtedly containing countless laboratories, factories and research establishments. The exploitation officials did not have long to wait.

Into the Reich

In the first few months of 1945, Western Allied forces began to seriously threaten, and then broke through, the German defences along the Rhine. By late March and early April, British, American, Canadian, French and Polish troops (among others) began pouring across the border of the Reich and surging eastwards. The Germany that they encountered was a land of desolation: cities bombed to rubble, a population living without adequate access to food, power or shelter and death in abundance. The Wehrmacht, by now well aware that victory or even an agreeable stalemate were out of the question, fought on regardless, fighting a war without real strategic considerations, but simply for its own sake, obeying the orders of a crumbling but still dangerous regime.[97] While the primary consideration of the Allied forces was obviously to bring the war to a swift end and secure a satisfactory peace in Europe, the exploitation initiative was not neglected. Rather, it too, like the war effort, was increased as the vast scientific, technological and industrial spoils of Germany became more accessible.

The T-Forces moved ahead with, and occasionally in front of, the main combat troops and seized targets on the Black List, as they had done in the campaign across France and the other occupied countries. CIOS teams were rapidly dispatched to investigate these targets, usually only a very short time after the fighting had ended. This was certainly true of Cologne, where investigators were already at work on the American-occupied left bank of the Rhine by late March 1945, while districts on the right bank remained in German hands until the middle of April. Lieutenant-Commander John N. Bradley of the Royal Naval Volunteer Reserve (RNVR), who was deployed on investigations in Cologne and the surrounding area, wrote home to his wife Margaret on 23 March and recounted the precarious nature of his situation. He described how it had been impossible to secure a guard from the Americans for the works they were visiting, so they had had to act as guards themselves. Fortunately, he also noted how obliging the local German people were, who would 'literally do anything you want, including putting up beds for us in the works manager's office'.[98]

Bradley also described the entry into Lübeck, following a rapid advance, and noted that 'never have I seen such chaotic and fantastic scenes'. He was among the first Allied troops to enter Kiel, an accolade that has been claimed variously by 30AU, T-Force and the Special Air Service (SAS), and it was to the latter that Bradley was attached. Despite these exciting and daring exploits, which Bradley related home with a mixture of relish and discontent, he soon found that 'my sphere of usefulness was over now that the ranking experts had arrived. I seemed to be doing the duties of a liaison officer and general runabout.' As soon as important sites were secured, as Bradley quickly learned, it was a matter of great urgency to get the appropriate experts involved.[99]

Of course, during these final months and weeks of the war, one of the targets of greatest importance was Berlin. Although many of the key scientific and industrial facilities had been evacuated out of the capital, either as a result of bombing or out of fear of Soviet capture, all the Allies were convinced that the heart of the Nazi regime would still contain some of the greatest spoils of war.[100] T-Force preparations for the assault on Berlin began in the summer of 1944 when a planned airlift of 119 specialist officers, to arrive on the day the city fell to the Allies, was deemed adequate. This even allowed for the inclusion of three 'specialist police officers' who were to be responsible for restructuring the German *Kriminalpolizei* and gathering its intelligence on international criminals, as well as the exploitation of 'all technical and scientific aids for the prevention and detection of crime'.[101] By November, expectations had drastically increased and it was estimated that, to be effective, the Berlin T-Force would need to number 'a minimum of one infantry regiment plus two battalions, a battalion of engineers, a military police company and signal troops'.[102]

In January 1945, plans had reached an even more advanced stage of development. Conventional military personnel would not be sufficient, it was felt, and special teams were thus deemed necessary. The form that these teams would take was laid down as follows:

> 8 Microfilm Teams, 8 Document Teams, 5 Interpreter Teams, 4 Interrogation Teams, 1 Safe Breaker Team, 4 individual females to search females held

at Detention Centre, and 30 individual guides who speak German and are familiar with the city. These Special teams and individuals will comprise approximately 40 officers and 160 other ranks.[103]

In addition to this, it was decided that some 500 'attached specialists will be provided by various interested intelligence agencies when needed. Some will accompany T-Force into the city; others will arrive later.' In all, the picture that emerges is one of a carefully planned operation on a vast scale and all this was borne on the assumption that T-Force troops would 'be used to seize and hold only *intelligence* targets, of which there are more than 1,300 now listed'.[104] However, the Soviet seizure of Berlin on 2 May 1945, after a period of vicious fighting, meant that these plans were made in vain and no exploitation airlift into the capital ever took place.

Less than a week after Berlin fell, the war in Europe ended. In the early hours of 9 May 1945, representatives of the four main Allies (Britain, USA, Soviet Union and France) and representatives of the three parts of the German armed forces (Army, Navy and Luftwaffe) signed a document of capitulation and brought hostilities to a conclusion.[105] At the Yalta Conference, held in Crimea in February 1945, Churchill, Stalin and Roosevelt had agreed on the division of a conquered Germany into three roughly equal zones (a fourth zone was later allocated to the French).[106] At the end of the war, the armies of the Western Allies had advanced further east than these prearranged zonal boundaries and it was July before this territory was duly ceded to the Soviets and all the Allies took formal control of their respective zones. As such, between April/May and July, particularly high priority was accorded to exploitation of those targets in areas of Germany which had been occupied by British or American forces, but which were very soon to be handed over to Soviet control, primarily in Saxony and Thuringia.[107] From July onwards, when the theoretical divisions became official borders, the politics of mutual access for foreign investigation teams to each zone added a new and complex dimension to exploitation which would persist throughout the period.[108]

Once the war had ended, it became clear that the exploitation activities conducted in the formerly Nazi-occupied territories had been a mere prelude. The German authorities had been reluctant to relocate many sites of scientific or technological significance outside of Germany, so the majority of important targets remained within the borders of the Reich. Even once these borders had been crossed by the Allies, accessing these targets had not always been easy while conflict continued, but in peacetime access and investigation swiftly became more viable. In addition, once the war was over, military targets no longer had such an exclusively high priority and interest grew in industrial and economic targets. This created issues for the current system of co-ordination, handled by Special Sections Sub-Division of G-2 SHAEF, which had been in place since the first Allied landings in Europe, but which was not sufficiently equipped to handle this sharply increased demand for exploitation.

As a result, it was decided that a more powerful organisation than Special Sections Sub-Division was needed and that this new body 'should have a technical staff and include a reference library and card-index of all relevant reports and enemy documents'. On 31 May 1945, therefore, a SHAEF Directive formally established the Field Information Agency, Technical (FIAT) – a combined Anglo-American effort under the auspices of G-2 SHAEF. Its commander was the British Brigadier Raymond J. Maunsell, who had spent much of the war in Cairo, where he had been the founder and head of Security Intelligence Middle East.[109] FIAT would become a central part of the exploitation machinery for almost its entire post-war duration.[110] Although it would later be split into separate, but complementary, British and American elements, it continued to serve as a guiding force throughout. Its stated purpose, in broad terms, was 'to provide for the seizure, freezing and exploitation of intelligence targets of scientific, technological and economic interest in enemy territory, to deal with which was outside the interest and beyond the competence of the troops and staffs of field formations'.[111]

Such a brief summary does not really reveal the true complexities of FIAT's role, which served to reflect the curiously convoluted and diverse nature of the exploitation programme, especially once it truly got off the ground. A central part of the FIAT mission statement was that 'it does not itself undertake intelligence investigations, but merely provides reference and administrative facilities for those who do undertake them.' It comprised six branches: Scientific and Technological, Industrial, Economic and Financial, Naval, Army, and Air, and there was also an Integration and Planning Branch which 'knits together the operations of the six technical branches, is responsible for the general co-ordination of technical operations and the allocation of technical priorities'. In order to fulfil these multiple functions, FIAT was outfitted with an Interpreters Pool, a Publications Branch (for printing and disseminating reports), a Records Branch (to keep records of all FIAT operations), an Enemy Documents Branch, an Administration Branch and a Control Branch (which handled transportation, movement of supplies and obtaining necessary clearances).[112]

At a meeting of CIOS in early June, deputy chairman R.P. Linstead commented that FIAT 'was designed to strengthen rather than in any way to disturb the existing CIOS machinery for obtaining intelligence'.[113] Certainly, with the FIAT structure in place, it was possible for CIOS to begin exploitation in earnest. In January 1945, during a typical fortnight, CIOS despatched fewer than a dozen investigators; by the end of May, a typical fortnight saw the despatch of more than 250. This was in addition to the 240 CAFT assessors who had been in the field continuously since late March. The 'CIOS Progress Report for 1945' noted that the investigations had borne a 'rich harvest', with results which were 'indeed better than those who planned the operations had dared to hope'. The report went on to assert that the role of CIOS had been twofold: 'to get technical experts to the most important places, and to get them there quickly'. CIOS had successfully achieved both of these aims and, had they not, it was suggested, 'much priceless material would have gone underground, never to become available again'.[114]

Nonetheless, irrespective of how successful CIOS had been in its eleven months of operation, it could not realistically be expected to last indefinitely. It was, after all, a product of the distinctly Anglo-American character of SHAEF and now that the war was over, it seemed likely that SHAEF would cease to exist and that CIOS would have to follow suit. The full programme of exploitation had barely begun so there was no doubt that some new organisation would need to take its place. Even as early as April, while the war was still being fought, the demise of SHAEF was foreseen and discussions about how exploitation would continue were conducted. The central proposal was that 'each controlling power will handle intelligence objectives in its own zone' and that they would each 'afford facilities for representatives from the other controlling powers to visit these objectives'.[115] Though this notion was couched in fairly generic terms, it did indeed form the basis for the future of exploitation in the post-war period.

Having served its purpose in co-ordinating the Allied military offensive on the western front in the European theatre, SHAEF was due to be dissolved on 13 July 1945. Brigadier-General T.J. Betts, the chairman of CIOS commented that 'this would affect CIOS in the sense that CIOS had been created to work with SHAEF and would raise the question of the continuation of CIOS after SHAEF. He felt that it was most desirable that CIOS should continue in the post-SHAEF phase.' On 22 June, Field Marshal Bernard Montgomery, Commander-in-Chief of the British Army of the Rhine (BAOR; the British occupying force in Germany) received a telegram from the Combined Chiefs of Staff informing him that CIOS would be terminated along with SHAEF, and that in order 'to provide interim arrangement for continuing exploitation of intelligence targets ... you authorise the US investigators to visit targets in your zone and that you exchange intelligence procured from such targets with the commander of US zone'.[116]

No such interim arrangement was ever necessary because, by the time SHAEF and CIOS were terminated, new organisations were already in place and prepared to shoulder the burden of exploitation. In fact, on 18 July 1945, only five days after SHAEF had been liquidated, Squadron Leader S.M. Harris, acting secretary of the Joint Intelligence Sub-Committee wrote to all British members of CIOS and informed them that 'the functions of CIOS, so far as the British interests are concerned, shall be continued by the British portion of the CIOS organisation under the title of British Intelligence Objectives Sub-Committee (BIOS).'[117] It was a remarkably smooth and primarily administrative, changeover and marked the transition from a period of joint Anglo-American exploitation to unilateral, national programmes, the relationship between which would vary between extremes of co-operation and competition.[118] A month and a half later, having only been in existence for three months in total, FIAT also split into two components; one American and one British.

CIOS may have ceased to operate as a central body, but investigators sent out under its auspices continued their work throughout the summer of 1945. Its final report, printed in September in the form of a book entitled *The Intelligence Exploitation of Germany*, noted that CIOS had despatched a total of 2,197 personnel, of whom 1,876 were investigators and 321 were CAFT assessors.[119] In the course of their investigations, these experts visited 3,377 different individual targets and filed 58 final reports.[120] These figures are impressive in their own right, but even more so when considered in context – prior to CIOS, 'no planned and co-ordinated exploitation of enemy technical intelligence had ever been attempted'; in addition, CIOS had only been operational for eleven months and for nine of those, the war with Germany was ongoing. Brigadier-General Betts and Professor Linstead wrote in their foreword to this report that 'the value of the scientific knowledge and "know-how" thus obtained cannot now be fully measured' but 'the benefits of this knowledge to British and US industry will be measured in terms of economic progress and well-being for many years to come.'[121]

As we have seen, the roots of the exploitation initiative were firmly entrenched in the Second World War and were nurtured both by the growing prominence of science within the strategic considerations of British military planners and by the changing nature of intelligence, which became increasingly preoccupied with the contents of foreign arsenals, especially those of potentially hostile nations. From these influential but initially inauspicious beginnings and guided by the experiences of its first iterations – 30 Assault Unit and Alsos – the programme grew rapidly and underwent a remarkable transformation throughout the last year of the war. Gone was the dominance of these small agencies, operating under military authority and racing from target to target in a desperate pursuit of the next piece of valuable technical intelligence, with their main aim being to facilitate a swifter and safer end to the war against Germany. Their place had been taken by much larger and more interconnected organisations, with control and oversight in the hands of the civil service as opposed to military authorities and thus subject to a greater number of regulations, but also with access to a much larger pool of resources, including investigators drawn from very specialised non-military backgrounds. Those responsible for this new phase of exploitation were not just looking towards the end of the war, but looking beyond it, and trying to find ways not to curtail their operations at the point of Germany's unconditional surrender, but to expand them.

This transitional period was also one of growth, both in terms of the scale of the endeavour's objectives and in terms of the machinery necessary to achieve them. It was during the last months of the war that T-Force began operating and, in so doing, developed a skillset and strategic approach which would continue to facilitate the smooth running of exploitation missions throughout the postwar period, as well as providing an outline eagerly adopted by FIAT upon its inception in May 1945. CIOS, too, were able to formulate a methodology which proved effective and would also persist after the war's end – it was no accident that BIOS appeared to be almost a carbon copy of its predecessor, albeit with exclusively British membership. Valuable though the operations in the formerly Nazi-occupied territories undoubtedly were, it would be fair to surmise that the experience gained once the borders of the Reich were breached was especially instructive. Once the officials and investigators were able to visit facilities on German soil and speak to German staff, they were not only able to ascertain quite how rich and tantalising the scientific and technical spoils on offer were, but they were also able to refine the strategy to best exploit them. Overall, while exploitation was indubitably a product of the Second World War as a whole, counting the

Bruneval Raid of 1942 and Alsos actions in Italy in 1943 among its antecedents, it was during the final year of the war, from D-Day to VE-Day, that it really came of age and showed its true potential – had it failed to do this by May 1945, there is a good chance that post-war exploitation proper would have failed to material- ise. In short, the successes of the exploitation initiative during this crucial period, as evidenced by the CIOS statistics mentioned above, augured well for the next phase, which could take place with the benefit of experience, on a grander scale, and in the immensely preferable conditions of peacetime.

Notes

1 TNA, WO 219/1669, Col. G. Vickers to Maj-Gen. K.W.D. Strong, 7 September 1944.
2 Michael S. Goodman, *The Official History of the JIC*, vol. I (Abingdon: Routledge, 2014), 28.
3 TNA, CAB 81/92, 'Minutes of 2nd Meeting of JIC', 11 January 1944.
4 TNA, FO 942/27, 'Post-Hostilities Equipment Policy', 29 March 1944.
5 TNA, CAB 81/92, 'Minutes of 21st Meeting of JIC', 16 May 1944.
6 TNA, FO 1032/35, 'The Investigation of Initial Stocks of War Material and War Material Factories', 5 August 1944.
7 TNA, FO 942/27, 'Post-Hostilities Equipment Policy', 29 March 1944.
8 TNA, CAB 81/92, 'Minutes of 22nd Meeting of JIC', 23 May 1944.
9 TNA, WO 219/1669, 'Proposed Function for Technical Sub-Division G-2 SHAEF during SCAEF Period of Occupation of Germany', 19 September 1944.
10 Vannevar Bush, *Pieces of the Action* (New York: Morrow, 1970), 115–16.
11 Süss, *Death from the Skies*, 539.
12 Norman Longmate, *Hitler's Rockets: The Story of the V-2s* (London: Hutchinson, 1985), 382.
13 Rankin, *Ian Fleming's Commandos*, 236.
14 TNA, CAB 81/47, 'Joint Committee on Research and Development Priorities', 15 November 1944.
15 McGovern, *Crossbow and Overcast*, 13.
16 Basil Collier, *The Battle of the V-weapons, 1944–45* (London: Hodder & Stoughton, 1964), 150–51.
17 TNA, FO 942/27, 29 March 1944.
18 TNA, CAB 81/47, 'Strategic Survey of the War', 24 May 1944.
19 Gimbel, *Science, Technology, and Reparations*, 20.
20 H.O. Hooper, 'Weeks, Ronald Morce, Baron Weeks (1890–1960)', *Oxford Dictionary of National Biography* (Oxford: Oxford University Press, 2004) [accessed online 28 April 2016, http://www.oxforddnb.com/view/article/36814].
21 TNA, FO 942/27, 29 March 1944.
22 Norman A. Graebner and Edward M. Bennett, *The Versailles Treaty and its Legacy: The Failure of the Wilsonian Vision* (Cambridge: Cambridge University Press, 2011), esp. 107–24.
23 Alec Cairncross, *The Price of War: British Policy on German Reparations, 1941–1949* (Oxford: Blackwell, 1986), 10.
24 TNA, FO 1050/67, 'Intelligence Division: formation of Scientific and Technical Intelligence Branch', 9 November 1946.
25 Ann and John Tusa, *The Nuremberg Trial* (New York: Skyhorse, 2010), 51.
26 See Chapter Ten.
27 TNA, FO 942/27, 29 March 1944.
28 TNA, CAB 158/2, 'Joint Intelligence Sub-Committee: Memorandum', December 1947.

29   Bower, *Paperclip Conspiracy*, 92.
30   TNA, FO 942/27, 29 March 1944.
31   TNA, WO 219/1669, 'Collection of Economic Intelligence: Policy and Organisation', 15 September 1944.
32   Gimbel, *Science, Technology, and Reparations*, 4. On the British total war effort see Stephen Broadberry and Peter Howlett, 'Blood, Sweat and Tears: British Mobilisation for World War II', in Roger Chickering et al., *A World at Total War: Global Conflict and the Politics of Destruction, 1937–1945* (Cambridge: Cambridge University Press, 2005), 157–76.
33   TNA, AVIA 12/191, G.W. Turner to C.W.G. Walker, 20 October 1943.
34   TNA, FO 942/8, 'Draft Armistice Terms 1: Article 21(a)', 14 July 1944.
35   TNA, FO 935/1, 'Research and Development Centres in Germany', 1944.
36   TNA, WO 219/1669, 19 September 1944.
37   TNA, CAB 81/24, 'Joint Technical Warfare Committee', 25 April 1944.
38   TNA, CAB 81/24, 'Joint Technical Warfare Committee', 25 April 1944.
39   TNA, FO 935/1, 1944.
40   TNA, CAB 81/24, 25 April 1944.
41   TNA, FO 1031/49, 'History of T-Force', July 1944.
42   TNA, ADM 223/500, '30 Assault Unit and 30 Commando: Papers', July 1944.
43   Gimbel, *Science, Technology, and Reparations*, 4.
44   TNA, FO 1031/49, 1946.
45   TNA, FO 1031/49, 1946.
46   TNA, WO 193/432, 'Combined Intelligence Priorities Committee', 12 June 1944.
47   Gimbel, *Science, Technology, and Reparations*, 4.
48   TNA, WO 193/432, 17 June 1944.
49   A.G.M. Barrett and D.H.R. Barton, 'Linstead, Sir (Reginald) Patrick (1902–1966)', *Oxford Dictionary of National Biography* (Oxford: Oxford University Press, 2004) [accessed online 21 July 2015, http://www.oxforddnb.com/view/article/34549].
50   Balmer, *Secrecy and Science*, 73.
51   IWM, CIOS Report, 'The Intelligence Exploitation of Germany', 15 September 1945, 9–10.
52   TNA, FO 1031/49, 1946.
53   TNA, FO 935/25, 'Minutes of a CIOS Sub-Committee Meeting', 2 February 1945.
54   TNA, FO 935/25, C.H. Noton to R.P. Linstead, 14 February 1945.
55   IWM, CIOS Report, 'The Intelligence Exploitation of Germany', 15 September 1945, 24.
56   William R. Sears, 'Project Paperclip' book review, *Bulletin of the Atomic Scientists*, 28:6 (1972), 55.
57   Klaus-Dietmar Henke, *Die amerikanische Besetzung Deutschlands* (Munich: Oldenbourg, 1995), 746.
58   Gimbel, 'US Policy and German Scientists: The Early Cold War', *Political Science Quarterly*, 101 (1986), 437.
59   TNA, FO 942/79, 'Combined Intelligence Objectives Sub-committee', 30 May 1944.
60   TNA, FO 1031/49, 1946.
61   TNA, FO 1032/470, 'Minutes of a Meeting on CIOS Field Teams', 28 December 1944.
62   Astrid M. Eckert, *The Struggle for the Files: The Western Allies and the Return of German Archives after the Second World War* (Cambridge: Cambridge University Press, 2012), 31.
63   TNA, FO 1032/475, 'Report of the Combined Intelligence Objectives Sub-committee for 1944', January 1945.
64   TNA, FO 1050/1419, 'Combined Intelligence Objectives Sub-Committee (CIOS) black list', 18 August 1944.
65   TNA, FO 1032/470, 'Minutes of a Meeting on CIOS Field Teams', 28 December 1944.

66 TNA, FO 935/25, 30 January 1945.
67 IWM, CIOS Report, 'The Intelligence Exploitation of Germany', 15 September 1945, 17.
68 TNA, FO 1050/1421, 'Combined Intelligence Objectives Sub-Committee Grey List', 9 March 1945.
69 TNA, FO 1032/470, 28 December 1944.
70 TNA, FO 935/25, R.P. Linstead to Lt-Col. White, 13 November 1944.
71 TNA, FO 935/25, A. Loewy to Brig. John G. Foster, 6 December 1944.
72 Liberman, 'Spoils of Conquest', 140–41; Liberman, *Does Conquest Pay?*, 53.
73 TNA, CAB 81/24, 25 April 1944.
74 See Chapter Ten.
75 TNA, FO 935/25 'Progress Report Paris Field Teams', 6 September 1944.
76 Churchill Archives Centre, Metals Society, METL.2, CIOS Report I-1, September 1944.
77 TNA, FO 1032/470, 28 December 1944.
78 TNA, WO 219/1987, 'Field Information Agency, Technical: Reports on T Force Operations and Activities', 24 January 1945.
79 IWM, CIOS Report, 'The Intelligence Exploitation of Germany', 15 September 1945, 22.
80 Churchill Archives Centre, Metals Society, METL.2, CIOS Report X-13, 1 November 1944.
81 TNA, FO 1031/49, 1946.
82 Longden, *T-Force*, 88.
83 TNA, FO 935/25, 'Comments on Difficulties Experienced as a Civilian Investigating Targets in or Near the Battle Zones', 13 December 1944.
84 TNA, WO 219/1986, Maj. E.D. Magnus to Brig. R.J. Maunsell, 10 April 1945.
85 Ulf Schmidt, *Justice at Nuremberg: Leo Alexander and the Nazi Doctors' Trial* (Basingstoke: Palgrave Macmillan, 2004), 76.
86 TNA, WO 219/1986, 'Field Information Agency, Technical : T Force Organisation', 3 March 1945.
87 TNA, WO 219/1630A, 'T-Force Planning', 7 April 1945.
88 McGovern, *Crossbow and Overcast*, 3–5.
89 TNA, FO 1031/49, 1946.
90 TNA, WO 219/1987, Wg Cdr T. Jackson to Air Cdre C.M. Grierson, 21 December 1944.
91 TNA, WO 219/1986, Flt-Lt. S.M. Harris to Brig-Gen. T.J. Betts, 26 December 1944.
92 TNA, FO 935/25, 10 November 1944.
93 TNA, FO 1032/475, January 1945.
94 TNA, FO 1031/49, 1946.
95 TNA, FO 1032/475, January 1945.
96 TNA, FO 1032/475, January 1945.
97 Bessel, *Germany 1945*, 12.
98 IWM, 10/7/1, Lt-Cdr John Bradley to Margaret Bradley, 23 March 1945.
99 IWM, 10/7/1, Lt-Cdr John Bradley to Margaret Bradley, 23 March 1945.
100 Tony Le Tissier, *Race for the Reichstag: The 1945 Battle for Berlin* (London: Frank Cass, 1999).
101 TNA, WO 219/1986, 'Field Information Agency, Technical: T Force Organisation', 11 September 1944.
102 TNA, WO 219/1630A, 'T-Force Planning', 7 November 1944.
103 TNA, WO 219/1986, 14 January 1945.
104 TNA, WO 219/1986, 14 January 1945. Emphasis in the original.
105 Bessel, *Germany 1945*, 131.
106 Deighton, *Impossible Peace*, 5.
107 Gimbel, 'US Policy and German Scientists', 437.
108 See Chapters Seven and Eight.

109 TNA, FO 936/39, Lt-Gen. Ronald Weeks to HQ, 21 Army, 13 August 1945; H.O. Dovey, 'Maunsell and Mure', *Intelligence and National Security*, 8 (1993), 60–77.

110 John Gimbel, 'Science, Technology & Reparations in Postwar Germany', in Jeffrey Diefendorf, Axel Frohn and Hermann-Josef Rupieper (eds.), *American Policy and the Reconstruction of West Germany, 1945–55* (Cambridge: Cambridge University Press, 1993), 178.

111 TNA, FO 1032/1459, 'Field Information Agency, Technical', 23 August 1945.

112 TNA, FO 1065/12, 'Proposed New Establishment of Field Information Agency Technical', 23 August 1945.

113 TNA, FO 1031/50, 'Minutes of 21st CIOS Meeting', 6 June 1945.

114 TNA, FO 1031/51, 'CIOS Progress Report for 1945', 4 June 1945.

115 TNA, FO 1032/475, 'Organisation and Functions of CIOS', 13 April 1945.

116 TNA, FO 1031/51, Chiefs of Staff to Field Marshal Montgomery, 22 June 1945.

117 TNA, FO 1032/177, S/Ldr. S.M. Harris to British members of CIOS, 18 July 1945.

118 See Chapter Seven.

119 IWM, CIOS Report, 'The Intelligence Exploitation of Germany', 15 September 1945.

120 Paul Maddrell, 'British-American Scientific Intelligence Collaboration during the Occupation of Germany', *Intelligence and National Security*, 15–2 (2008), 79.

121 IWM, CIOS Report, 'The Intelligence Exploitation of Germany', 15 September 1945.

3   Exploitation in earnest

With the war over, the character of exploitation changed substantially. Speed was no longer as essential as it had been before. The risk of targets being damaged by fighting or bombing, or sabotaged by Nazis, was greatly diminished and even the potential for valuable material to be destroyed or removed by displaced persons or over-zealous Allied troops was gradually being lessened. Some semblance of order was being imposed by the occupying powers on Germany and this allowed exploitation to expand and become more thorough.[1] With no risk of stumbling into open combat, the investigators could now be selected from a much wider pool – they did not need to have any military connection and civilian experts of all stripes were now drawn wholesale from private industry and elsewhere. The experiences of these men and women offer a valuable insight into the every-day realities of the investigators and exploiters on the ground in Germany. The new post-war dynamic also necessitated a change in the exploitation machinery; 30 Assault Unit (30AU), for instance, had been well suited to daring raids and frontline activity, but their gung-ho *modus operandi* was deemed 'hardly suit-able for taking a team of middle-aged metallurgists across the plains of northern Germany'.[2] The same was true of T-Force, which had been the forward combat echelon of the exploitation initiative during the war, but now began evolving into a logistics provider for the increasing numbers of investigators who were travel-ling from Britain to Germany.[3]

This greater volume of exploitation agents raised its own set of problems. It was not feasible that all these operatives, many of whom had different but over-lapping remits, would be able to work in harmony. The sheer value and finite amount of the scientific and technical spoils on offer in Germany at the end of the war meant that fierce competition was to be expected. In addition, exploitation was not the only objective for British and Allied occupation forces; the exploita-tion teams were joined in the field by a plethora of other investigators – those looking into Nazi war crimes, the impact of Allied bombing strategies or the sal-vage of Europe's cultural and documentary treasures, to name just three. To fully understand exploitation, and particularly to appreciate the importance afforded it, viewing its internal and external relationships is essential. The overall purpose of this chapter is, therefore, to study how exploitation was conducted in its most comprehensive phase; the months and years immediately following the capitulation, when Germany lay completely at the mercy of its Allied occupiers.

Committees and agencies

The end of the war and the relative stability of the occupation period saw the demise of the opportunist, cavalier approach to exploitation, as the emphasis shifted from enterprising field-based units, such as Alsos and 30AU, to the more considered work of committees back in Britain. Foremost among these in the post-war era was the newly formed British Intelligence Objectives Sub-Committee (BIOS), based in the same office in Bryanston Square, London, as its predecessor CIOS (Combined Intelligence Objectives Sub-Committee), and initially chaired by CIOS Deputy Chairman Professor R.P. Linstead. In addition, it retained the same basic composition as the British half of CIOS, but also expanded it somewhat, so that it comprised representatives from 'the Admiralty, the War Office, the Air Ministry, the Foreign Office, the Ministry of Supply, the Ministry of Aircraft Production, the Board of Trade, the Ministry of Fuel and Power, the Department for Scientific and Industrial Research and the Government of the Dominion of Canada'.[4]

The primary remit of BIOS was to handle 'all requests of British Government departments for intelligence of military, political, industrial or economic significance which may be available in Germany and in European countries lately under German occupation'. BIOS officials would then compile target lists, liaise with organisations responsible for logistics and, upon completion of the mission, 'arrange for the appropriate dissemination of the resulting information to the British departments concerned'. In this way, BIOS largely mirrored the actions of CIOS, but now the approach was unilateral – British investigators, working to British target lists, and preparing reports for British circulation. Nonetheless, careful liaison and arrangements for mutual information exchange with the Americans became another central function of BIOS.[5] Crucially, there was already an awareness among those involved in Britain's exploitation planning that other nations were pressing ahead and that, through delay or poor execution, 'we should damage our own interests while the other Allies were helping themselves'.[6]

While the experts were marshalled and prepared for their missions by BIOS in Britain, once they got to Germany, their interactions with the machinery for exploitation would have been primarily with FIAT (Field Information Agency, Technical) and with T-Forces. These organisations were the workhorses of the project and there is no doubt that without them it could never have gone ahead and certainly not on the scale which was eventually achieved.[7] Exploring the relationships between BIOS investigators and the agencies which were responsible for facilitating their efforts reveals two interesting elements of the initiative as a whole: the first is the sheer enormity, complexity and diversity of the programme, which utilised a great multitude of committees, teams and special units, each with their own identity and terms of reference, and with their various interrelationships governed by careful and strict administration; and the second is the continuation of that fusion of the civilian and military spheres which had proved so essential during the war and now allowed for the specialist knowledge of civil industry and science to be effectively paired with the unique logistical capabilities of the armed forces.[8]

T-Force, initially the true spearhead of the exploitation programme, had numbered some 5,000 personnel at the time of the crossing of the Rhine but was reduced to 3,000 in November 1945. At this time, it also took on responsibility for handling reparations teams and British business owners visiting their properties in Germany, alongside its exploitation commitments and, by June 1946, oversight for its activities was switched from HQ, British Army of the Rhine, to the Control Commission for Germany (British Element) (CCG(BE)).[9] This was clearly representative of its shift from a direct military purpose to a more general logistical one, as was the (unheeded) suggestion by the Chief Administrative Officer that the name T-Force be changed to Technical Travel Agency.[10] Its duties ranged from provision of accommodation to the evacuation of equipment, documents and personnel.[11] In information provided to BIOS investigators before they arrived in Germany, they were advised to 'contact T-Force for all their requirements while out at their targets', while the general responsibilities of T-Force were elucidated as follows:

> T-Force arrange your clearance into the British zone, supply transport to your targets from limited resources, run a chain of messes from the Ruhr to Kiel, provide interpreters, escorts, fuel supplies *en route*, valuable intelligence re: targets, and arrange the evacuation of documents and equipment.[12]

To give some sense of scope, T-Force operated 15 Transit Messes throughout the British zone and, in the period June 1945 to October 1946, they handled the visits of 6,084 BIOS investigators, as well as 1,400 Reparations/Restitution Teams, and facilitated visits to 7,300 separate targets in total.[13]

As the situation in Germany became more stable and the multiple threats which had abounded immediately after the end of the war subsided, the military nature of T-Force was called into question. The majority of targets which were being examined were peaceful industrial concerns and demobilisation across the British armed forces was the order of the day.[14] It was confidently considered 'that all T-Force functions could efficiently be carried out by civilians' but those assessing the organisation's future felt it was 'doubtful if wholesale civilianisation would be in the interests of efficiency or economy'. There was even discussion surrounding the possibility of allowing British personnel to return home by replacing them with German citizens. Though this may have offered many benefits to the operation of T-Force, security concerns, among other things, precluded it from going ahead.[15] This distinguishes exploitation from other spheres of the occupation, where the British were quite content to hand over the day-to-day administration to German officials; a form of indirect control adopted wholesale from Britain's colonial experience. Interestingly, when implementing this 'imperial' mode of administration, British occupation officials were reminded that the German people were 'highly developed in most spheres' and should therefore be treated with 'patience and tact' – presumably a different approach to that employed in Britain's other, less 'developed' colonies.[16]

Alongside T-Force, FIAT was the other element which helped to bear the burden of the exploitation programme's logistical requirements. With the split of the combined FIAT into its separate British and American components at the end of August 1945, FIAT (Br) was placed under the administration of the British Control Commission, and its costs were to be borne on the Commission's establishment. Its purpose was formally stated thus:

> The Field Information Agency, Technical (British) will co-ordinate, integrate and direct the activities of the various missions and agencies interested in examining, appraising and exploiting all information pertaining to German economy other than direct military intelligence. It will provide centralised information services and facilities covering this technical intelligence field. It will not, however, be responsible for final collation of such information or its exploitation.[17]

Although this set-up was adopted so as to provide the greatest possible assistance to BIOS investigators when navigating the difficulties of carrying out their work in the chaos of post-war Germany, it did, as complicated bureaucratic structures so often do, generate almost as many problems as it solved. Linstead complained that 'the clearing of British investigators for the British zone which had to be done through FIAT' was causing considerable delays and that, in some cases, 'FIAT did not pass on the clearance for several days and then they cleared so many trips that RAF Transport Command could not cope with the requests for air bookings'.[18]

Perhaps the greatest attribute which FIAT possessed was its role as an Anglo-American co-ordinating body, a happy hangover from its origins as a combined unit. Even after the two national elements were split, both retained their headquarters in the same building in Frankfurt and continued working in the spirit of 'closest co-operation'. Moreover, FIAT was 'the only British unit in the American zone of Germany' and the goodwill which it had built up with the Americans (and French) was 'of immense value to the Control Commission, and to Ministries and Departments in the UK'.[19] This was important as FIAT (US) was responsible for granting passes and permits for British investigators to visit targets in the American zone, where many sites of interest were located, and the arrangement was duly reciprocal. FIAT (Br) was also the agency to which all British investigators had to report upon entering, and before leaving, the American zone. It helped to provide the facilities which T-Force offered in the British zone – transport, accommodation, library access and the evacuation of documents or material, as long as clearance could be obtained from the Americans.

Despite the structures put in place by FIAT and T-Force on the ground in Germany, problems still arose, often due to poor administration or a lack of proper briefings offered by BIOS in London. At one BIOS meeting in December 1945,

Brigadier Pennycook, of T-Force HQ, gave a scathing assessment of the serious shortcomings in the system: 'in many cases teams arrived without having seen the reports available in London and totally unfamiliar with the targets they were scheduled to visit or even why they had come out at all.' This in turn fed into another problem which was the saturation of teams and the increased pressure which this put on the limited resources available. The issue regarding ground transport was one of severe scarcity, which was only exacerbated in bad weather, when 'trips were taking an average of three weeks instead of ten days' and one-fifth of vehicles were off the road at any one time for maintenance.[20] In addition, some of the trips which were contributing to this congestion were 'obviously not of a serious nature', as a result of the sponsoring ministries 'not showing the sense of responsibility that they showed during the CIOS period'. Often these teams were larger than they needed to be (thus claiming unnecessary extra transport) and were revisiting targets which had already been completely exploited, causing a state of affairs which 'created a lack of enthusiasm on the part of the staff working in London and in the field'.[21]

It is worth noting here that the Whitehall administrative structure to which BIOS was answerable did little to simplify matters. The origins of exploitation lay within the realms of espionage and, therefore, the initial responsibility for its co-ordination lay with the Joint Intelligence Sub-Committee (JIC). This arrangement was able to operate, without much-imposed change, until the end of the war in Europe. However, during this immediate post-war period, it was monitored closely not only by the JIC but also by the Deputy Chiefs of Staff (DCOS), who were responsible for co-ordinating scientific and military co-operation and for whom exploitation became one of the most frequently discussed subjects at their fortnightly meetings. In the summer of 1945, questions began to be asked as to whether the JIC was still the body which should have main control over exploitation activities. In July, it was suggested that BIOS was no longer a conventional intelligence operation and should therefore 'transfer [its] whole allegiance' from JIC to DCOS, while still maintaining some input from the former.[22]

In September, responding to the belief that future BIOS work would likely be 'of greater value to industry than to defence', Norman Bottomley, the Deputy Chief of Air Staff, suggested that BIOS be responsible to the DCOS committee on technical military matters, the JIC on intelligence matters and the Board of Trade on industrial matters.[23] One year later, responsibility for BIOS was fully transferred to the Board of Trade, with the exception of a few issues in which the DCOS committee retained a particular interest.[24] As this shows, exploitation was not closeted away in some obscure corner of Whitehall bureaucracy, but rather was open to observation and interference by numerous governmental parties. Indeed, the decisions taken by these varied committees and agencies, combined with the peculiar conditions of post-war occupied Germany, directly created the operational reality in which the investigators and exploiters had to conduct their examinations of German science and technology. Their experiences in the field can therefore shed a unique light on exploitation at the grassroots level and give some measure of the success of the frameworks and procedures put in place to facilitate, and regulate, this endeavour.

Investigators and exploiters

The investigators who were despatched to Germany by CIOS and BIOS from 1945 onwards were a diverse collection of people – men and women, of all ages, from across Britain (and occasionally beyond) and drawn from all variety of scientific, technical and industrial backgrounds. While some had had military experience during the war, many were civilians who had not seen active service and were therefore unused to any form of military culture and who would find the British forces of occupation as unfamiliar to them as Germany itself. There were also a great number of them, travelling back and forth across the British zone and beyond, with very little day-to-day oversight or co-ordination. Moreover, the investigators were active during a period of great diplomatic uncertainty, with Britain, the USA, the Soviet Union and Germany, among many others, all trying to find their place in the new world order. This led to a serious risk that exploitation, a contentious issue from inception to termination, and the investigators who personified it, could jeopardise that fragile peace. Therefore, in order to enact a successful exploitation policy, in which all the desired information and material was gathered efficiently and which did not too severely disrupt the general occupation of Germany and the broader geopolitical circumstances, it was considered necessary to establish a solid set of procedural guidelines and governing regulations for the investigators on the ground. These rules, whether rigidly adhered to or brazenly contravened, played a key role in shaping the experience of exploitation teams on the ground.

Some of these policies were almost exclusively bureaucratic, including the three different officials (the BIOS Administrative Officer, Technical Liaison Officer and a member of the Economic Division of the CCG) that investigators had to report to upon arrival in Germany, but this was deemed necessary on account of the sheer volume of visitors who arrived in, and travelled through, the British zone every day in the year or so after the war's end. BIOS teams were often also needlessly large to the point of being unwieldy – in the interests of fairness and avoiding favouritism, membership of each team was drawn from across the relevant field of industry, so that all major firms were represented, even though it was felt that efficiency was diminished if they numbered more than 3–4 members.[25] Furthermore, there was a perceived risk of unofficial individuals visiting German factories and obtaining technical intelligence for the exclusive benefit of themselves or their employers, which necessitated the repeated statement of the importance of possessing the right credentials and always obtaining permission to visit targets from the local Military Government detachment.[26] Indeed, the issue was significant enough to warrant Lieutenant-General Brian Robertson, the Deputy Military Governor of the British zone, sending a memo to the Trade and Industry Division (the agency responsible for the control and revival of German industry) warning that 'unauthorised visits are most undesirable, may lead to severe abuses and in any case waste the time of the Factory Manager or his staff', and instructing that a Visitors Book be made compulsory at all factories and similar sites.[27]

In general, although many of the investigators were sourced from private industry, during their time on BIOS trips they were considered to be official representatives of the British government. Once they had signed the Official Secrets Act, they were told 'you become a temporary Government Servant and remain one until your final report is published'. What this entailed was made abundantly clear to all involved:

a   All information that you obtain, even if it is outside the actual scope of your investigation, is the property of HM Government and is a Government secret. You are bound to report it fully and accurately to the proper authority. Until your final report is published, you must not discuss this information with anyone who is not a Government Servant.

b   You are an official representative of Great Britain, wearing British Armed Forces uniform and you are regarded as such by our Allies and by the Germans. Consider carefully in this light everything you do and say.

c   You are NOT permitted to conduct business of a private nature whilst you are in Germany.[28]

Nonetheless, in autumn 1946, a decision was made to allow individual British firms to send teams of their own investigators to Germany, under the BIOS banner, 'to secure, in the national interest, any significant technical information which had been omitted from BIOS reports'. This came about in spite of 'a very real possibility that the technical experts sent out by a particular firm to report on the commercial value of the process in question might, employing duress (and that is to some extent what the wearing of uniform involves), extract from the German firm all the necessary know-how before purchase was agreed by either side.'[29]

One safeguard which was put in place came in the form of the limited scope of operations. Investigators were strictly informed that 'BIOS terms of reference limit your investigations to scientific and technical developments and do NOT cover current and future production and commercial practices.' In short, technical exploitation was allowed but economic exploitation was not. All BIOS investigators were also cautioned on matters of security; not just their own, which was largely guaranteed by the significant British military presence anyway, but also that of the German citizens they encountered, who could possibly be at risk of reprisals by a handful of diehard Nazi fanatics who saw them as treacherous collaborators.[30] Investigators were sternly reminded of 'the necessity for withholding from the Press and any other unauthorised persons the names of German individuals or organisations which have co-operated with you or assisted you in any way'. However, other elements of the programme were far less restricted; for instance, BIOS teams were also told that 'by inter-Allied agreement, any plant situated in Germany and in operation during the war is a legitimate target. It does not make any difference if it is partly or wholly owned by a member of the United Nations.'[31] This is indicative of the difficulty which BIOS faced throughout the period, of wanting to ensure that no stone was left unturned and the maximum benefit was gleaned from their examinations of German science and industry, so that they were not left behind by their Allied rivals, while still ensuring that Germany and the German people did not suffer disproportionately as a result.

Although the Black Lists were used as comprehensive guides of the targets which investigators should look into, the members of BIOS were aware that, as a result of the secrecy which characterised the Nazi regime, there would be many targets which they were not previously aware of which might be of great value. As such, experts who had travelled to Germany were allowed to rely on their own knowledge and instincts and make their way to additional 'targets of opportunity' for their examination. This was an essential part of the programme, but it did generate its own set of problems, particularly in terms of opportunity targets in Berlin. Major Baukham of FIAT complained to T-Force in July 1946 that 'quite a lot of these investigators, on receiving clearance to Berlin, do a thorough job of sight-seeing and then return to carry on with their other targets'.[32] His complaints fell on deaf ears however; T-Force acknowledged the issue, saying that sight-seeing occurred not only in Berlin but in many other major cities and famous locations, but then was dismissive of any action to curtail it: 'It is felt that any attempt to completely control all investigators' movements would not only raise great administrative difficulties but would also cause widespread resentment amongst all concerned.'[33]

In some cases, the rules which exploitation personnel had to negotiate were much the same as the general controls which applied to all British personnel in Germany during this period. One such area was accommodation and the provision of basic supplies. The end of the war was not a time of plenty and with so many people heading over to Germany from Britain in an official capacity, the authorities could not be responsible for providing them with all the essentials during their trip. Investigators were instructed to take with them sheets and a pillowcase (but not a blanket or bedroll), knife, fork and spoon, a water bottle, a torch, a towel and soap, as well as 'sufficient supplies of cigarettes, razor blades, shoe-cleaning material, soap (including laundry soap), toothpaste, chocolate etc. to last you for the whole trip'. They were advised that canteen facilities would be limited in the British zone and non-existent elsewhere.[34] It became even more restrictive if they travelled to the FIAT mess at Höchst, near Frankfurt, in the American zone, where, for instance, the standing orders dictated that visitors should provide their own blanket; they could be loaned one but if they were to 'purloin' it, they were to be charged 'through the normal Army Accounting Channels'.[35] These rules seemed to be 'rather brusque' to many investigators, and Major Hughes of T-Force headquarters wrote to his counterpart at FIAT, to comment that 'life seems very Spartan at Höchst', before adding 'I am surprised that you find Standing Orders necessary. We used to have similar instructions to investigators in all our bedrooms and we found that they merely annoyed people unnecessarily and did very little good.'[36]

Another problematic issue, which proved very difficult to navigate for all concerned, was that of non-fraternisation. The standard orders given to regular servicemen on this matter were clear; in the information booklet issued to all members of the British armed forces as they prepared to cross the Rhine in 1944, they were told, in no uncertain terms, to 'keep Germans at a distance, even those with whom you have official dealings'.[37] In principle, this was something that British troops found easier than the Americans, who had suffered no direct German attack on their homeland but, in practice, both nation's troops found it to be largely unworkable.[38] By the end of September 1945, all the official British non-fraternisation measures had been abolished, with the exception of marriage to German citizens and the billeting of troops in German homes. This marked a shift in favour of common sense policy as key occupation initiatives such as reconstruction and re-education were hardly feasible if no civil contact was permitted between the British officials and the German people.[39] This was particularly true in exploitation, where experts could often only elicit the best information through lengthy and in-depth conversations with their German counterparts, many of whom they knew on a personal level from before the war, having mixed in the same professional circles and even collaborated on some projects.[40]

Although with hindsight this litany of restrictions may appear to be cumbersome and inflexible, the programme of exploitation was a mammoth one and it was only through this meticulous preparation and careful governance that it was able to proceed at all. Studying the policies imposed from above is, therefore, essential to understanding exploitation but it does not present the complete picture; for that, it is important to examine how the process was viewed by those directly involved – the men and women who served as investigators in the field. One such individual was Gilbert Hunter, of the coking firm Stewarts and Lloyds of Corby, who travelled to Germany in January 1946 as a member of BIOS Trip 1539, tasked with investigating the German coking industry. He recognised from the outset that 'one cannot defeat the system, so a wise man should learn it and adopt it with as good a grace as possible.' It did, in fact, have many benefits to offer too; he marvelled that after a mere half an hour in the office of Major Peterson at T-Force HQ, 'we'd virtually got the freedom of the country', and commended the whole arrangement: 'Hats off to the British Army for a very fine job of organisation.'[41]

Monica Maurice, of the family-run Wolf Safety Lamp Company of Sheffield, who travelled to Germany on a BIOS trip in April–May 1947 to investigate German lamp manufacturers, was not quite as enamoured with the military organisational style and complained that 'the Army is unable to move, think or decide anything unless it has been written down on a piece of paper.' One area where such comprehensive administration proved helpful though was in terms of transport, which Ms Maurice noted was essential as without it 'one cannot move at all as messes are sometimes or mostly several miles away from HQ or map-room or whatever else one wants.'[42] Despite the considerable shortages described in the preceding section, both Hunter and Maurice commended the transport and drivers which they were given – Hunter described his team's Humber cars as 'comfortable and cosy' and the drivers as 'quiet, reliable fellows',[43] while Maurice and her companion formed such a bond with the driver of their brand new Austin 12 that they took turns driving from Berlin to Hanover to let him rest up after drinking rather too much the night before.[44] Eddie Aspden, however, who visited Germany to investigate engine factories in autumn 1945, described covering some 2,500 miles in five weeks, and how they were spent mostly in 'acute discomfort'.[45]

To a certain extent, these differences in experience can be explained by the timing of the trips. Those who travelled to Germany in 1945, whether during the war or immediately after it, found resources short and comfort lacking but, by mid-1946, conditions had improved considerably and visiting experts tended to have a less miserable time of it, and nowhere was this more clearly reflected than in comments on accommodation and provisions. When Lieutenant-Commander John Bradley was making his way through northern Germany in the last weeks of the war, he and his colleagues mostly slept in barns or out in the open or would on occasion 'calmly dispossess' a farmer of his house, during their stay in which 'we may drink his schnapps and use his utensils but we do not do useless smashing'.[46] Matters were little improved five months later when Aspden and his team visited Nuremberg; accommodation was hard to come by 'because the city had been almost completely destroyed' and they were only able to obtain meals 'in a transient mess … [which] in quality were nothing more than mediocre'.[47]

Within a further four months though, in February 1946, Gilbert Hunter felt it only fair to 'accord T-Force a very hearty vote of thanks for their hospitality', especially on account of 'the seriousness of housing conditions in Germany'. Perhaps his perceptions were improved by his team's experience of dining at Villa Hügel, the 269-room mansion near Essen, formerly the home of the Krupp family of industrialists and requisitioned after the war by the Allied North German Coal Control agency, of which he wrote this glowing review:

> Dinner was something to be remembered for many, many days to come and should be written down in capital letters … The dinner was good, the wines excellent, the silver, the china, the glass, the napery, the room, the service, etc., etc. were all as nearly perfect as makes no matter, but the real attraction was the atmosphere which was so perfectly natural that one felt it might have been rehearsed many times.[48]

Michael Howard, who served as an Intelligence Officer with No. 1 T-Force in 1946–47, also spoke fondly of Villa Hügel, especially the squash courts set up in the old indoor riding school adjacent to the house, of which he made use frequently throughout 1946.[49] Elsewhere, in April 1947, Monica Maurice recorded visiting the Landeshaus Club in Düsseldorf which was exclusively for the use of Control Commission personnel and which she described as 'a marvellous and beautifully furnished club with dining room, games room, lounge, and dance floor, on a corner overlooking the river'.[50] This image of sophisticated socialising seems unrecognisable when placed alongside such earlier accounts of shortage, discomfort and extreme improvisation, or against the privations and hardships suffered by most ordinary German citizens in this period.[51]

Another factor which shaped investigators' experiences of their time in Germany was the zone in which they were based. Though BIOS teams spent the majority of their time in the British zone, trips to the American zone were considered quite the treat, as Eddie Aspden recorded:

> Conditions in the American zone were much more comfortable than those in the British, where we had to use our sleeping kit the whole of the time, and to wash and shave in cold water. A bath was almost out of the question. Laundry too was difficult, though in both zones. Rations for the road were very good in the American zone, where it was possible to buy the American Army K-rations, a carefully balanced meal packed into a sealed carton and graded breakfast, dinner and supper. We consumed quite a number of these. In the British zone, however, we were provided with bully beef sandwiches, frequently wrapped in newspaper.[52]

These comments on the general conditions of each zone were actually indicative of a larger truth – that the American zone was far better-appointed than those of the other Allies, on account of its economic strength, which affected the lives not only of US soldiers and British visitors but the German civilian population too.

One element which all accounts seem to agree on, though one which contradicts wartime expectations for the experience of investigators on the ground, was the nature of their interactions with German citizens. Lieutenant-Commander Bradley dealt frequently with German naval personnel and admitted to his wife that 'I admire the discipline and bearing of the great majority of them and I should find it difficult to hate them.' As such, he accurately predicted that non-fraternisation measures would prove futile and difficult to enforce.[53] Maurice described the German individuals she encountered as 'willing and anxious to talk and apparently to show us everything' and those who were not could be easily coerced, as with one engineering expert who was 'obviously pleased to be presented with ½ lb coffee'.[54] There were understandably some German people who resented the outcome of the war and the imposition of having investigators visit and root out their industrial secrets, but Hunter noted that even those who were 'far from happy' still 'received us with dignity and as naturally as present circumstances (for them) could permit'.[55] There were, of course, exceptions. Brigadier W.P.T. Roberts, who led a team to investigate a cartridge case factory in Karlsruhe in May 1945, described a major-general he met as having 'the courage still to maintain his Nazi sympathies'.[56] Gilbert Hunter, meanwhile, looked upon lingering political loyalties with less sympathy, writing of a visit to a hydrogenation plant that 'our guide was a typical Nazi and every one of us would have enjoyed kicking him in the slats!'[57]

More generally, as part of the relatively small cohort of British civilians who visited Germany directly after the war, the investigators' records of their time there, and their interactions with ordinary German people, offer a rare insight into how the British perceived their former enemy. The impacts of Allied bombing on German cities struck a particular chord. Michael Howard described a journey he made along the Route 1 autobahn through the heavily targeted Ruhr area, along which there was 'hardly a street for fifty miles which did not have at least one in four of its houses bombed or bomb-damaged'.[58] Eddie Aspden wrote that the 'awful damage' in Cologne 'made one feel sick' and lamented that 'people in England had no conception of the scale of destruction done to Germany through bombing and shelling – it is a great pity that they can never know.'[59] Monica Maurice felt similarly, writing that 'we were mad about Coventry, Manchester, Sheffield and others at home. It has been more than avenged over here to a point where one wonders how anyone can take up any sort of life again.'[60] Indeed, Gilbert Hunter felt that 'most of the Germans look a very beaten lot' but believed 'it is their womenfolk who will restore their pride and self-respect if it is to be restored. Hats off to the German womenfolk. They're clean and alive-looking and as smart as it is possible for them to be under very difficult circumstances.'[61] It is interesting to note that the sentiments displayed here are primarily those of sympathy, pity and even remorse, and that exploitation – a policy forcibly enacted by the victors over the vanquished – may have actually served, at the grassroots level at least, to foster better understanding between the two nations and peoples.

Overall, accounts by investigators of their time in Germany could vary enormously, particularly in terms of whether they viewed it positively or negatively. For example, Eddie Aspden had a generally dismal impression of his experience as an agent of exploitation, describing his tour as 'nerve-wracking', 'arduous and exacting' and requiring 'great patience'.[62] On the other hand, Michael Howard felt great pride in his work and wrote home to his parents that T-Force 'is the only unit in Germany which is not a liability to the taxpayer in that the consequences of the work have a considerable and direct bearing on our economic recovery. This does help one feel that one is doing a good job of work.'[63]

As many of these accounts have shown, exploitation was far from a smooth process. Though many operatives were enthusiastic about their missions and satisfied, even impressed, with the structures in place to facilitate them, all of them found themselves butting up against administrative roadblocks, paralysed by seemingly counterintuitive regulations or hamstrung by a paucity of key resources. Part of the blame for this unfortunate state of affairs lies with the fact that British exploitation was not a linear process under the jurisdiction of a single organisation, but rather the product of myriad agencies and departments working in varying degrees of co-operation and competition, both on the ground and in London. Even though all these elements were striving towards roughly the same aim – to attain the greatest spoils of German science and technology for the betterment of Great Britain – their relationships were not always smooth. To fully understand the process of exploitation, therefore, it is necessary to examine the interactions which the various representatives of the exploitation initiative had with each other and with agencies pursuing different aims, whether they were of British or foreign origin, in the field in Germany.

Competition and co-operation

As has been discussed, the situation in post-war Germany was chaotic, with numerous military and civilian organisations all striving to contribute towards creating the best possible post-war world – some were concerned with gathering evidence of war crimes, others wished to assess the merits of the Allied strategic bombing campaign, while others were interested in securing Germany's documents and archives. How these other operations related to the exploitation programme, and how different strands of that same programme related to one another, is critical to understanding the context within which exploitation took place. On the ground in Germany, these endeavours faced countless occasions of logistical limitation, shortage of resources and conflated objectives, prompting either competition or co-operation among the agents in the field. In many cases, the actions taken by these units and individuals had to be carried out with great speed in response to rapidly changing circumstances and thus had very little recourse to official policy as dictated from above. In short, the rigid departmental line was often incompatible with the pragmatism necessary on the ground.

The first example of true competition emerged in the dying months of the war when the earliest exploitation teams were racing across Europe, trying to keep up with the advancing Allied armies, desperate to be the first to seize the most valuable scientific and technical targets. One CIOS assessor complained about other services running 'semi-piratical expeditions', the operatives of which were nicknamed 'witches', because 'they flew over our heads on broomsticks'.[64] Elsewhere, the Economic Sections of Supreme Headquarters Allied Expeditionary Force (SHAEF) complained that Ministry of Supply parties were causing confusion by gathering economic intelligence which did not fall under their remit, and this then had to be referred to CIOS for adjudication.[65] The confusion worsened once individual countries began sending their own unilateral expeditions to Europe too – in February 1945, Colonel Geoffrey Vickers of the Ministry of Economic Warfare warned his colleague C.H. Noton that 'a new mission called "The US Technical Mission for Europe" had been formed which looked as if it would be another knight errant in the CIOS field'. Vickers hoped that there would be some way that this new mission could somehow be absorbed into the existing CIOS machinery.[66] In fact, Vickers may well have been referring to the US Technical Oil Mission which was already active in Germany and was responsible to both CIOS and the American Technical Industrial Intelligence Committee (TIIC).[67]

In many cases, it seemed that the exploitation agencies active in late 1944 and early 1945 were more concerned with keeping their activities secret from one another than they were with securing the best spoils of Nazi Germany. As time passed, the number of different organisations involved increased exponentially and by 1946, British FIAT recorded that there were 52 other agencies with which they had to liaise in the course of their regular operations.[68] However, many of these other units were tasked, at least in part, with rendering assistance to the larger organs of the exploitation machinery. The Scientific Intelligence Advisory Service (SIAS), for instance, was established under G-2 SHAEF to 'find out, or to create opportunities for others to find out, what scientific discoveries of fundamental importance have been made in Germany since 1939'. They believed that they could offer the greatest assistance by being granted the opportunity to handle 'that part of the procurement of German scientific intelligence which necessitates lengthy contact between German and Allied persons of extreme intellectual distinction'.[69]

However, in some cases, bureaucratic clutter meant that the utilisation of these supplementary services did not run at maximum efficiency. For example, the German Economic Department, part of the Ministry of Economic Warfare, which was in possession of 'a considerable body of information regarding German industry, its location, rate of production and technical developments' and which had knowledgeable staff available for consultation,[70] complained that there were 'far too many agencies making demands upon them' and requested that a uniform procedure be adopted to allow them to 'deal more satisfactorily and expeditiously' with these demands.[71] Similarly, the Inter-Services Topographical Department (ISTD), which, as its name suggests, supplied topographical intelligence to all branches of the British armed forces and which felt it had much of value to offer to CIOS and others, complained of being:

> insulated from all knowledge of, or contact with, the working parties and their chairmen. This insulation was carried to the degree that, not only were we forbidden all contact except via SHAEF (and this channel did not work), but we knew none of the personnel who, it was suggested, were all 'greybeards'.[72]

Here, though the will for co-operation was present, the lack of adequate means by which to facilitate it meant that much progress was hindered. Lessons were certainly learnt from this chaotic arrangement and led, in many ways, to the formation of the Joint Intelligence Bureau in 1948 – this office, part of the Ministry of Defence, was responsible for co-ordinating economic, scientific, industrial and atomic intelligence throughout the early Cold War, and even incorporated some existing agencies, such as the ISTD, to ensure smooth, well-informed operations.[73]

Once the war ended, so too did the most urgent phase of the race for the spoils. Now, with all of Germany open to exploitation, it was no longer a case of which agency could reach a certain target first, but rather which could cover the most targets at any one time; the most valuable attribute was no longer speed, but scale. This, in turn, meant that logistical concerns now came to the fore and opened up new fertile territory for both competition and co-operation. CIOS and its successor agencies, with their ability to command the support of troops, transport and storage facilities, became the gatekeepers for any civilian agency wishing to conduct exploitation in Germany, even for entities as powerful as the Foreign Office or the US State Department.[74] On the ground itself, similar roles were fulfilled by FIAT and the T-Forces.

As the post-war period wore on and the number of agencies active in Germany grew to unprecedented levels, the scope for competition increased exponentially. One group which proved especially problematic was the Monuments, Fine Art and Archives (MFAA) initiative. Nicknamed the 'Monuments Men', this organisation was one of the more unusual operating in this period; led by a motley band of archaeologists, art historians and museum curators, they were tasked with protecting the cultural and architectural treasures of Europe from damage during the fighting, as well as being responsible for the restitution of works or art or other precious possessions which had been stolen by the Nazis or hidden for safe-keeping.[75] As their remit included archives, those exploitation investigators who were seeking valuable documents on German science and technology were often brought into contact with MFAA officials. No top-down policy as to how the two operations should interact was enacted here, rather it was declared that the 'precise limits of responsibility between the two will be decided by … officers [from both sides] in conference at the appropriate level'. This was a clear example of how a pragmatic approach by the men on the ground was considered to be the best way to ensure healthy relations between concurrent programmes. In fact, it could even lead to positive collaboration, as the specialist MFAA agents offered 'advice and assistance on all technical matters concerning German archives' which could be of great use to uninitiated exploitation investigators seeking specific documents.[76]

More significant problems arose when departments in Britain dispatched teams to Germany outside of the proper BIOS channels. Here there was a great risk of duplication or redundancy of effort as well as competition for certain targets. The Air Ministry's Assistant Directorate for Intelligence (Science), under the leadership of R.V. Jones, sent its own investigators to continental Europe two days after D-Day. Jones proudly recalled how these men were enthusiastic enough to often be ahead of the frontline troops and were soon sending back a steady stream of information, documents and equipment. He notes that there were indeed 'other overseas parties besides ours', but does not record any clashes or conflicts of interest.[77] What does become apparent, however, is that the left hand did not always know what the right was doing. In an Air Ministry report from October 1945, it notes that many prominent German atomic scientists, including Professors Hain (misspelt – Hahn) and von Lane (again misspelt – von Laue) are 'reported to be in America', when in fact they were being held at Farm Hall, just outside Cambridge, a fact which was known by ordinary members of the public but which had apparently eluded the Air Ministry. In the same report, they also assert that the Soviet Union is 'by no means a bad fourth' in the race for the spoils of Germany which, even by this stage, was a gross miscalculation.[78]

Similarly, the Department of Scientific and Industrial Research (DSIR) also dispatched its own experts to Germany at this time in order to procure laboratory equipment, such as microscopes, centrifuges and x-ray sets, for research establishments and private industry in Britain, despite the Board of Trade and Research Branch both insisting that such removals could only be carried out legitimately through the existing BIOS channels.[79] These two examples barely scratch the surface of the true range of different exploitation programmes which coexisted in this period. By late 1947, often in conjunction with British authorities, the US Army was operating six different projects, through military intelligence channels, to tap the talents of German specialists, and the focus was not just on scientific and technical expertise. For example, Operation Pajamas was concerned with European political trends, Birchwood with economics, Dwindle with cryptography and codebreaking and Panhandle with operational military expertise.[80]

Countless agencies roamed through Germany, determined to exploit the most useful aspects of the defeated nation's human resources for their own benefit.

Teams working on utilisation of other aspects of Germany's expertise were not the only other investigators that scientific and technical exploitation units encountered in the field. Another major operation which was undertaken at this time was designed to uncover the impact of Allied bombing on Germany and to try to ascertain which techniques were the most successful in order to prepare the air forces of Britain and America for any future war. After all, the Second World War was the first conflict to employ strategic bombing on such a scale and both the RAF and USAAF were convinced that the next war would do much the same, but to an even greater extent.[81] In order to conduct effective assessments of their bombing efforts, both Britain and America created investigative agencies, which were tasked with sending teams to Germany and compiling reports of their findings; respectively, these were the British Bombing Research Mission (BBRM; later also the BBSU – British Bombing Survey Unit) and the United States Strategic Bombing Survey (USSBS).[82]

Both these enterprises soon discovered how important it was to be the first on the scene, in order to secure valuable documents before they were seized by another party; an endeavour in which they actually benefited from their small size, as they went unnoticed while they bent rules and succeeded in 'slipping through the net of restrictive regulations' which was in place as the war ended.[83] The British especially lacked the manpower to send multiple teams out into the field so instead hoped that collaboration would offset the problems of other agencies reaching key sites first. They hoped to work closely with CIOS, perhaps even to the extent where they would be able to attach an operative of their own to a CIOS team, thus facilitating a sharing of expertise.[84] They also liaised with FIAT to ensure that relevant German individuals would be relocated from the highly-restricted Dustbin detention centre at Schloss Kransberg to the more accessible Combined Services Detailed Interrogation Centre at Bad Nenndorf when only the BBRM retained an interest in them.[85] There was also international co-operation, with the BBRM utilising a lot of the data collected by the USSBS, which had the downside of leading to the British drawing many of the same conclusions as the Americans, despite the fact that the strategies of the RAF and USAAF had differed notably during the war.[86] It is also worth noting that Britain was largely disinterested in the results of their bombing survey and swiftly swept it under the carpet after publication, fearing that it might harm their 'hearts and minds' campaign in Germany and draw unfavourable comparisons with the very war crimes for which they were planning to prosecute senior Nazis.[87]

In fact, the pursuit of post-war justice, specifically the prosecution of war crimes or the newly designated crimes against humanity, was an initiative which seriously threatened to clash, or even present a genuine conflict of interests, with the exploitation scheme. During the war, as word of the atrocities committed by the Nazis filtered back to Britain, it became clear that bringing the perpetrators to justice would have to be a main priority of the post-war period and, in October 1942, Churchill stated that 'retribution for these crimes must henceforth take its

place among the major purposes of the war'.[88] This was in itself a British commitment to an earlier declaration, signed in January by representatives of nine
European governments-in-exile in London, which both condemned Nazi atrocities and asserted that justice would be sought with determination at war's end.[89]
The best-known manifestation of this was the International Military Tribunal,
held in Nuremberg, which began in November 1945 and tried 24 members of
the Third Reich's leadership and which was conducted jointly by all four occupying powers.[90] After this, the Americans conducted further war crimes trials at
Nuremberg, dealing with senior figures from various sections of Nazi society,
such as doctors, lawyers and industrialists.[91] Meanwhile, the bulk of the other trials were held on a unilateral basis, usually confined to each nation's own zone of
occupation (and often to specific concentration camps, such as Bergen-Belsen or
Ravensbrück) and Britain's policy in this respect was inconsistent and beset by
numerous difficulties throughout the immediate post-war period.[92]

Perhaps foremost among these difficulties were problems in gathering evidence
and detaining the accused individuals and this presented fertile ground for conflict
with the concomitant exploitation initiative. This was particularly true in the field
of medicine, where the Nazis had often relied on brutal human experimentation
in the pursuit of progress – means which were rarely, if ever, justified by the
ends.[93] Some historians have claimed that the Allies faced a stark choice in this
matter, between exploiting German know-how and prosecuting its criminality,
and there is some evidence to that effect.[94] At a meeting of the Joint Intelligence
Sub-Committee in September 1945, it was stated that 'War Crimes trials had priority over Intelligence investigations ... and that the loss of Intelligence from
this source would have to be accepted'. R.A. Clyde, of the British War Crimes
Executive, responded to this with surprise, commenting that his organisation had
'always considered the preparation of cases against war criminals to be of secondary importance to our intelligence requirements'.[95]

However, interpretations such as these were exceptions rather than the rule.
Any notion of a larger conflict of interest only emerges through creative retrospective analysis and would not have appeared as such to the agents on the ground.[96]
The two initiatives coexisted but co-operation was more likely than competition
in most cases. This was mostly a result of the individuals involved in exploitation
who, being human beings in possession of a moral compass, were incensed by the
horrors of which they found evidence during their investigations. For example,
Leo Alexander, an Austrian-Jewish doctor who immigrated to the USA in 1933
and would later become a senior medical adviser to the Nazi Doctors' Trial at
Nuremberg in 1946, began his post-war career as a CIOS investigator, tasked
with examining German aviation medicine.[97] In several of his official CIOS
reports, which were always supposed to be objective and factual, Alexander criti-
cises the inhumanity of the experiments, describing one set as 'a callous waste
of unnecessarily large numbers of human lives'.[98] In another, he unequivocally
recommended that the German doctors involved 'should definitely be tried as
war criminals for these forced experiments on human beings', citing not only
the 'unnecessary infliction of pain, suffering and death' on the subjects but also

the fact that the results added nothing new to what had been learnt from previous animal experiments.[99] Elsewhere, when Ministry of Supply investigators encountered Otto Ambros, the IG Farben nerve agent specialist who was accused of authorising cruel human experiments and overseeing slave labour at Auschwitz, they 'commented adversely on the friendly treatment being given to this man who is suspected of war criminality'.[100]

It was not just morally indignant individuals who created the links between exploitation and war crimes investigation; there were also connections made via unofficial networks operating through official channels. For example, on 15 May 1946, Brigadier R.J. Maunsell, the chief of FIAT, chaired a meeting, consisting of nine Brits, four Americans and two Frenchmen, the purpose of which was to establish a policy for exploitation staff to handle any material pertaining to war crimes.[101] In his opening statement, Maunsell explained that, in the course of their regular work, FIAT officials 'had accumulated some material which bore on the commission of war crimes by German scientists' but that FIAT 'could not deal with this question since it was not within their terms of reference and no investigational organisation of this character was available within FIAT'. Instead, he pledged that all possible assistance would be rendered by FIAT to the war crimes agencies, including the loaning of scientific experts to aid legal investigators and interrogators.[102] This was especially essential as the organisations created to investigate war crimes were often short-staffed and under-resourced, particularly when compared to the well-equipped behemoth of exploitation, and is yet another example of co-operation and collaboration trumping any competitive urge.[103]

In all, the relations which the exploitation agencies had with other organisations operating in Germany at the same time were mixed. Initially, they followed an instinct of competition, which was reflected on an international scale, too, driven by the desperation to secure the best parts of Germany's finite human resources (as well as documents and other material) for themselves. As time passed, however, it became clear that their interests were best served by co-operation, not least because they shared a common purpose – ensuring Germany never again posed a threat to world peace – and later, a common enemy – the Soviet Union. This co-operation was, in many ways, the key to the exploitation programme's procedural successes as it allowed for a sharing of scarce resources and access to expertise in a very broad range of fields.

This chapter has explored the programme of exploitation during its most active phase. In the immediate aftermath of the war, with Germany at the mercy of its occupiers, the process to gather its scientific spoils both gathered speed and increased in scale. The administrative and logistical machinery necessary to carry it out was quickly established and, though often complicated and occasionally obtuse, largely served its purpose, bringing a huge number of investigators into Germany, directing and transporting them to the most pertinent targets and then shipping back their confiscated materials and circulating their findings. To say it ran like clockwork would be an exaggeration, but it was perhaps the best that could have been hoped for, given the problematic post-war context. It should be noted that the real strength often lay in the organisations in the field in Germany, which operated flexibly and pragmatically, rather than in the committees back in Britain which, distanced from the realities on the ground, were often too rigid and unable to adjust to changing circumstances.

A testament to the operation's successes can perhaps be best seen in the recorded experiences of the exploitation personnel themselves, many of whom praised the facilities and support on offer, even if the quantity and availability did not always meet their expectations. Accounts which paint a more negative picture perhaps tell us more about the individual personality of the author than they do about the actual practices of British exploitation in Germany. Perhaps the area where the programme faced the greatest difficulties was in its relationships with other, parallel organisations and initiatives, though here, once again, the potential conflicts which periodically threatened to flare up in Whitehall were often ameliorated by the genuine collaborative spirit among operatives on the ground. There were, of course, larger issues at play in this than a simple competition over resources and priorities, especially in terms of the moral predicaments posed when particularly valuable material or individuals were tainted by their association with the darkest elements of the Third Reich, which is a topic that will be explored in greater depth elsewhere in this book. In any case, it is neither appropriate nor effective to judge exploitation on the merit of its procedure and implementation alone; a much more relevant assessment can only be achieved by also considering the fruits of its labours, in the various forms which these took.

Notes

1 Bessel, *Germany 1945*, 169–70.
2 Longden, *T-Force*, 95.
3 TNA, FO 1032/177, 'Notes for Investigators in the British Zone', 5 September 1945.
4 TNA, FO 1032/177, 'BIOS: Organisation', 18 July 1945.
5 TNA, FO 1032/177, 'Draft Terms of Reference for BIOS', 18 July 1945.
6 TNA, CAB 122/342, 'Exploitation of German Science and Technology', 12 December 1945.
7 Longden, *T-Force*, 277.
8 Edgerton, *Britain's War Machine*, 273ff.
9 TNA, FO 1031/4, 'Statement from the Administrative Staff of the Deputy Military Governor Regarding T-Force', 10 October 1947.
10 TNA, FO 1065/12, 'Organisation and Future of T-Force', 5 December 1946.
11 TNA, FO 1031/4, 10 October 1947. For more on this distinction, see Chapter Six.
12 TNA, FO 1032/177, 'Notes for Investigators in the British Zone', 5 September 1945.
13 TNA, FO 1065/12, 'Investigation of T-Force: Report No. 47', 12 December 1946.
14 Rex Pope, 'British Demobilization after the Second World War', *Journal of Contemporary History*, 30 (1995), 65–81; Alan Allport, *Demobbed: Coming Home after the Second World War* (New Haven, CT: Yale University Press, 2010); David French, *Army, Empire and Cold War: The British Army and Military Policy, 1945–1971* (Oxford: Oxford University Press, 2012), 36–7.
15 TNA, FO 1065/12, 'Investigation of T-Force: Report No. 47', 12 December 1946.
16 Graham-Dixon, *Allied Occupation of Germany*, 84.

17 TNA, FO 1065/12, 'Establishment of Field Information Agency, Technical (FIAT), British Component', 23 August 1945.
18 TNA, FO 1031/50, 'Minutes of 2nd BIOS Meeting (1945)', 29 August 1945.
19 TNA, FO 1032/1459, 'Field Information Agency Technical (FIAT)', 10 December 1946.
20 TNA, FO 1031/50, 'Minutes of 10th BIOS Meeting', 20 December 1945.
21 TNA, FO 1031/50, 'Minutes of 7th BIOS Meeting', 7 November 1945.
22 TNA, CAB 81/93, 'Minutes of 49th Meeting of JIC', 24 July 1945.
23 TNA, CAB 82/3, 'Minutes of 11th Meeting of DCOS', 6 September 1945.
24 TNA, CAB 82/8, 'Minutes of 26th Meeting of DCOS', 4 September 1946.
25 Douglas O'Reagan, 'Science, Technology, and Know-How: Exploitation of German Science and the Challenges of Technology Transfer in the Postwar World', Ph.D. dissertation, University of California, Berkeley (2014), 27.
26 TNA, FO 1032/177, 'Notes for Investigators in the British Zone', 5 September 1945.
27 TNA, FO 1031/7, Deputy Military Governor to T&I Division, 25 November 1946.
28 TNA, FO 1031/7, 'Advance Notes for Investigators', 2 September 1946.
29 TNA, BT 211/541, 'Minutes of BIOS/Board of Trade Meeting on "Facilities for Technical Investigations in Germany by Individual British Firms"', 25 October 1946.
30 Perry Biddiscombe, *Werwolf! The History of the National Socialist Guerrilla Movement, 1944–46* (Toronto: University of Toronto Press, 1998), 50.
31 TNA, FO 1031/7, 'Advance Notes for Investigators', 2 September 1946.
32 TNA, FO 1031/7, 'Opportunity Targets – Berlin', 31 July 1946.
33 TNA, FO 1031/7, 10 August 1947.
34 TNA, FO 1031/7, 'Advance Notes for Investigators', 2 September 1946.
35 TNA, FO 1031/7, 'British Investigators Billets Standing Orders', 22 July 1946.
36 TNA, FO 1031/7, Maj. Hughes, HQ T-Force to Maj. Baukham, FIAT, 1 August 1946.
37 Foreign Office, *Instructions for British Servicemen in Germany 1944* (London: Foreign Office, 1944), 51.
38 Noel Annan, *Changing Enemies: The Defeat and Regeneration of Germany* (London: Harper Collins, 1995), 148.
39 Atina Grossmann, *Jews, Germans and Allies: Close Encounters in Occupied Germany* (Princeton, NJ: Princeton University Press, 2009), 33.
40 Eckert, *Struggle for the Files*, 32; Schmidt, *Justice at Nuremberg*, 99.
41 IWM, 09/21/1, Private Papers of Mr Gilbert A. Hunter, January 1946.
42 IWM, 99/76/1, Private Papers of Monica Maurice, 16–17 April 1947.
43 IWM, 09/21/1.
44 IWM, 99/76/1, 25 May 1947.
45 IWM, 05/48/1, Private Papers of Edward C. Aspden, September–October 1945.
46 IWM, 10/7/1, Lt-Cdr John Bradley to Margaret Bradley, May 1945.
47 IWM, 05/48/1.
48 IWM, 09/21/1.
49 Howard, *Otherwise Occupied*, 56–57.
50 IWM, 99/76/1, 21 April 1947.
51 Bessel, Germany 1945, 333–37; Hoffmann, 'Germany is No More', 604–05.
52 IWM, 05/48/1.
53 IWM, 10/7/1.
54 IWM, 99/76/1, 25 April 1947.
55 IWM, 09/21/1.
56 IWM, 99/36/1, Private Papers of Brigadier W.P.T. Roberts CBE, May 1945.
57 IWM, 09/21/1.
58 Michael Howard, *Otherwise Occupied: Letters Home from the Ruins of Nazi Germany* (Tiverton, Devon: Old Street, 2010), 57.
59 IWM, 05/48/1.
60 IWM, 99/76/1, 19 April 1947.
61 IWM, 09/21/1.

62 IWM, 05/48/1.
63 Howard, *Otherwise Occupied*, 90.
64 TNA, FO 1031/49, 'History of T-Force', 1945.
65 TNA, FO 935/25, C.H. Noton to Lt-Col. E. Ewart-Williams, 30 January 1945.
66 TNA, FO 935/25, Col. C.G. Vickers to C.H. Noton, 19 February 1945.
67 Arnold Krammer, 'Technology Transfer as War Booty: The US Technical Oil Mission to Germany, 1945', *Technology and Culture*, 22 (1981), 68–103.
68 TNA, FO 1031/68, 'List of British and American Agencies', 11 September 1946.
69 TNA, FO 1031/51, 'Scientific Intelligence Advisory Section of G-2 SHAEF', 7 April 1945.
70 TNA, FO 1031/51, 'Facilities Offered to BIOS by German Economic Department', 4 October 1945.
71 TNA, BT 211/116, R.G. Somervell to Sqn. Ldr. Harris, 14 November 1945.
72 TNA, FO 935/25, Maj. G.W.H. Andrews to C.H. Noton, 30 January 1945.
73 Dylan, *Defence Intelligence*, 1–3.
74 Eckert, *Struggle for the Files*, 33.
75 Michael J. Kurtz, *America and the Return of Nazi Contraband: The Recovery of Europe's Cultural Treasures* (Cambridge: Cambridge University Press, 2006).
76 TNA, FO 1032/179, 'The Handling of German Documents and Archives', 6 September 1945.
77 Jones, *Most Secret War*, 608–10.
78 TNA, AIR 20/1715, 'German Scientists', 18 October 1945.
79 TNA, FO 942/425, C.F.C. Spedding to C.A. Spencer, 29 March 1946.
80 Simpson, *Blowback*, 73.
81 Overy, *Bombing War*. See also Süss, *Death from the Skies*; Mark Connelly, *Reaching for the Stars: A History of Bomber Command* (London: I.B. Tauris, 2014).
82 British Bombing Survey Unit, *The Strategic Air War against Germany* (London: Frank Cass, 1988); David MacIsaac, *Strategic Bombing in World War Two: The Story of the USSBS* (New York: Garland, 1976).
83 TNA, AIR 19/434, Air Cdre C.B.N. Pelly to N.H. Bottomley, 23 April 1945.
84 TNA, AIR 20/4818, 'Minutes of 1st BBSU Advisory Committee Meeting', 6 June 1945.
85 TNA, FO 935/140, 'British Bombing Survey Unit', 30 July 1945.
86 Sebastian Cox, 'Introduction: An Unwanted Child – The Struggle to Establish a British Bombing Survey', in BBSU, *The Strategic Air War against Germany*, xix.
87 Connelly, *Reaching for the Stars*, 158–59.
88 Tom Bower, *Blind Eye to Murder* (London: Andre Deutsch, 1981), 37; see also Richard Breitman, *Official Secrets: What the Nazis Planned, What the British and Americans Knew* (New York: Hill & Wang, 1998).
89 Jay Winter, 'From War Talk to Rights Talk: War Aims and Human Rights in the Second World War', in David Welch and Jo Fox (eds.), *Justifying War: Propaganda, Politics and the Modern Age* (Basingstoke: Palgrave Macmillan, 2012), 242.
90 Tusa and Tusa, *Nuremberg Trial*.
91 Kim C. Priemel and Alexa Stiller (eds.), *Reassessing the Nuremberg Military Tribunals: Transitional Justice, Trial Narratives, and Historiography* (Oxford: Berghahn, 2012).
92 Donald Bloxham, 'British War Crimes Policy in Germany, 1945-1959: Implementation and Collapse', *Journal of British Studies*, 42 (2003), 91-118; Anthony Glees, 'The Making of British Policy on War Crimes: History as Politics in the UK', *Contemporary European History*, 1 (1992), 171–97.
93 Ulf Schmidt, *Karl Brandt, the Nazi Doctor: Medicine and Power in the Third Reich* (London: Continuum, 2007), 255–96; Ulf Schmidt, 'Scars of Ravensbrück: Medical Experiments and British War Crimes Policy, 1945–1950', *German History*, 23 (2005), 20–49; Robert Proctor, 'Nazi Doctors, Racial Medicine, and Human

Experimentation', in George J. Annas and Michael A. Grodin (eds.), *The Nazi Doctors and the Nuremberg Code: Human Rights in Human Experimentation* (Oxford: Oxford University Press, 1992), 17–31.

94 Weindling, *Nazi Medicine and the Nuremberg Trials*, 3.
95 TNA, CAB 81/93, 'Minutes of 64th Meeting of JIC', 18 September 1945.
96 Schmidt, *Justice at Nuremberg*, 111.
97 On Leo Alexander, see Schmidt, *Justice at Nuremberg*.
98 IWM, CIOS Report XXVI–37, 'The Treatment of Shock from Prolonged Exposure to Cold, Especially in Water'.
99 IWM, CIOS Report XXIX–21, 'Miscellaneous Aviation Medical Matters', 22 August 1945; Proctor, 'Nazi Doctors', 36.
100 TNA, FO 1031/86, 'Poison Gas: Interrogation and Reports', 4 September 1945.
101 Schmidt, *Justice at Nuremberg*, 124–25; Weindling, *Nazi Medicine and the Nuremberg Trials*, 111–15.
102 TNA, FO 1031/74 'Scientific and Technological Branch Policy on Unethical Medicine and Medical War Crimes', 15 May 1946.
103 For more on this, see Charlie Hall, 'A Conflict of Interests? British Efforts in the Pursuit of Post-War Justice and Technical Intelligence in Occupied Germany', M.A. dissertation, University of Kent (2013).

4   The spoils of war

At its core, the exploitation programme was driven by a desire to use German science and technology to benefit the armouries and industries of Britain. This began, as we have seen, by sending teams of expert personnel into the ruins of the Third Reich to investigate facilities, examine equipment, and interview relevant members of staff. This was only the first half of the scheme, however, and results were not truly felt until the findings and seizures made by these teams began to flow back to Britain. Alongside the German scientists and technicians who were detained, interrogated and, in some cases, recruited by the British during this period, the scientific and technological spoils of war came in several different manifestations, which this chapter will explore. First, there were documents and archives which, once discovered in their often secretive locations in Germany, were shipped to Britain for comprehensive analysis and assessment. Second, equipment and materiel were also transported back, either to fill shortages from which post-war Britain was suffering, as in the case of machine tools, or to be deconstructed and studied to reveal their technological secrets, as in the case of V-2 rockets or nerve gas shells. Thirdly, the final reports which every Combined Intelligence Objectives Sub-Committee (CIOS) and British Intelligence Objectives Sub-Committee (BIOS) team was told was 'the real object' of their trip to Germany, were compiled upon their return and all those on non-military topics were then mass-produced, thus becoming widely available to the public and to private industry across the country.[1]

Finally, there was something of a reverse in flow in the exploitation process. For a number of reasons, the focus changed from bringing materials and intelligence back to Britain and, instead, with the increased stability which had been achieved in occupied Germany, it became more common to facilitate the visits of all manner of private individuals to German factories, laboratories and other sites of interest, and to transform this process from one of intensive government oversight to a more widely accessible form of scientific, industrial and commercial 'tourism'. In addition, this chapter will look at three areas of German scientific and technological endeavour which became the subject of particularly intense focus from the British exploitation staff – chemical and biological warfare, rocketry and aeronautics. These case studies provide a closer look at the motivations and mechanisms of exploitation, the development of special missions to handle particularly pertinent subjects, and the wider successes and failures experienced therein.

Material spoils

Documents and archives were among the most important prizes of exploitation, whether these constituted blueprints, technical specifications, instruction manuals or scientific research articles. The bulk of those brought over to Britain were handled at the Halstead Exploiting Centre, near Sevenoaks in Kent, where a number of German POWs and civilians were engaged in translation.[2] Once translated, these documents could provide clear and comprehensive information on anything from industrial processes to records of development and experimentation and, especially in the case of blueprints or technical drawings, could allow British laboratories or businesses to very quickly replicate German practices after the war. They were described as having 'a considerable intrinsic value. At the least, this may be the cost of raw materials and man-hours but, in actual value, the cost would be much higher.'[3] As such, it was considered desirable to implement a clear policy on them as soon as post-war exploitation got underway. In September 1945, the British Element of the Control Council issued a directive entitled 'The Handling of German Documents and Archives', which intended to:

> ensure the opportunity of access by all exploiting agencies to such German documents as are essential to their researches and to prevent the researches of one agency from impeding those of another; and to provide means for the collection and dissemination of information concerning the location, movement and content of German documents and archives.

What this meant in practice was the establishment of Documents Centres at the headquarters of the British Army of the Rhine and in each of its Corps Districts, which were to be responsible for the protection of the files, the maintenance of a register of information on all of the contents and the circulation of this register to all interested bodies.[4] Unfortunately, as the staffing of these Documents Centres was mostly provided by ordinary T-Force personnel, proper archival procedure was not always followed and archivists often had to explain in exasperation that if finding aids were lost or destroyed, a professional, organised archive lost the great majority of its value.[5]

Evacuation of documents from the British zone back to Britain itself also became a highly contentious issue. In December 1945, the Economic Division released a memo which stated that 'sufficient emphasis is not given to the commercial value of technical drawings and specifications', especially when compared to that given to prototype machines and similar items, and that, as a result, they were being removed without regard to 'proper safeguards'.[6] There were also considerable delays in making documents accessible to interested parties in Britain as a result of a number of factors, including investigators not following instructions, the necessity of relying on sea transport to evacuate files and the lack of any kind of reproduction service operated by the British (there was a German service but it was perpetually short on supplies).[7] All of this was being discussed and debated at the same time as pressure was mounting to destroy large numbers of documents and drawings, as the teams involved had 'not the time nor the staff' to assess them all, even though there were warnings that 'by so doing we may miss something useful'.[8]

The procedure for seizure and evacuation of equipment was equally beset by problems. Demand was high, whether for 'the purpose of equipping defence establishments' or for 'research and intelligence purposes', but was also restricted, most notably by the fact that it had to be accounted for in terms of the total reparations which Britain was allowed to claim.[9] Moreover, if equipment was removed from a target by one team, it could substantially prejudice the investigations of another team, if they were to visit that same target subsequently.[10] This, of course, took on an additional degree of complexity if the desired equipment was located in the American zone where, unless it was considered 'reparationable', the only way of obtaining it was by way of a 'straight purchase'.[11] It is worth noting also that this was not a small-scale issue – by the end of 1946, T-Force reported that '6,590 tons of equipment have been shipped to the UK since June 1945 and a further 11,182 tons have been earmarked and are awaiting shipment'. The type of equipment this included ranged from delicate scientific apparatus and optical instruments to tanks and major machine tools.[12] Sometimes, the confiscated objects were even larger, as in the case of *U-1407*, an experimental high-speed U-boat, equipped with a prototype high test peroxide engine designed by the German submarine expert Hellmuth Walter, which was transported to Britain, refitted at the Vickers shipyard in Barrow-in-Furness under Walter's direct supervision and then recommissioned into the Royal Navy as HMS *Meteorite* on 25 September 1945, to be used for trials and testing purposes.[13]

In a few cases, Britain – and British companies – considered going beyond the removal of machinery or finished products and instead aiming to utilise entire German factories or production centres. The most prominent example of this concerned the Volkswagen plant in Wolfsburg (a new town which had been constructed by the Nazis to support the production of their so-called *KdF-Wagen*, or 'Strength Through Joy Car') which lay just inside the British zone of occupation. In 1945, British administrators, led by Major Ivan Hirst of the Royal Electrical and Mechanical Engineers Corps, took the plant under 'trusteeship', a technique of overseas rule more commonly seen in Britain's imperial colonies, and set about manufacturing vehicles to be used by the occupation authorities. During this time, members of the British automobile industry visited Wolfsburg to inspect the plant and the cars and to decide whether it should be relocated to Britain. While they were impressed with the scale and potential efficiency of the Volkswagen facilities, they believed the car itself (later to enjoy enormous success as the Volkswagen Beetle) would not meet the high standards expected by the British public and were concerned that the introduction of such a huge factory to Britain would throw domestic motor manufacturing into turmoil.[14] While the decision not to seize the Volkswagen factory has been judged in retrospect to have been a mistake, the reality is that at the time it was a considered and seemingly prudent conclusion to reach. On the whole, Britain and its industries preferred to pursue a more gradual, piecemeal approach to exploitation.

In fact, it would be fair to surmise that the main product of exploitation was always expected to be a sizeable catalogue of Final Reports, prepared by CIOS and BIOS investigators and covering every topic of scientific, technological or industrial interest in Germany. Exploitation staff were told in no uncertain terms that 'the information you gain is valueless unless it is fully and clearly set out in this report', and they were notified that it would 'have a wide distribution which includes UK, USA, France and other Allied countries'.[15] By November 1946, there were already 1,039 BIOS Reports available to interested parties, divided into categories such as Communications, Transport, Nutrition and Heavy Industry.[16] This figure does not account for any military topics, which were given a 'Top Secret' classification and thus remained out of the public domain.[17] The number also continued to grow after 1946 and, though a precise final figure is difficult to come by, estimates of around 3,000 seem reliable.[18] Some of the topics covered in these reports were incredibly specific and sometimes bordered on the bizarre – for example, BIOS Report 227 was concerned with 'Manufacture of Harmonicas (mouth organs) and Accordions' while BIOS Report 275 contained a section entitled 'Equipment for Mechanical Cheese Making'. Others focused on the manufacture of fountain pens, chocolate, gramophone needles, crocodile-skin leather and toilet paper.[19] The sheer breadth and variety of topics covered reveal a remarkable appetite for these scientific and commercial spoils of war in Britain as well as a willingness within the official exploitation machinery to facilitate such investigations.

In Britain, the publication of all reports was handled by His Majesty's Stationery Office (HMSO), from whence all non-classified reports were then sent to libraries, trade associations and HMSO's own sales offices.[20] By July 1947, the Board of Trade noted that some 2 million copies of the 2,000 different reports prepared by British and American agencies had been sold or distributed.[21] For those looking to buy reports outright, the costs varied hugely, from 2 pence (roughly 30 pence today) for 'Technical Developments in the German Margarine Industry' to 42 shillings (roughly £75) for 'A Survey of the German Can Industry during the Second World War'.[22] However, direct sales of the reports were never intended to represent the bulk of circulation; instead, the greater part of the burden was to be shouldered by libraries. In early 1946, the Board of Trade developed a template letter to be sent to all public libraries across the country, which began by saying that the government was 'faced with the problem of making the Technical Intelligence that our industrialists have obtained from Germany available to the business community of England'. The letter continued by asserting that libraries could 'play a conspicuous part in bringing this information to the notice of the small man'.[23] By the spring of 1948, 66 libraries and eleven Chambers of Commerce held a collection of reports for reference; these were located across the country, from Aberdeen to Plymouth and from Ipswich to Swansea, with twelve locations in London alone.[24] It is interesting to note that the security classification of certain material could vary depending on its eventual use – two reports on the same piece of technology or scientific research could be classified completely differently, for example as Top Secret for a military application, but as unrestricted for any peaceful, civilian use.[25]

While the quantity, dissemination and scope of the Final Reports was ample, their quality and value did not always meet such high standards. In the 4 July 1946 edition of *World's Press News*, an article appeared entitled 'Criticisms of BIOS Final Report No. 343' which lambasted this report (which detailed the German diesel engine industry), focusing particular ire on the fact that 'practically all the photographic reproductions are so blurred that it is quite impossible to distinguish any detail whatever, so that they have no value to the industry at all' and describing most of the drawings as 'appalling and a disgrace to any publication'.[26] In addition to complaints about production quality such as these, reports were also criticised, including by senior trade association figures, for containing information of nothing novel and for being poorly written, very limited on detail and impossible to follow up.[27] That said, BIOS reports were considered far superior to the equivalents produced by the USA (a judgement shared by American exploitation officials), which was attributed to the broad composition of the British teams, comprising representatives from across the relevant industry. The Soviet Union were so impressed with the BIOS reports that they spent approximately $400,000 per year purchasing every single one.[28]

Nonetheless, Britain's business community remained dissatisfied with simply reading these dry and ultimately limited reports and, as such, it is really no wonder that the form which exploitation took shifted so as to represent a reversed flow. As well as general dissatisfaction with the quality of the Final Reports, there are several other reasons why facilitating the travel to Germany of private individuals became the most viable way for exploitation to continue after the initial post-war rush. As mentioned above, documents and blueprints were far too numerous and labour-intensive to be effectively utilised and all equipment which was shipped to Britain had to be deducted from the national reparations allowance, which was obviously both finite and highly dependent on problematic assessments of value. Moreover the removals of any form of physical material began to be wound down from mid-1947, as part of multilateral international agreements, though the British found workarounds to continue this to some extent, in contradiction of announced policy.[29] In November 1947, the BIOS Secretariat was absorbed into the new Technical Intelligence and Documents Unit (TIDU), but T-Force continued to exist almost a year beyond this point.[30]

The role of T-Force however changed dramatically in this period. The number of BIOS teams they handled was expected to drop to a nominal figure, while the number of commercial buyers was expected to reach an average of some 200 a month, visiting businessmen to stay at roughly 60 a month and official Reparations and Restitution Teams were expected to rise from 150 to 600 a month.[31] This shifting onus of exploitation from official ministry-sponsored teams to private individuals and groups was, to a certain degree, an inevitable outcome of the close involvement which business and industry had had in the programme from the start. This was itself an extension of the innumerable contributions British firms had made to the war effort and which had proved so invaluable to Britain's capacity to fight in such a technologically advanced conflict.[32] Initially, as we have seen, this post-war collaboration was characterised by industrial concerns supplying the experts to participate in BIOS missions to Germany, and working closely with the various branches on any matters relating to the procurement of information from Germany.[33] This was a productive relationship because, as BIOS investigator Monica Maurice remarked, there were many important pieces of information which 'you can only detect by knowing the job'.[34]

As time moved on though, it became clear that it would perhaps be more mutually beneficial if representatives of industry were able to conduct investigations unrestrained by the regulation and administration of government. It was acknowledged as early as the summer of 1945 that 'the question of British industrial and commercial visitors who are now on private rather than official business' would be subject to 'very great ... political pressure in favour of permitting such visits'. However, no such allowance could be made at the time as resources were so scarce that 'no [German] industry can be started that is not vital either to the needs of the occupying forces, or for maintaining the standard of the civil population at the minimum necessary to prevent disease and unrest'.[35] There were, of course, loopholes to be exploited. In October 1946, Margarete Steiff GmbH, a stuffed toy manufacturer based near Heidenheim, famed for its invention of the teddy bear, complained to US occupation authorities that representatives of rival British toy firms had visited their factory, taken photographs of special machinery and demanded samples. The company argued, justifiably, that the manufacture of toys was the most peaceful of industries and the actions of the British amounted to no less than unrestricted commercial espionage.[36] However, by early 1948, these cases ceased to be unsavoury exceptions, as policy changed to facilitate this private exploitation on a larger scale.

One crucial organisation in this period was the Joint Export Import Agency (JEIA), an Anglo-American body tasked with certain aspects of Germany's post-war economic recovery, including facilitating trips of Western businessmen to Germany and German businessmen abroad.[37] In February 1948, at a meeting of the JEIA, it was asserted as essential that 'the facilities afforded to businessmen [in Germany] be as nearly as possible equivalent to those in other European countries today'. The first step in the process towards achieving this end was the reactivation of the German hotel industry which, by this stage, had been successfully carried out in nine hotels in the American zone but not at all in the British zone. The idea was that these hotels, usually de-requisitioned T-Force messes, would 'lead an entirely normal life; bars would be provided and stocked, prices would be on a commercial basis and visitors would not have to conform to the Military Government regulations normally to be found in Transit Hotels'. However, they would also be helped with the initial set-up by JEIA, who would, for instance, help source furniture and buy food in bulk.[38] This coincided with a remarkable desire among the German population to revitalise their own domestic tourism industry, as a method of returning to pre-war 'normality'.[39]

An attempt to replicate this American success in the British zone was made on a purely experimental basis, where a hotel was handed over to German ownership but provisioned by the British Army. The verdict was not wholly promising:

> The experiment had not been a complete success owing to the extreme demands of the employees, suspected black market activities in food and a reluctance on the part of a German manager to continue under these conditions. It had also been found necessary to employ a British supervisor to deal with complaints and to supervise the general running of the hotel.[40]

Discouraged, the British authorities began to consider alternative options, such as the outsourcing of all private and civilian travel to the commercial travel agent, Thomas Cook, but not only did the company 'refuse to undertake such a commitment, it was also established that it would be uneconomical to outsource this task anyway'. In the end, as the disbandment of T-Force loomed ever closer, the chosen course of action was to hand over responsibility for these hotels from T-Force to the Regional Administrative Offices of the Control Council.[41] By 23 May 1949, the issue had become moot as the occupation zones of Britain, the USA and France ceased to exist and were replaced by the Federal Republic of Germany, an essentially sovereign nation.

It is worth noting as an aside that, after the war, Germany was not the only nation which was subject to exploitation. BIOS was also responsible for co-ordinating a similar programme in Japan, although the bulk of the investigations there were conducted by American personnel, as they constituted the majority of the occupying force.[42] In August 1945, the BIOS Committee was informed that although Japanese research had been fairly advanced, they had generally failed to apply science to war on a substantial scale and that, as a result, any investigations in Japan should focus on research rather than development and on laboratories rather than plants.[43] There were certain areas of interest though, such as biological warfare, in which the Japanese pursued fairly advanced research.[44] By November, BIOS had been instructed to liaise with the British Staff Office in Tokyo and perhaps create a similar Black List as the one used in Europe, albeit smaller.[45] The initial steps which BIOS took were uncertain and poorly governed, while the USA was pressing ahead and flooding Japan with scientific intelligence missions.[46] In the end, it was decided that 'BIOS had no mandate for securing Japanese intelligence, only of receiving reports' and their role would only be to help in the translation of documents and the 'channelling' of reports into Britain.[47] An interesting epilogue to this narrative is that CIOS and BIOS reports on German science and technology were made available for consultation in Japan from 1949 onwards, which was considered of great value to Japanese businesses and research institutions and helped to close the technology gap between these two former Axis allies.[48]

In Germany, the story which the exploited materials tell is one of an effort to obtain the best results for British science and industry by amassing a vast quantity of documents, equipment and Final Reports compiled by expert investigators. Certainly, these spoils of war had their benefits, but they could not unravel the whole, or even the greater part, of the secrets of German science and technology. Within a month of the war's end, CIOS had recognised that 'some of the most spectacular results have tended to be associated with information obtained from men rather than from places' and it was not long before British policy changed to reflect these realisations. Greater emphasis in exploitation was therefore placed on German scientists, technicians and other knowledgeable personnel, who not only had the potential for longer-term benefit to Britain but were also less subject to the strict regulations surrounding reparations allocations. In the meantime, however, it was concluded that 'optimum results will be obtained when men, equipment and records bearing on a single problem are examined concurrently at the same place' and three examples of this approach are explored in the following section.[49]

Chemical and biological warfare

There were several areas of German military research and development that particularly piqued the interest of the British exploitation planners towards the end of the war, largely because they offered the most striking instances where the scientific capabilities of Nazi Germany had seemingly exceeded that of Britain and its allies. For example, the technology of the Luftwaffe, which had helped secure Nazi mastery over much of continental Europe and delivered the horrors of the Blitz to London, generated avid curiosity about the Third Reich's aeronautical facilities. Meanwhile, rocketry – particularly in the form of the V-2 ballistic missile, which was deployed enthusiastically (albeit briefly and not particularly effectively) by the German military against targets in Britain, Belgium, the Netherlands and France – seemed to herald the dawn of a whole new era of warfare. Chemical warfare, on the other hand, was not afforded a particularly high priority, largely because it had not been actively employed in the European theatre during the war. Accordingly, sites relating to chemical warfare research and development proved to be the very definition of 'targets of opportunity', as unsuspecting British soldiers uncovered a whole new class of war gases – the nerve agents – by chance. Therefore, studying the way in which chemical warfare was investigated by exploitation teams sheds light on the adaptability and flexibility which was integral to the programme as a whole.

Despite the ubiquity of gas masks in wartime Britain, carried by men, women and children as they went about their daily lives, no belligerent decided to employ chemical weapons during the conflict. This was probably because a mutual fear of retaliation existed and no country wished to be the first to unleash weapons of such horror, lest they be visited on their own civilian population in return. Germany, however, was in a stronger position than its military leaders or scientific experts realised. In 1936, scientists at Anorgana, a subsidiary of industrial giant IG Farben which focused on chemical products, had, while researching new insecticides, developed an agent which could inflict great harm on the human nervous system, sometimes resulting in death, which gave it great potentiality as a chemical weapon. It was given the designation Tabun and, through further research, two even more potent derivatives were developed, named Sarin and

Soman. These have been described as being as great an advance over the chemical weapons of the First World War as the machine gun was over the musket.[50] Neither Britain nor any of its allies had a chemical weapon that even came close to the destructive power of these new agents but, due to the endemic secrecy of wartime research, Germany remained as oblivious of the Allies' vulnerability as the Allies did of Germany's superiority.[51]

As such, it is unsurprising that the exploitation programme did not at first count chemical warfare as a category of any particular importance in their preparation for entry into Germany. The first indication that it was an area of any interest at all came on 6 April 1945, when British troops came across truckloads of strangely marked shells at a rail-yard in Espelkamp, seven miles north of Lübbecke.[52] In addition, the discovery of a wealth of information, including documents and manufacturing equipment, at a chemical warfare experimental station at Raubkammer and munitions storage facilities at nearby Munster-East and Oerrel proved that this topic warranted further investigation.[53] Fortunately, procedures for the discovery of enemy possession of new chemical weapons had been in place during the war and samples were immediately sent to the British Chemical Defence Experimental Establishment (CDEE) at Porton Down, Wiltshire, where their contents could be examined by experts.[54] So alien to these experts were the chemicals contained within that they all but dismissed them, reporting that 'apart from novelty, [it is] not clear that this charging has any advantage over other well-known chargings.'[55] In short, the initial reaction was that this chemical agent was roughly equivalent in potency to Britain's own obsolescent agent, PF-3; in reality, it was up to ten times more dangerous.[56]

This illusion of relative impotency did not last long and, within days, scientists at Porton had realised that, if deployed on the battlefield, these new nerve agents could have an absolutely devastating effect.[57] Just over two weeks after the nerve gas shells had been discovered, a 19-man CIOS team, including 9 experts from Porton Down, plus 5 other British members, 4 Americans and a Canadian, travelled to Raubkammer and the surrounding sites to conduct their investigations. On arrival they discovered that a great deal of equipment and documents had been transferred there from the main German chemical warfare establishment at Spandau Citadel, near Berlin, during the war, to avoid it being captured by the advancing Red Army. The CIOS team was on site for 6 weeks, at the end of which they filed a 482-page report 'based upon an examination of the range, laboratories, plant and equipment, upon a preliminary examination of a mass of documents and samples, and upon a thorough questioning of all available witnesses'.[58] This was an early example of the type of 'concurrent examination' which was quickly being recognised as the most effective method of exploitation.

Even this extensive assessment was not considered to be sufficient so, in the summer of 1945, Britain despatched a unilateral investigation team, known as Porton Group No. 1, to Raubkammer to conduct a three-month study of German chemical warfare.[59] One of the core elements of this examination was the use of field trials – there were 26 in total, many of which were carried out by German technicians, with members of the Porton group simply acting as observers, in order to understand German technique as well as the nature of the weapons them-selves.[60] This observation-led method would later be replicated in Operation Backfire, for the study of V-2 launching procedure. All in all, the comprehensive investigation into German chemical weapons allowed Britain and the United States to add nerve agents to their arsenals within a few months of their discovery and it also ensured that chemical warfare retained its place in British military doctrine.[61] It also guaranteed the survival of the CDEE at Porton Down, which had been considered for closure due to the supposed eclipse of chemical weapons by the new atomic bomb. Instead, Porton Down became one of the world's leading nerve agent research centres throughout the Cold War, a status it retains to this day. Scientists there embarked on large programmes of experimentation with these toxic substances in the post-war period, as they attempted to learn more about them and develop increasingly lethal variants. Many of these tests involved human subjects, who had not always given their full and informed consent before participating – an issue which became especially problematic following the death of Leading Aircraftsman Ronald Maddison, a National Serviceman who died after being exposed to Sarin in May 1953.[62] Porton Down also oversaw the construction of a nerve gas production facility at Nancekuke in Cornwall, which produced around 20 tons of Sarin by 1955.[63] In this way, Britain's sizeable chemical warfare arsenal was a direct product of the exploitation process.

Porton Down was also responsible for Britain's biological warfare programme, and the energetic wartime work in this field, both defensive and offensive, carried out there was driven by the need to be able to protect the country from, and retaliate in kind against, any form of German bacteriological attack which might occur.[64] This policy was based on a widely held assumption, which predated the war, that Nazi Germany was actively developing biological weapons and possessed an arsenal which far outstripped that of the Allies. In reality, though, this assumption was almost completely false. Though the Third Reich had possessed a biological warfare programme, it had been rendered wholly dysfunctional by the bureaucratic quagmires and departmental infighting which typified so much of the Nazi regime's activity and had not enjoyed any support from the leadership, with Hitler himself expressly forbidding any offensive research.[65] In addition, Kurt Blome, the civilian director of Germany's biological warfare research, revealed in a post-war interrogation that 'all the leading bacteriologists in Germany consider B[iological] W[arfare] to be impracticable and not worthy of any serious study'.[66] Alsos investigators, who counted chemical and biological warfare as part of their remit along with atomic research, recognised, before the war was even over, the truth of the matter and, in March 1945, the War Office decreed that the amount of work entailed in the biological warfare aspect of German disarmament would not even 'justify the services of a full time Technical Officer'.[67]

Nonetheless, exploitation officials still conducted a relatively thorough examination of facilities, documents and personnel with any link to biological warfare in Germany and Alsos produced several reports on their investigations into this field. They also encountered many of the same issues as did their counterparts working on other topics, such as problems in locating senior German experts – even the wife of the Wehrmacht biological warfare expert, Heinrich Kliewe, did not know where he was when interrogated in April 1945 (he later turned up in a temporary German hospital)[68] – and difficulty in securing the relevant files – in late May, Alsos operatives reported locating around 55 large chests, containing 'Top Secret' German biological warfare information which had been evacuated from the Surgeon-General of the Army's office in Berlin, hidden in the cellar of a monastery in the village of Niederviehbach, near Landshut in Bavaria.[69]

It is also worth noting that, despite the overwhelming evidence that Germany posed no biological warfare threat whatsoever, the genuine fear of biological warfare among the Allied officials made them unwilling to ignore any possible lead on the subject, which in turn made them susceptible to being misled, as happened in the series of incidents collectively codenamed as 'Mayfly'. This essentially comprised a plot by a small group of low-ranking German officials to influence the occupation policies of the Western Allies by offering the British authorities exclusive access to the details of a German 'BW weapon capable of destroying the Anglo-Saxon states quickly', which might otherwise end up in Russian hands, if the British did not comply. When the British displayed their clear scepticism, the German plotters switched to bare-faced extortion, threatening to leak documents which 'proved' that Britain was negotiating unilaterally for access to this new biological weapon to Britain's allies, with the aim, presumably, of destabilising the already tense relations between the occupying powers. In return for not disclosing this 'evidence' to the Americans and Soviets, the plotters demanded, among other things, 'the immediate release of POWs and internees, the limiting of denazification and the cessation of the dismantling of German industry', as well as the empowerment of a German cabinet to be headed, almost certainly without his knowledge or consent, by future West German chancellor, Konrad Adenauer. Despite the fact that the British seemed at no point to fall for this almost farcical attempt at blackmail, the 'Mayfly' case rumbled on for six months, from September 1946 to March 1947.[70] Exceptional incidents such as this aside, biological warfare only formed a very minor part of the exploitation scheme, for obvious reasons. Despite this, the British biological warfare programme did not disappear after 1945, with the threat of the Soviet Union simply supplanting that of Nazi Germany as a *raison d'être*.[71] However, as with chemical weapons, though to a far greater degree, any strategic advantages conferred by biological weapons were, by this stage, adopting a secondary role behind the atomic bomb.[72]

Rocketry and aeronautics

By contrast, the potential future use of atomic weapons was one of the many reasons why a good understanding of rocketry was considered so important in the aftermath of the Second World War. Wernher von Braun, German pioneer of rocket technology and one of the most widely desired prizes of exploitation, wrote a report for CIOS when in Allied custody after the war in which he saw 'possibilities in the combination of … the harnessing of atomic energy together with the development of rockets, the consequence of which cannot yet be fully predicted'.[73] Major-General A.M. Cameron, of the Allied Special Projectile Operations Group, shared von Braun's vision of the future, noting that if a V-2 could be fitted with an atomic warhead, 'its destructive ability will be colossal'. Cameron also hypothesised optimistically about a piloted rocket which could be used as a mail service and would be able to cross the Atlantic in forty minutes, something he felt 'might be of more value than a weapon of war'.[74] More peaceful applications aside, it was commonly believed that rockets would change the face of warfare in the future, with all manner of long-range, and potentially intercontinental, ballistic missiles possible.[75] For Britain, this was an especially acute fear, knowing that the traditional defensive value of the English Channel had been eroded by the development of long-range weapons, particularly in the form of guided missiles.[76]

Whatever its eventual uses were to be, there was no doubt that a good understanding of rocket technology was essential for any nation which desired to exist as a world power after the war. The key to unlocking this was undoubtedly the V-2, also known as the A.4, the results of research into which were largely considered to be 'applicable to all … rocket propulsion problems'.[77] Here it was felt that Britain had a certain advantage, partly because its cities had been the primary target for these new weapons and partly because of the efforts of Duncan Sandys, Financial Secretary to the War Office and later Minister of Works, sometime son-in-law of Prime Minister Winston Churchill, and a staunch advocate of investigation into the German long-range rocket programme, long before the first V-weapons fell on London. Not only had Sandys instigated a major bombing raid on the German rocket development site at Peenemünde in August 1943 but he also chaired the aforementioned wartime Crossbow committee – a subsidiary of the War Cabinet which handled all matters relating to defence against flying bombs and rockets.[78] As a result, by the end of the war, Britain had unilaterally amassed a great quantity of intelligence on German rocketry – Major Robert Staver, Chief of the Jet Propulsion Section within the Research and Intelligence Branch of the US Army Ordnance Corps, admitted that British rocket experts had given him 90 per cent of his target intelligence, including all the information they had gathered on Peenemünde during the Crossbow investigations.[79]

As soon as the war ended, exploitation teams rushed to sites of interest across Germany – not only to Peenemünde, which had been captured by the Red Army in May 1945, but also to storage dumps under British and American control and to the Mittelwerk, the enormous underground missile factory, located near Nordhausen in the Harz Mountains, which was occupied and stripped of everything of value by the Americans before being handed over to the Soviets as part of their zone of occupation. CIOS filed several reports on the subject, including one based on the interrogation of key figures who were held at Garmisch-Partenkirchen in southern Bavaria and were led by both Professor von Braun and the military head of the V-weapon programme, General Walter Dornberger. Both these men had 'the attitude that if they can convince the British and Americans of the value of their work, there is a chance that facilities may be offered in England or America for continuing it'.[80] For the British, though, these investigations were mere preludes to what would prove to be the most comprehensive evaluation of German guided missile technology conducted in the post-war period – Operation Backfire.

On 22 June 1945, General Eisenhower instructed Major-General Alexander Cameron, head of Special Projectile Operations Group, to conduct an operation, the primary object of which was to ascertain the German technique of launching long-range rockets. This was to conclude by actually conducting several launches, in order to prove this method, as well as offering 'opportunities to study certain subsidiary matters such as the preparation of the rocket and ancillary equipment, the handling of fuels, and control in flight'. Three weeks later SHAEF was disbanded, but the operation, now known as Backfire, continued regardless, with the British shouldering the majority of the burden and command for it assumed by the War Office. The chosen site for this operation was Cuxhaven on the North Sea coast, fifty miles north of Bremen, within the British zone.[81] As no 'complete, undamaged and serviceable' rockets had been found, the idea was to assemble them from various parts which had been acquired 'from fields, from ditches, from railway yards, from canals, from factories' during the initial rush of exploitation activity in Germany. The assembly, preparation and firing were all to be conducted by German personnel, with the British experts acting simply as technical officers and observers.[82]

The operation was afforded 'overriding priority' by the British authorities so that it 'should not be handicapped through non-availability of the necessary technical personnel'.[83] Certainly, the most valuable component of Backfire was the assembled group of German technical experts, who brought with them the benefit not only of their accomplishments but also of their mistakes – 'the real ingredient of experience'. One of the experts, Dieter Huzel, pithily described the project as the British effort 'to become familiar with the other end of a trajectory', but felt it was well-conducted and considered the British treatment of the German personnel to have been 'generous'.[84] On the whole, Operation Backfire was a success. Despite initial expectations that it would require 30 rockets,[85] in the end, only three were launched. The first took place on 2 October 1945 and 'the behaviour of the rocket from the moment of take-off to the point of fall was perfect'; the second was far less successful and crashed into the sea almost immediately after take-off; and the third, taking place on 15 October, was designed to be 'a demonstration to representatives from the United States, Russia, France, the Dominions, Whitehall, and the Press'.[86] Hinting at future discord and underhand tactics, the Soviet Union sent six observers despite the fact that only three had been invited, meaning the other three had to linger beyond the secure area and try to catch a glimpse of the rocket in flight.[87] On the whole, Major-General Cameron was suitably impressed by the conduct of his operation, noting in the conclusion to his official report that, in relation to German rocket technology, 'it is believed that all is known and that it now remains for others to make use of that knowledge'.[88]

The success was even celebrated in the public domain, at least in Britain. Edmund Townshend, the *Daily Telegraph*'s special correspondent who attended the third launch, wrote an article which conveyed his awe at the rocket, which he described as 'a pencil on a spear of flame as long as itself'.[89] Members of the

British technical team who had supervised the operation were awarded a trophy in the shape of a V-2 and were invited to attend lectures on the subject, screenings of the official Backfire film and a celebratory dinner which included, for dessert, the mysterious 'A.4 Special'.[90] The far-reaching importance of Backfire was not lost on those involved either and Cameron added a grave warning to the end of his report that, 'for the sake of their very existence, Britain and the United States must be masters of this weapon of the future'.[91] Certainly it can be argued that the most significant weapons technology of the Cold War, after the atomic bomb, was the rocketry which would most likely be used to deliver it – as a result, ballistic missile defence became both a technically challenging and politically controversial field.[92] Despite the successes of Backfire, British attempts to employ German rocket specialists on longer-term contracts had minimal success and, a year after the test-firings, only ten had been successfully recruited for the British government's Rocket Propulsion Establishment at Westcott in Buckinghamshire.[93] Part of the problem was an inability to compete with more appetising offers from the USA. This was already becoming clear to those involved, as noted by American Major Robert Staver, who, having admitted how much his country had relied on British V-weapon intelligence during the war, now recognised the irony that in the post-war missile race between the world powers, it was the British who would finish last.[94]

In contrast, in aeronautics, Britain remained a leading global force throughout the Second World War and beyond. Jet engines, for example, had been developed separately but at roughly the same time in both Britain and Germany, and both the RAF and Luftwaffe had deployed jet-powered aircraft towards the very end of the war.[95] However, like rockets (and even, arguably, the atomic bomb) their impact was more as a showcase for what a future war might look like, rather than any direct effect on the course of the present conflict. On the whole, it was clear that aerial warfare, and the doctrine of air power, was here to stay and those nations that wished to secure a place at the top table of global politics after 1945 would need a strong and technologically advanced air force at their disposal. In addition, developments in aviation technology, especially the jet engine, had obvious and major implications in a civilian context and Britain was keen to use its edge in this field to build a strong post-war export market.[96] As a result, British government and military authorities, and private companies, were keen to get a firm grasp on German wartime developments in order to maintain, or even extend, their global lead.

It is, therefore, perhaps unsurprising that one of the first concerted exploitation efforts launched by the British after the end of the war was concerned with aeronautics. The Fedden Mission – named for its leader, the eminent aircraft engineer Roy Fedden – was established by the Ministry of Aircraft Production (MAP) and travelled to Germany on 12 June 1945 for three weeks.[97] It was tasked with investigating various aeronautical topics, including fuel injection, ignition for aero-engines, gas turbines, jet engines and variable pitch propellers, as well as scouring 'universities, research departments and engineering works in Germany … to earmark plant, equipment, books, instruments etc., suitable for the new College of Aeronautics [at Cranfield] which is now being set up in England'. Travelling through the ruins of Germany, the members of the Mission often found themselves without accommodation, telephone connections or proper food, coming to rely very heavily on American K-rations, but they were impressed by the whole-hearted co-operation of most of the German scientists and technicians whom they interviewed and by the 'superabundance and extravagance of the instruments and subsidiary tools and checking equipment to be found at every factory and labora-tory'.[98] They shipped back to Britain some 2,000 tons of equipment, including jet engines, rocket motors and turbine blades, as well as a large number of blueprints and other documents.[99] The conclusion of the Fedden Mission's final report stated that its members were 'greatly interested with what they saw in Germany' and that they felt 'British industry will be well advised to learn all the lessons possible from German experience and research work', recommending both further interro-gation of the relevant experts (though in Germany, alongside their materials, and not in Britain) and the evacuation of the more elaborate research equipment.[100]

Trends in the investigation of German aeronautics mirrored those in exploi-tation more broadly as the dust settled on post-war Germany – small, fast-moving operations with a cavalier approach (such as the Fedden Mission) fell from favour, while larger, more comprehensive undertakings came to the fore. Operation Surgeon is a prime example of the latter – beginning in July 1945, it was a wide-ranging and well-resourced Air Ministry initiative to discover as much as possible about wartime German aeronautical science, jointly conducted by the Ministry of Aircraft Production and Ministry of Supply.[101] It took place at a num-ber of 'Surgeon stations' across the British zone of Germany, which included the Kaiser Wilhelm Institute at Göttingen, AVA (*Aerodynamische Versuchsanstalt –* Aerodynamic Research Institute) Reyershausen, and the Focke-Wulf facility at Detmold, but by far the most significant was LFA (*Luftfahrtforschungsanstalt –* Aeronautical Research Institute) Völkenrode, located near Braunschweig. British investigators described Völkenrode as having a 'magnificence in layout, struc-ture and furnishing that beggars the imagination of anyone who has seen similar institutions in the UK'.[102] Indeed, it can be argued that the facilities available at these sites for designing and testing aircraft (such as wind tunnels), unrivalled by anything similar in Britain, were prized much more highly than the actual aircraft developed there.

Operation Surgeon consisted of two phases – exploration and removal. During exploration, British investigators examined plant and equipment and ran it for 'calibration, testing and the collection of essential data'.[103] In addition, the British sought 'to pick the brains of German aeronautical scientists by setting them to write monographs of their research work in recent years'.[104] This work entailed the use of 180 German experts and by November 1946 it was reported that LFA Völkenrode alone would produce 252 separate monographs, each of which would be reproduced by a dedicated on-site press and printing department, with a print run of 6 in German and 200 in English.[105] In order to circumvent the restrictions on warlike research laid out in Allied Control Council Law No. 22 (which the British had helped to formulate and were supposed to enforce), the writing of monographs was designated as interrogation and not research.[106] During the sub-sequent removal phase, large quantities of the plant and equipment, which had been studied so closely earlier on, were now dismantled, shipped to Britain and then reassembled. The total amount identified for removal to Britain was 14,000 tons and this included specialist 6,000-horsepower Siemens electric motors for supersonic wind tunnels and advanced ancillary optical equipment to be used for flow visualisation therein.[107] In addition, subsidiary efforts, codenamed Operations Medico and Gold Dust, were mounted to secure scientific equipment from other sources, including some 25 private German companies associated with aircraft and instrument production.[108] Back in Britain, these spoils were shared out among at least 35 public, semi-public and private bodies, alongside the national defence research establishments.[109] Curiously, a section of one wind tunnel evacuated to Britain after the war later made its way back to Germany and is now on display at the Deutsche Museum in Munich.[110]

Perhaps the most significant part of the removal phase, though, was the recruitment of German scientists and technicians who could potentially contribute to British research and development in aeronautics. Not only did Britain hope to gain from the services of these experts, it also sought to deny them to the Soviet Union, fearing that if the Soviets were able to utilise this expertise, it would 'allow them to achieve a long-range bomber force superior to any other in the world in numbers and speed'.[111] Interestingly, in terms of jet engines, German work was actually discovered to be largely inferior to its British equivalent, and both the Soviets and the Americans copied British rather than German designs in the post-war period.[112] The recruitment element of Operation Surgeon ran up against the same bureaucratic roadblocks and moral handwringing as the broader British attempts to employ German scientists and technicians after the war.[113] However, the Surgeon efforts were generally more successful than other parallel attempts and, by the end of 1948, some 87 German aeronautical experts had been transferred to British government defence research establishments, including 26 hired directly from Völkenrode following their participation in Operation Surgeon.[114] The contribution which Operation Surgeon, and the scientists recruited therein, made to subsequent British aeronautical development has been much disputed. Writing in 1957, Roy Fedden lamented that, unlike the USA or the Soviet Union, Britain apparently 'could not be bothered to appreciate the implications of the new aeronautical techniques which Germany had assimilated in such a remarkable way'.[115] More recent scholarship has disputed that and, in any case, though the results may have ultimately been less than were hoped for, this cannot be attributed to any lack of effort on the British part.[116]

Without a doubt, the exploitation process was at its most extensive and comprehensive in the period immediately following the collapse of the Third Reich and the end of the war in Europe. This chaotic situation provided the ideal circumstances for all the victorious powers to pursue their policies of exploitation with maximum vigour and enthusiasm. Documents, prototypes, entire laboratories and other valuable material and plant were all purloined wholesale, often in circumstances of dubious legality. These physical spoils were supplemented with the expertise gained by government-sponsored investigators who conducted thorough examinations of facilities and processes and compiled equally thorough reports on their findings (even if the quality or utility of these reports was sometimes found to be lacking). Combined, these various fruits of exploitation allowed the Allies to learn much about German science and technology – across a remarkably diverse range of topics – and improve their own armouries and industries as a result. Unsurprisingly, the areas which received the greatest attention were those with direct military applications, especially those where German progress had outstripped that of Britain and which looked likely to define, or at least feature heavily, in future conflicts, such as chemical warfare and rocketry. Even in cases where German science had not advanced further than its equivalent in Britain, such as aeronautics, the British were still able to make use of German research facilities to help them continue to innovate in the future. Nonetheless, it soon became clear to the British exploitation authorities that the most promising spoils of war were not files, factories or Final Reports, but rather the German scientists and technicians themselves. As a result, the focus of the exploitation scheme shifted from documents and machines to the men and women who had created and operated them. By utilising this expertise, through interrogation and recruitment, it was believed that greater and more lasting benefits could be obtained for Britain.

Notes

1 TNA, FO 1031/7, 'Advance Notes for Investigators', 2 September 1946.
2 TNA, AVIA 54/1404, 'Halstead Exploiting Centre: Review of Activities'.
3 TNA, FO 1032/179, Maj. D.E. Evans to Brig. Spedding, 14 January 1946.
4 TNA, FO 1032/179, 'Intelligence Directive No. 7: The Handling of German Documents and Archives', 6 September 1945.
5 Eckert, *Struggle for the Files*, 16.
6 TNA, FO 1032/179, 'Evacuation of Documents from British Zone', 11 December 1945.
7 TNA, FO 1031/50, 'Minutes of 9th BIOS Meeting', 19 June 1946.
8 TNA, FO 1032/179, 'Disposal of Drawings', November 1945.
9 Reparations are covered in Chapter Ten.
10 TNA, FO 1031/50, 'Minutes of 4th BIOS Meeting', 26 September 1945.
11 TNA, FO 1031/50, 'Minutes of 8th BIOS Meeting', 29 May 1946.
12 TNA, FO 1065/12, 'Investigation of T-Force: Report No. 47', 12 December 1946.
13 Longden, *T-Force*, 265.
14 Bernhard Rieger, *The People's Car: A Global History of the Volkswagen Beetle* (Cambridge, MA: Harvard University Press, 2013), 97–106; Ian Turner, 'British Occupation Policy and its Effects on the Town of Wolfsburg and the *Volkswagenwerk*: 1945–1949', PhD dissertation, University of Manchester, 1984.
15 TNA, FO 1031/7, 'Advance Notes for Investigators', 2 September 1946.
16 TNA, BT 211/541, 'Survey of Results of BIOS Investigations, 25 November 1946.
17 Balmer, *Secrecy and Science*, 59.
18 Graham Simons, *Operation Lusty: The Race for Hitler's Secret Technology* (Barnsley: Pen & Sword, 2016), 201.
19 Churchill Archives Centre, Metals Society, METL.1/METL.2, CIOS and BIOS reports.
20 TNA, BT 211/23, 'CIOS, BIOS and FIAT Reports', 1946.

21 TNA, AVIA 54/1403, 'The Exploitation by British Industry of German Scientific and Industrial Knowledge', 29 July 1947.
22 IWM, 'Reports on German and Japanese Industry – Classified List No. 18', 31 March 1948. Price adjustments calculated on TNA Currency Converter [accessed online 23 August 2018, http://www.nationalarchives.gov.uk/currency-converter/].
23 TNA, BT 211/23, 'CIOS, BIOS and FIAT Reports', 1946.
24 IWM, 'Reports on German and Japanese Industry'.
25 Balmer, *Secrecy and Science*, 66.
26 'Criticisms of BIOS Final Report No. 343', *World's Press News*, 4 July 1946.
27 Bower, *Paperclip Conspiracy*, 215.
28 O'Reagan, 'Science, Technology, and Know-How', 28–29.
29 Gimbel, *Science, Technology, and Reparations*, 132.
30 TNA, FO 1031/9, 'BIOS papers: general', 22 October 1947.
31 TNA, FO 1065/12, 12 December 1946.
32 Edgerton, *Britain's War Machine*, 244.
33 TNA, FO 1032/177, 'Relations between Industry and Econ. Div. Branches', 27 July 1945.
34 IWM, 99/76/1, Private Papers of Mrs Monica Maurice, 26 April 1947.
35 TNA, FO 1031/10, 'Civilian Travel to Germany', 8 September 1945.
36 Gimbel, *Science, Technology, and Reparations*, 165–6.
37 John Backer, *Priming the German Economy: American Occupational Policies, 1945–48* (Durham, NC: Duke University Press, 1971), 143.
38 TNA, FO 1031/2, 'Minutes of JEIA Meeting', 24 February 1948.
39 Alon Confino, 'Dissonance, Normality, and the Historical Method: Why Did Some Germans Think of Tourism After May 8, 1945?' in Bessel and Schumann, *Life after Death*, 323–48.
40 TNA, FO 1031/2, 'Minutes of JEIA Meeting', 24 February 1948.
41 TNA, FO 1031/4, 'Termination and Transfer of T-Force Commitments', May 1948.
42 Michael Schaller, *The American Occupation of Japan: The Origins of the Cold War in Asia* (Oxford: Oxford University Press, 1985).
43 TNA, FO 1031/50, 'Minutes of 2nd BIOS Meeting', 29 August 1945.
44 Sheldon H. Harris, *Factories of Death: Japanese Biological Warfare, 1932–1945, and the American Cover-Up* (Abingdon: Routledge, 2002).
45 TNA, FO 1031/50, 'Minutes of 8th BIOS Meeting (1945)', 21 November 1945.
46 R.W. Home and Morris F. Low, 'Postwar Scientific Intelligence Missions to Japan', *Isis*, 84 (1993), 528.
47 TNA, FO 1031/50, 'Minutes of 7th BIOS Meeting', 8 May 1946.
48 Yuki Nakajima, 'The Allied Forces and the Spread of German Industrial Technology in Post-War Japan', in Pierre-Yves Donzé & Shigehiro Nishimura (eds.), *Organizing Global Technology Flows: Institutions, Actors, and Processes* (Abingdon: Routledge, 2014), 197–212.
49 TNA, FO 1031/51, 'CIOS Progress Report for 1945', 4 June 1945.
50 Robert Harris and Jeremy Paxman, *A Higher Form of Killing: The Secret History of Chemical and Biological Warfare* (London: Arrow, 2002), 54.
51 Florian Schmaltz, *Kampfstoff-Forschung im Nationalsozialismus: zur Kooperation von Kaiser-Wilhelm-Instituten, Militär und Industrie* (Göttingen: Wallstein, 2005); Jonathan B. Tucker, *War of Nerves: Chemical Warfare from World War I to Al-Qaeda* (New York: Anchor, 2006), 31–40.
52 Schmidt, *Secret Science: A Century of Poison Warfare and Human Experiments* (Oxford: Oxford University Press, 2015), 158.
53 Tucker, *War of Nerves*, 85.
54 Schmidt, *Secret Science*, 157.
55 TNA, WO 208/2183, 'Reports on Phosphorus and Nitrogen Compounds Tabun and Sarin', April 1945. In this context, charging means the active contents inside a chemical weapon, i.e. payload.

56 Schmidt, *Secret Science*, 159.
57 Rob Evans, *Gassed: British Chemical Warfare Experiments on Humans at Porton Down* (London: Stratus, 2000), 115.
58 IWM, CIOS Report XXXI-86, 'Chemical Warfare Installations in the Munsterlager Area', 3 June 1945.
59 Schmidt, *Secret Science*, 165–75.
60 TNA, WO 208/2174, 'Field Technical Assessments by Porton Group', 2 November 1945.
61 Gradon P. Carter, *Porton Down: 75 Years of Chemical & Biological Research* (London: HMSO, 1992), 55.
62 Schmidt, *Secret Science*, 218ff.
63 Tucker, *War of Nerves*, 154.
64 Balmer, *Britain and Biological Warfare*, 19.
65 Erhard Giessler, 'Biological Warfare Activities in Germany, 1923–45', in Erhard Giessler and John Ellis van Courtland Moon (eds.), *Biological and Toxin Weapons: Research, Development and Use from the Middle Ages to 1945* (Oxford: Oxford University Press, 2009), 91.
66 TNA, WO 208/3974, 'Interrogation of Blome', 30 July 1945.
67 TNA, FO 1032/247, Lt.Col. Wansbrough-Jones to Brig. Spedding, 11 March 1945.
68 TNA, WO 208/4280, 'Report on Potential BW targets Visited Between 11–14 April 1945', 26 April 1945.
69 TNA, FO 1031/83, 'Official German Documents and Records on BW', 24 May 1945.
70 TNA, FO 1032/247, 'Mayfly – Situation as at 5 March 1947', 6 March 1947.
71 Brian Balmer, 'The UK Biological Weapons Program', in Mark Wheelis, Lajos Rosza and Malcolm Dando (eds.), *Deadly Cultures: Biological Weapons since 1945* (Cambridge, MA: Harvard University Press, 2006), 83.
72 Gradon P. Carter and Graham S. Pearson, 'British Biological Warfare and Biological Defence, 1925–45', in Giessler and Moon, *Biological and Toxin Weapons*, 188.
73 IWM, CIOS Report XXVIII-56, 'Rockets and Guided Missiles'.
74 IWM, Misc. 21/382, 'Report on Operation Backfire', 7 November 1945.
75 Collier, *Battle of the V-weapons, 1944–5*, 150.
76 Robert Hathaway, *Great Britain and the United States: Special Relations since World War II* (Boston, MA: Twayne's, 1990), 11.
77 IWM, CIOS Report XXXII–125, 'German Guided Missile Research'.
78 McGovern, *Crossbow and Overcast*, 13.
79 Ibid., 98.
80 IWM, CIOS Report XXVIII–56, 'Rockets and Guided Missiles'.
81 McGovern, *Crossbow and Overcast*, 200–1.
82 IWM, Misc. 21/382.
83 TNA, CAB 82/6, 'Minutes of DCOS Meeting', 16 July 1945.
84 Dieter K. Huzel, *From Peenemünde to Canaveral* (Westport: Greenwood, 1962), 200.
85 TNA, WO 219/2165, 'Operation Backfire', 6 June 1945.
86 IWM, Misc. 21/382.
87 Simons, *Operation Lusty*, 133.
88 IWM, Misc. 21/382.
89 'German Fires V2 under British Control', *Daily Telegraph*, 16 October 1945.
90 IWM, 06/27/1, Private Papers of Major P.A. Chittenden, 23 October 1945.
91 IWM, Misc. 21/382.
92 Jeremy Stocker, *Britain and Ballistic Missile Defence, 1942-2002* (London: Frank Cass, 2004), 33.
93 'German Technicians for Britain', *The Times*, 1 November 1946, 4.
94 McGovern, *Crossbow and Overcast*, 98.
95 Hermione Giffard, *Making Jet Engines in World War II: Britain, Germany, and the United States* (Chicago: Unversity of Chicago Press, 2016).

96   Jeffrey A. Engel, '"We are not concerned who the buyer is": Engine Sales and Anglo-American Security at the Dawn of the Jet Age', *History and Technology*, 17 (2000), 45–6.

97   Simons, *Operation Lusty*, 120–7.

98   TNA, AVIA 9/83, 'The Fedden Mission to Germany', 4 July 1945.

99   Simons, *Operation Lusty*, 127.

100   TNA, AVIA 9/83, 'The Fedden Mission to Germany', 4 July 1945. See also: John Christopher, *The Race for Hitler's X-Planes* (Stroud: The History Press, 2012).

101   Matthew Uttley, 'Operation Surgeon and Britain's post-war exploitation of Nazi German aeronautics', *Intelligence and National Security*, 17 (2002), 1–26.

102   TNA, AVIA 12/82, 'Operation Surgeon: Memorandum', 23 November 1946.

103   Uttley, 'Operation Surgeon', 6.

104   TNA, AVIA 12/82, 'Operation Surgeon: Memorandum', 23 November 1946.

105   TNA, AVIA 12/82, 'Operation Surgeon: Memorandum', 23 November 1946.

106   TNA, FO 942/426, 'MAP and Ministry of Supply Participation in RAF Operation Surgeon', 15 February 1946.

107   Andrew Nahum, '"I believe the Americans have not yet taken them all!": the Exploitation of German Aeronautical Science in Postwar Britain', in Helmuth Trischler and Stefan Zeilinger (eds.), *Tackling Transport* (London: Science Museum, 2003), 107–8.

108   Uttley, 'Operation Surgeon', 10.

109   Carl Glatt, 'Reparations and the Transfer of Scientific and Industrial Technology from Germany: A Case Study of the Roots of British Industrial Policy and of Aspects of British Occupation Policy in Germany between Post-World War II Reconstruction and the Korean War, 1943–1951', Ph.D. dissertation, European University Institute (1994), 246–7.

110   Neufeld, 'German Aerospace Exodus', 60.

111   TNA, AVIA 54/1403, 19 March 1947.

112   Giffard, 'Engines of Desperation', 843.

113   See Chapter Six.

114   Uttley, 'Operation Surgeon', 9.

115   Roy Fedden, *Britain's Air Survival: An Appraisement and Strategy for Success* (London: Cassell, 1957), 32.

116   Uttley, 'Operation Surgeon', 21.

5 Exploiting expertise

The exploitation of German personnel by the victorious Allies is arguably the best-known element of the post-war exploitation programme, largely because of the controversy which it often created. Certain cases have even attained relatively prominent positions in the public consciousness, such as that of the rocket expert Wernher von Braun, interest in whom can be attributed to his subsequent fame and the moral ambiguity of his Nazi past. However, von Braun's story was not unique; he was just one of hundreds of German scientists and technicians who were treated as targets of exploitation by the Allies after the war and his experience of detainment, interrogation and then employment was a familiar one.[1] In intelligence circles, the use of enemy personnel as assets was not a new approach, as shown by the wiretaps placed on senior German prisoners of war during their imprisonment in Britain, in the hope that they would reveal important military secrets which the Allies could utilise.[2]

Once the war ended, German expertise in a number of fields was sought out by the victorious powers, and other foreign countries, to be deployed to their benefit, resulting in a broad diaspora of former Nazi personnel. For example, the Americans tasked SS-*Hauptsturmführer* Klaus Barbie, the so-called 'Butcher of Lyon', with tackling the German Communist Party in Bavaria, and also pulled Nazi collaborators from French prisons, setting them to gather intelligence on Communists in France.[3] The French Foreign Legion, meanwhile, employed SS anti-partisan specialists to hunt down guerrillas in the jungles of Indochina while the post-war independent Syrian government hired a number of German military advisors (some directly out of POW camps), as well as SS men who had been involved in the Holocaust, to provide training in counter-terrorism and torture techniques.[4] Even the present-day German intelligence agency, the *Bundesnachrichtendienst* (BND), began life in 1946 as the Gehlen Organisation, named for its founder Reinhard Gehlen, a former Wehrmacht intelligence officer tasked by the American occupation authorities with monitoring the Soviet threat in Germany and Eastern Europe.[5]

However, it was German scientific experts that were often considered the most desirable prizes of this personnel exploitation initiative. As we have seen, continuing a trend from the First World War, science and scientists became even more influential during the Second World War and began to exert an influence over military strategy.[6] This, coupled with the fact that scientific intelligence as a field had truly proved its worth during the war, ensured that at its conclusion much attention was focused on accessing the expertise of formerly enemy scientists and technicians. While some accounts have described Britain as a junior partner to the USA in this endeavour,[7] asserting that they 'preferred the inventions to the inventors' on account of lacking the financial resources to support personnel in detention and interrogation,[8] the truth is not quite so clear-cut. Indeed, the British certainly displayed no lack of effort or excess of scruples in their pursuit of these specialists.[9] In principle at least, there is little doubt that Britain was as enthusiastic about the opportunity of exploiting German scientists and technicians as its wartime allies, if not more so, and that it therefore formed a major part of its postwar exploitation efforts on the whole.

Despite this enthusiasm, British efforts in the exploitation of German expert personnel began slowly. During the development of this risky and controversial policy, it was felt that the direct recruitment of German specialists was too politically unpalatable, at least in the short-term, so instead the process was limited to detention and interrogation on German soil. One of the first sites established for this work was Schloss Kransberg, a medieval castle repurposed as a detention centre for some of the most high-profile German experts and given the euphemistic codename Dustbin. Running in parallel to this was the much smaller Operation Epsilon, which saw ten of Germany's most accomplished nuclear physicists interned and interviewed at a country house in Cambridgeshire. Soon, the personnel exploitation scheme was extended and expanded and saw a greater number of German specialists brought to Britain for more comprehensive and detailed debriefings. Overall, in these early stages, Britain took the first significant steps towards a substantive personnel exploitation scheme, building upon the far simpler policy for the acquisition of material spoils and moving towards the longer-term and more complex recruitment of German experts in Britain, which came afterwards and had much further-reaching consequences.

Dustbin and Epsilon

Schloss Kransberg was first constructed in 1170, near the village which shares its name, in the Taunus Mountains, some twenty-five miles north of Frankfurt. It changed hands among the German nobility a number of times throughout the centuries and underwent substantial renovation and redesign along the way. In 1939, it was appropriated by the Nazi government and Hitler's chief architect, Albert Speer, set about transforming it into a military command headquarters, with the designation *Adlerhorst* (eagle's eyrie), installing an underground bunker network and various bombproofing measures. Hitler did not favour it, however, and it later became a military rehabilitation centre and a personal retreat for Hermann Göring. Hitler did reside there briefly, between 11 December 1944 and 16 January 1945, when he used it as a base to oversee his armed forces' last-ditch offensive against the Western Allies in the Ardennes region.[10] The castle was captured by American forces on 30 March and taken over by Field Information Agency,

Technical (FIAT (Field Information Agency, Technical) in June, at which point it became the site of the largest scientific and technical detention centre in the western zones of Germany, replacing a similar facility which FIAT had operated in Chesnay, near Versailles, in the last months of the war. Run on a joint Anglo-American basis, Dustbin became home to a number of senior German scientists and specialists, who were interrogated at length and instructed to compile comprehensive reports on their wartime work.[11] Leading experts in practically all the subjects in which Germany was judged to be ahead of the Allies were held there – topics covered in interrogation included chemical and biological weapons, radar, electromagnetic radiation, steel manufacture, submarine design and guided missiles, alongside many others.[12] Among the internees were the Third Reich's Economics Minister Hjalmar Schacht, automobile designer Ferdinand Porsche, and Wernher von Braun and it was the status of these figures which helps to explain the relaxed and relatively comfortable conditions at Dustbin – prisoners could converse freely with one another, walk at leisure in the gardens and arrange evening entertainment.[13]

Dustbin's most notable resident, however, was Albert Speer – architect, Nazi Minister of Armaments and War Production and close confidant of Hitler – who, in a twist of not uncommon post-war irony, ended up a prisoner in summer 1945 at a property which he had helped redesign only six years previous. In his memoirs, Speer recalls his time at Dustbin relatively favourably: the excellent views from the unbarred windows of his top-floor room, the sizeable US Army rations and the entertainment organised by the other internees, including comic cabaret scenes on account of which, in Speer's own words, 'tears of laughter ran down our faces at the tumble we had taken'.[14] As with many of Speer's recollections, Gitta Sereny takes issue with this particular account – her interviews with some of Speer's fellow detainees reveal that he largely isolated himself while at Kransberg, interacting only with his secretary, Annemarie Kempf, and not participating in any of the musical or sports events which the others used to pass the time. Sereny also describes how Speer became quite despondent while at Dustbin and how the Allied authorities encouraged him to write reports on various technical and political topics in order to lift his mood and thus make him more amenable to their purposes. This was most successful when he worked with a British intelligence officer, Captain Hoeffding, to prepare profiles of other senior Nazis, a process which he evidently found cathartic.[15] In September 1945, Speer was taken to Nuremberg to stand trial before the International Military Tribunal, and thus the scientific informant became a war criminal.[16]

As the main site for detaining and interrogating the Third Reich's scientific and technical elite, Dustbin's value was undeniably high, which often made it a contentious subject between the British and Americans. When SHAEF (Supreme Headquarters Allied Expeditionary Force) was disbanded in July 1945, General Walter Bedell Smith, Eisenhower's Chief of Staff, issued a policy memo which stated that 'the appropriate US and British agencies will have equal facilities and responsibilities for the intelligence exploitation of Dustbin, and each will receive copies of all reports resulting from this exploitation'.[17] Dustbin was operated on this combined Anglo-American basis until August 1946, when it was replaced with an informal system of mutual co-operation between the two national components of FIAT, which in turn was removed in the following November, when the Americans assumed complete control and excluded the British officers from any and all policy decisions. At the end of 1946, the Americans closed Dustbin, an outcome which the British considered deeply unsatisfactory.[18]

The source of this British dissatisfaction lay in the unique nature of Dustbin, as it featured, in the opinion of Air Commodore Victor Bennett, the chief of British FIAT, 'certain facilities which cannot be obtained in other detention centres'. First and foremost, it had 'an atmosphere most likely to induce [the detainees] maximum co-operation and hence facilitate their exploitation by US and British agencies'. The usual atmosphere of an ordinary detention centre, Bennett continued, was not suitable for dealing with these German experts 'who are often temperamental and who only respond favourably to gentle and careful treatment'. In addition, Dustbin had ample space to accommodate new internees at short notice, which was important as speed was often of the essence in ensuring desirable German targets were not snapped up by rival powers.[19] Furthermore, the security at Dustbin was high; all the guards, and even the drivers, were armed and no members of the press were to be admitted, as they could end up 'seriously prejudicing interrogations' being conducted there. One reason for these tight restrictions was that Dustbin also served, in the words of one British official, as 'a convenient place where persons, in whom other powers have shown interest, can "disappear"'.[20] In other words, certain German experts could be kept hidden away at Dustbin to prevent the Soviets from getting to them, foreshadowing a wider policy of denial which developed in the later occupation period.

While the operations at Dustbin were leading the way in the detention and interrogation of German scientists and technicians in Germany, another initiative had already begun paving the way for similar efforts to be carried out on British soil. The exploitation of enemy intelligence assets in Britain had a firm antecedent in the interrogation and covert observation of German prisoners of war who were interned in Britain, most notably at Trent Park in north London.[21] Here, so-called 'M' ('Miked') rooms were fitted with listening devices and allowed interrogators to eavesdrop on their more senior internees, including several Wehrmacht generals.[22] This process was eagerly adopted for the purposes of scientific intelligence within two months of the end of the war in Europe. In the same way that the actions of Alsos did much to shape the future activity of exploitation teams on the ground in Germany, it was investigations into the Nazi atomic bomb project which laid the template for a wider programme of exploitation of scientific and technical personnel back in Britain. From July until December 1945, Operation Epsilon was conducted, in which ten senior German atomic physicists who had been detained as part of the Alsos operations across Europe, including Werner Heisenberg, Otto Hahn and Max von Laue, were interned at Farm Hall, Godmanchester, 15 miles northwest of Cambridge.[23]

Here, they were secretly wire-tapped in the hope that something of significance about the German bomb project, which the scientists would not reveal in interrogation, would be overheard, although by this stage Britain and the US were fairly confident that their own bomb project had been considerably further advanced than the German equivalent. The German physicists did not necessarily share this view and were quite shocked when news reached them of the atomic bombing of Hiroshima and Nagasaki in August. The internees were also completely oblivious of the eavesdropping going on, with Heisenberg recorded brazenly laughing off the concerns of his colleagues and saying 'Oh no, they're not as cute as all that. I don't think they know the real Gestapo methods; they're a bit old-fashioned in that respect.'[24] The secret recordings also revealed some other interesting opinions held by this unique group of experts. Many of them failed to acknowledge the horrors the Third Reich had perpetrated, particularly in terms of the Holocaust, and some even maintained lingering traces of Nazi racial ideology – Erich Bagge, for instance, expressed his dismay that French Moroccan soldiers were being billeted in his house.[25] Looking ahead, they feared returning to Germany, lest they be deemed traitors for denying Germany the nuclear weapons with which it could have won the war, but they considered the idea of working for the Soviet Union even less appealing.[26] This was ironic as one purpose of the Farm Hall detention was to keep these valuable human assets out of the hands of the Soviets and thus contribute towards the retarding of the USSR's atomic bomb project.[27] The Farm Hall scientists soon realised that their fates would likely be tied up with the Western Allies in the future; however, this was something they saw not as a stroke of good fortune but as an unpleasant necessity – they viewed the British and Americans with ambivalence and resented them for their ignominious imprisonment.[28]

Indeed, in September 1945, the internees started to become restive and Heisenberg even made suggestions that he would escape from Farm Hall and 'try to get in touch with some of his British scientific friends in order to ask them to make public the fact that these German scientists are being kept in this country'.[29] In his biography of Heisenberg, David Cassidy noted that the major effort which the Allies expended to secure Heisenberg reinforced his greatly inflated sense of self-worth (though this was dealt quite a blow by the news of the atomic bombs dropped on Japan) and encouraged him to complain often about what he considered to be poor treatment.[30] This situation was exacerbated when Otto Hahn was awarded the Nobel Prize for Chemistry in December but could not attend the ceremony due to his secret internment (he only learnt of his award by reading about it in the *Daily Telegraph*), which prompted the British authorities to release a statement on the premise that 'if the secrecy, and therefore the sensationalism, is whittled away before the matter leaks out, it will be less embarrassing if the German scientists turn nasty'.[31]

Nonetheless, Members of Parliament (MPs) did face complaints from their constituents, mostly scientists, about the government's poor treatment of their fellow men of science – in February 1946, William Proctor, Labour MP for Eccles, received a letter arguing that the secrecy surrounding the fate of the German atomic physicists was 'not only fettering scientific progress, but bedevilling international relations'.[32] The civil servants who fielded these complaints evidently found it ironic that British scientists were prepared to make a row 'unless these people [German experts], who are living comfortably near Cambridge, are sent back to the discomforts of Germany'.[33] In reality, by the time these impassioned critiques of the poor treatment of esteemed German physicists began arriving in Parliamentary pigeon-holes, their period of incarceration was at an end. Despite their earlier fears and predictions for the future, all the men chose to return to Germany (though Paul Harteck emigrated to the United States in 1951) and many enjoyed highly successful post-war careers, including Hahn, who became the first president of the reconstituted Max Planck Society, and Heisenberg, who became Director of the Max Planck Institute for Physics.[34] On the whole, as case studies, both Dustbin and Operation Epsilon offer intriguing insights into the early processes of personnel exploitation, in Germany and in Britain, but they cannot tell the whole story. They were precursors to, and microcosms of, the much broader and more complex programme which emerged from 1945 onwards.

Detention and interrogation in Germany

As the personnel exploitation scheme grew in scale and scope, so too did the apparatus needed to enact it, which was responsible for, among other things, tracking down the targeted individuals, detaining them safely and securely and facilitating useful interrogations. Not only were these roles considered to be part of the remit of T-Forces and FIAT, but a separate sub-division of the latter was also established to focus exclusively on this element of the programme. The Enemy Personnel Exploitation Section (EPES) was formed on 1 May 1945 and, as with FIAT, was split into national components shortly thereafter. The main objective of EPES was to build up and maintain a comprehensive set of records on all German scientists and technicians of note and then to provide this information to interested agencies on request and to enable exploitation by establishing appropriate contact between the target and relevant agency. Within a year of operation, EPES had compiled an index of 18,000 personality cards and 400 persona dossiers and was considered to be 'the only place where a general picture of exploitation of German scientists and technicians can be obtained'. By December 1946, EPES was handling 'approximately 130 requests every month on behalf of British Ministries and Agencies who wish to trace and locate German scientists and technicians with a view to exploiting them'.[35]

Such extensive record-keeping was only possible because of the quality of the staff employed by EPES, who were 'well trained in intelligence duties' and therefore adequately prepared to handle the 'very considerable number of difficult and delicate problems [which] arise in the course of the work'.[36] Before being despatched to conduct their activities with FIAT Forward in Berlin, EPES operatives were sternly instructed to adhere to a number of security regulations, including not entering the Soviet zone, not talking to 'any person except those with whom it is necessary for you to converse' and always employing 'German agents to contact unknown persons at addresses which are not known to you outside the office'.[37] EPES was, however, only a fairly small organisation and the bulk of its workload was administrative, so the legwork involved was usually delegated to the greater manpower of T-Force and FIAT, as well as support being rendered by the other intelligence agencies of the occupying powers.[38]

As with the removal and transportation of documents and equipment and the facilitation of investigative trips to Germany, T-Force acted primarily in the British zone while FIAT was largely responsible for parallel work in the USA and, less often, French zones. Perhaps the most critical part played by T-Force in the execution of personnel exploitation was the locating of the scientist or technician in question, 'often on the scantiest of information'. This sometimes involved obtaining the address from the Regional Research Officer, or the Education Branch at Regional level, or any number of other administrative bodies operating at this time.[39] Once the German expert was located, it was necessary to secure clearance from the relevant Military Government officials before moving him to the desired location, either elsewhere in Germany or over to Britain for interrogation or employment. Such a sensitive issue as this unsurprisingly entailed a certain amount of bureaucratic excess as shown by the number of agencies with which EPES and FIAT were in regular contact – a staggering 52, including 20 British, 13 American, 14 Combined and 1 French.[40] This can, to some extent, be accounted for by the international character of FIAT, which was responsible for 'the location of German scientists and technical personnel in the US and French zones of Germany and Austria'.[41]

FIAT was also responsible for Dustbin but, while the facility at Schloss Kransberg possessed many unique qualities, it was not the only scientific and technical detention centre established after the war. For example, in the late summer of 1945, the British occupation authorities established an all-German 'Works Centre' in the buildings owned by Rheinmetall Borsig, considered to be 'the most advanced technically of the German armament firms', at Unterlüß, near Celle.[42] At its peak, 150 German specialists in a range of fields, especially munitions and ballistics, were billeted and administered at the Unterlüß Works Centre, 'under good conditions in order to counteract any tendency on their part to migrate to other zones' and tasked with compiling reports on their wartime work.[43] Information provided by the Unterlüß internees also enabled the Technical and Personnel Administration (TPA) of the Ministry of Supply to produce a comprehensive report in October 1948, entitled 'German Organisation and Personalities Engaged in Research and Development of Armaments during the Second World War'.[44] When the Works Centre was closed on 20 August 1948, the German scientists and technicians who had worked with limited compensation for the British for up to three years were not guaranteed employment in Britain but were simply moved to other secure facilities in Germany so that they would not be 'made liable to consider offers from the East or be made to suffer undue hardship'.[45]

This fate of uncertainty was a common experience shared by many, if not most, of the German scientists and technicians who were targeted by the British. Unwilling to let them slip away to take up employment in the Soviet zone or the USSR, but unable to offer them any serious prospects themselves, the British preferred to keep their German detainees in a state of limbo, preparing reports or subject to interrogation, with only a limited hope of future remuneration. This process began almost immediately after the war ended, as a result of a policy which was euphemistically known as 'freezing'. This process involved the British and American occupying forces keeping close tabs on all German individuals of interest, preventing them from relocating or even travelling too far from their homes, so that should they be required for any further investigation they could swiftly be rounded up and delivered. By September 1945, T-Force reported that there were 5,000 German scientists and technicians who had been 'frozen', the vast majority of whom remained 'in both British and US zones without any employment'. They urged the British departments to find work for these specialists soon, as it was untenable to sustain this 'freezing' process for much longer.[46] However, in January 1946, as attention moved from exclusively military personnel to those in civil industry, the British informed the Americans that they 'welcomed' the current 'freezing' and would like for it to continue.[47]

Obviously, such a policy was not at all favourable in terms of the treatment of the German specialists – in March 1947, the Scientific and Technical Research Board (STRB) reported that a group of 100 German aeronautical experts who had been tasked with writing monographs with little chance of future British employment were 'intensely dissatisfied with their lot'.[48] Some German specialists even founded protest groups to register their discontent with the occupation authorities.[49] This issue of mistreatment was highlighted especially clearly in the case of Bad Gandersheim, a small market town approximately 40 miles south of Hannover, which, in June 1945, became the temporary home of around 90 German scientists, technicians and their families, when SHAEF ordered that they be evacuated there from Magdeburg, before the British handed that city over to the Soviets.[50] However, this hasty evacuation was not swiftly followed by a multitude of job offers. Rather, the German specialists remained somewhat stranded at Gandersheim, being fed and housed by the British Army, accumulating considerable debts and with practically no prospects for the future.

Major Evans, of Research Branch, Control Commission for Germany (British Element) (CCG(BE)), complained that the 'Gandersheim Germans' were 'one of the major headaches left us by SHAEF', yet it was acknowledged that the British were 'to some extent morally responsible' as these people had been moved 'forcedly' at their behest.[51] The issue was raised at a British Intelligence Objectives Sub-Committee (BIOS) meeting on 3 March 1946, where it was stated that the 'position regarding the dependents of these scientists was not satisfactory', as well as concerns that if word of such poor treatment got out, it could seriously hinder British recruitment attempts in Germany.[52] The decision made was that the Control Commission was 'to make every effort to find suitable employment in the British zone for those scientists who are not offered contracts in Britain', and to provide assistance for the scientists' families when they had to move.[53] In reality, though, this did not materialise quite as promised and, by October, many of the men were still at Gandersheim and the Military Government continued to gripe about the costs of maintaining them there.[54]

Problem cases such as this exposed clear weaknesses in the British personnel exploitation endeavour. These shortcomings were thrown into even starker relief when the Americans began to limit British access to Dustbin. It is not surprising therefore that, when Dustbin closed at the end of 1946, Air Commodore Bennett, the Chief of British FIAT, considered the establishment of an exclusively British camp along similar lines to be a matter of utmost urgency. This needed an immediate solution, he argued, due to the steadily increasing number of individuals 'whom it is desired to house in conditions of security because they have been threatened with kidnapping ... by potentially hostile powers'.[55] Moreover, German experts who had been interrogated, often in Britain, but were now back in Germany awaiting a contract for long-term employment, would be dispersed and the Ministry of Supply complained that the 'problem of contacting personnel when required was considerable': an issue which could be largely offset by the use of a central detention camp.[56] A partial solution came in the form of a series of transit hotels, operated by T-Force, under the auspices of Operation Matchbox.[57] These hotels were oversubscribed, comparatively under-resourced and occasionally poorly run but, while they were far from an ideal solution, they did allow Britain to continue its unilateral personnel exploitation policy on the ground in Germany once Dustbin closed. In the meantime, however, an alternative approach was sought through the process of short-term detention and interrogation back in Britain.

Detention and interrogation in Britain

While Operation Epsilon was the first occasion where German scientists were brought to Britain for exploitation, it cannot be considered representative of the programme as a whole, not least because of its small scale and the eminence of the men targeted. The experience of other German experts who were brought to Britain to be detained and questioned, therefore, rarely mirrored that of the Farm Hall group. For instance, while Werner Heisenberg had dismissed the very thought of the British using so-called 'Gestapo tactics', others were less convinced. In August 1946, Ernest Bearder, the Controller General of the Chemical Industries Branch, wrote to the Secretariat of his branch's parent organisation, the Trade and Industry Division (part of the British Control Commission) complaining about the process for the removal of German specialists from Germany, which he outlined thus:

> Usually an NCO arrives without notice at the house or office of the German and warns him that he will be required. He does not give him any details of the reasons, nor does he present his own credentials. Some time later the German is 'seized' (often in the middle of the night) and removed under guard.

Bearder felt that this procedure 'savours very much of the Gestapo methods and ... is bound to create feelings of alarm and insecurity', which he did not think it was the intention to foster.[58]

Just over two weeks later, Bearder received a reply from Brigadier W.E.H. Grylls, the chief of T-Force, which curtly noted that, other than from Bearder himself, 'only one complaint has been received, although over 1,000 Germans have been evacuated through T-Force' and added that Bearder's Chemical Industries Branch was also the only division of the Control Commission which received special advance notification of any German who was to be taken. Grylls went on to say that his office was not 'aware of any Control Commission law, order or instruction that requires a British officer, NCO or soldier to present his credentials to a German under any circumstances whatever unless it be to a civil policeman on duty'. He concluded by making 'a strong protest' against Bearder's tone and suggesting that he 'may now wish to withdraw the letter'.[59] On the whole, the evidence suggests that Bearder's account of 'cloak and dagger methods' was exaggerated and far from accurate and, in reality, this was simply another minor chapter in the ongoing saga of dispute between those tasked with rebuilding German industry and those tasked with exploiting it.[60] Nonetheless, it is worth examining the process utilised when transporting German experts from Germany to Britain, as this was a crucial element in the satisfactory operation of the exploitation programme.

The responsibility for handling the experts in transit was left mainly to T-Force, who had to locate the individual specialist, obtain security clearance for their movement and then escort them from their home all the way to a specified reception point in Britain. This involved a considerable degree of administrative work, especially as by mid-1946, approximately 20 German scientists were making this journey every week.[61] Even contacting all the relevant agencies was no small task; T-Force reported that 'four Agencies in the UK and eight Agencies in Germany are concerned with the move of every German'.[62] Contrary to Bearder's sensationalised account, the procedure for each scientist usually involved giving them seven days' notice where possible, collecting them from their homes and taking them to the British Army of the Rhine (BAOR) HQ at Bad Oeynhausen for documentation and then transporting them by train and boat, usually in parties of three or four escorted by a British military or civilian officer, to Britain. During the journey, the scientists shared the officers' messing and accommodation facilities.[63]

Escorting officers were told that the men they had custody of were not prisoners of war and 'unless instructions to the contrary have been issued, you may assume them to be peaceable and co-operative'. Furthermore, 'as the value of their information depends to a certain extent upon their goodwill, they should be treated with reasonable consideration and should be adequately fed *en route*'.[64] However, these good intentions did not always easily manifest themselves, as related in the case of a 'prominent German scientist' who was returning to Germany from Britain after interrogation. When he boarded the ship at Harwich, 'the Captain insisted that the German should go below and he was taken to that part of the hold reserved for military prisoners returning to Germany under arrest'.[65] Nevertheless, not all German scientists who travelled to and from Britain had such a bad experience. In January 1947, the Board of Trade proudly circulated an extract from a censorship report which included comments from the wife of an unnamed scientist who had been brought to Britain. She said, 'They have shown great concern for my husband. [They] took him to London personally and will bring him back personally for Christmas.' She also commented that, in his absence, their home was protected from being requisitioned and they had been 'nobly looked after' with 'heavy-labour ration cards, ample fuel for the whole winter and a monthly remittance of RM 400 from the German Bank'.[66]

Once in Britain, in an effort to avoid potential public criticism, the German specialists were often housed out of view, in special interrogation centres, many of which had also served as POW camps during the war. The primary centre of this nature used by the British for scientific and technical personnel was Inkpot, based at the Beltane School in Wimbledon, directly south-west of London.[67] Here, German experts could be housed and then visited by experts from various government departments who had an interest in their particular area of expertise – staff from the Ministries of Aircraft Production and Supply, for instance, conducted interrogations on a vast range of topics, including radio control in guided projectiles, gas turbines, rocket fuels and parachute design.[68] Wernher von Braun and five of his Peenemünde colleagues were held in Wimbledon for ten days during the summer of 1945 and were interviewed by Sir Alwyn Crow, the head of the British rocket programme. Von Braun described these interviews as 'friendly shop talk', part of an effort by Crow to get the German rocketeers to sign contracts to work for the British government, an endeavour which failed in the face of more attractive offers from the United States.[69] Inkpot ultimately provided only a temporary solution and, at the end of 1946, the Beltane School site had to be relinquished to allow for an extension to the nearby Southlands teacher training college and to replace it a dedicated BIOS Reception Centre was created instead.

The site chosen for this was Spedan Towers, in Hampstead; 'a very large, modern private house in its own grounds' and formerly the home of retail magnate John Lewis. German scientists who were to stay there, known euphemistically as 'Visitors', were informed that:

> This is run on the lines of a hostel and is administered by a small unarmed military staff. There are no guards or barbed wire fences and there are no restrictions on the amount of mail either sent or received. The Visitors are accommodated in single-tier beds in rooms holding two or three each. In addition the visitors have at their disposal a dining room, a large well-furnished lounge and a library. The number of Visitors living at Spedan Towers at any one time varies between 25 and 30.

The lifestyle enjoyed by the Visitors at Spedan Towers was a largely pleasant one. They did not have complete freedom of movement, but board, lodgings and medical care were free, they were given a weekly cash allowance of 10 shillings (roughly £18 today) and 6 shillings (£10) worth of chocolate, cigarettes and similar items, as well as having access to a swimming pool, regular film shows, lectures on British culture and occasional tickets for concerts.[70]

Not only were the experts themselves well looked-after and protected from dismissal by their German employer during their period of interrogation, but their families also received numerous amenities, including an ample financial allowance, as well as increased rations and fuel allocations. The importance of this cannot be understated during a time when malnutrition and starvation were very real threats to the majority of ordinary German citizens.[71] Some German experts attempted to take advantage of this system, however, and secure these benefits for friends and distant relatives too, which resulted in BIOS issuing definitions of who exactly constituted dependants – wives, children, 'aged parents, sick relatives or any such members of the family who cannot fend for themselves', or a nominated housekeeper in the case of a widower scientist with children.[72] BIOS estimated that the total payment to each individual, including 'pocket money', the allowance to his family and his own board and lodging, was about £6 a week (over £200 today).[73] In return for this largely favourable treatment, the German experts were expected to co-operate whole-heartedly with all interrogations conducted at the Reception Centre, as well as being prepared to travel (escorted, naturally) for short spells to other locations around Britain, for interviews *in situ* at various private firms and establishments.

Despite this fairly comfortable arrangement, the British authorities decided to pre-empt any complaints the German experts might have about their accommodation in a pamphlet issued to all Reception Centre Visitors, which explained, with a hint of accusation, that 'the housing shortage due to air-attacks during 6 years of war makes it impossible to provide better accommodation'.[74] This pamphlet was not sufficient to deflect criticism by some Visitors – Friedrich Uhlmann, who, during the war, had owned a metallurgical research and production factory, was brought to 'austere' Spedan Towers in early 1946 and left in May, 'highly incensed' by what he described as 'miserly' and 'niggardly' treatment. He felt that in return for his considerable contributions to furthering the British hard metal industry, the amount of money paid to him and his family was 'nothing less than an insult'.[75] Uhlmann's case, however, appears to be an exception rather than the rule. The BIOS Reception Centre at Spedan Towers continued to operate at a steady rate throughout 1947 but, by the end of that year, the importance of short-term interrogation was considerably diminished, while the programme of recruitment for longer-term employment was in the ascendancy, and facilities were forced to change to reflect the shifting needs of the exploitation initiative as a whole.

Overall, it is clear why many of those involved saw the detention and interrogation of German scientists and technicians as the most valuable phase of the post-war exploitation of science and technology. While visits to facilities in Germany and the examination of confiscated documents and equipment had loose antecedents in the conventional looting and plundering which traditionally accompanied the occupation of enemy territory at the end of a conflict, the detainment and interrogation of individuals with specialist knowledge were all but unprecedented. This new development was the product of many factors: an intelligence network which had expanded considerably during the war and which understood the benefits of human intelligence, the contributions made by very talented individuals or small groups in terms of developing critical new innovations for the Allied arsenals (such as radar and the atomic bomb) and, perhaps most importantly, the looming spectre of some future war, in which it was widely assumed that a nation's scientists and technicians would be as significant as its soldiers, sailors and airmen.

As has been shown in this chapter, the process for detaining and interrogating German experts was far from a simple one and was certainly more complicated and problematic than the exploitation of documents and equipment. Difficulties arose in many areas: the location of the targeted specialists; the establishment and maintenance of appropriate detention facilities in which to hold them; dissatisfaction among the detainees about their treatment; complaints from the public, MPs and bureaucrats about the nature of the scheme; the secrecy within which it was necessary to shroud certain elements of the programme; and the mounting costs of supporting these men, and their families, for lengthy periods. However, personnel exploitation was judged to be so successful and vital to British interests that not only was it continued regardless of these many obstacles, it was later expanded and extended, moving from the phase of short-term detention and interrogation described here to the much further-reaching recruitment initiative, which saw German experts given full employment contracts in government research establishments and private industry in Britain. Ultimately, the story of the British detention and interrogation of German experts reinforces many of the characteristics attributed to the exploitation programme more widely – persistence, ingenuity and adaptability. In the years immediately following the end of the war, it became clear that the British were increasingly determined to secure Germany's best and brightest for their own use and were always willing to find new and creative ways in which to do so.

Notes

1 On von Braun, see Wayne Biddle, *Dark Side of the Moon: Wernher von Braun, the Third Reich and the Space Race* (London: W.W. Norton, 2009); Michael Neufeld, *Von Braun: Dreamer of Space, Engineer of War* (New York: A.A. Knopf, 2007).
2 Sönke Neitzel (ed.), *Tapping Hitler's Generals: Transcripts of Secret Conversations, 1942–45*, trans. Geoffrey Brooks (London: Frontline Books, 2013).
3 Richard J. Aldrich, *The Hidden Hand: Britain, America and Cold War Secret Intelligence* (London: John Murray, 2001), 181; Susan McCall Perlman, 'US Intelligence and Communist Plots in Post-war France', *Intelligence and National Security*, 33 (2018), 385.
4 Chern Chen, 'Former Nazi Officers in the Near East: German Military Advisors in Syria, 1949–56', *The International History Review*, 40 (2018), 732–51; Adam Chandler, 'Eichmann's Best Man Lived and Died in Syria', *The Atlantic*, 1 December 2014 [accessed online 27 June 2018, https://www.theatlantic.com/international/archive/2014/12/eichmanns-best-man-lived-and-died-in-syria/383296/].
5 Norman J.W. Goda, 'The Gehlen Organisation and the Heinz Felfe Case', in David A. Messenger and Katrin Paehler (eds.), *The Nazi Past: Recasting German Identity in Postwar Europe* (Lexington, KY: University Press of Kentucky, 2015), 273ff; Michael Wala, 'The Value of Knowledge: Western Intelligence Agencies and Former Members

of the SS, Gestapo and Wehrmacht during the Early Cold War', in Camilo Erlichman and Christopher Knowles (eds.), *Transforming Occupation in the Western Zones of Germany: Power Politics, Everyday Life and Social Interactions, 1945–55* (London: Bloomsbury, 2018), 271–80.

6   Bud and Gummett (eds.), *Cold War, Hot Science*, 6.

7   Bar-Zohar, *Hunt for the German Scientists*, 132.

8   Giles MacDonogh, *After the Reich: From the Liberation of Vienna to the Berlin Airlift* (London: John Murray, 2007), 390.

9   Gimbel, *Science, Technology, and Reparations*; Lasby, *Project Paperclip*, 110.

10 Annie Jacobsen, *Operation Paperclip: The Secret Intelligence Program that Brought Nazi Scientists to America* (Paris: Hachette, 2014), 13; Mark Felton, 'Adlerhorst – The Führer's Secret Castle', 20 March 2016 [accessed online 28 June 2018, http://markfelton.co.uk/publishedbooks/adlerhorst-hitlers-forgotten-headquarters/].

11 Gimbel, *Science, Technology, and Reparations*, 82; Tusa and Tusa, *Nuremberg Trial*, 42; Schmidt, *Justice at Nuremberg*, 131.

12 Maddrell, *Spying on Science*, 17–21.

13 Richard Overy, '"Instructive for the future": the Interrogation of the Major War Criminals in Germany, 1945', in Christopher Andrew and Simona Tobia (eds.), *Interrogation in War and Conflict: A Comparative and Interdisciplinary Analysis* (Abingdon: Routledge, 2014), 96.

14 Albert Speer, *Inside the Third Reich* (London: Sphere, 1971), 673–74.

15 Gitta Sereny, *Albert Speer: His Battle with Truth* (London: Picador, 1996), 559–61.

16 Weindling, *Nazi Medicine*, 43.

17 TNA, FO 1031/69, Walter Bedell Smith to Head, CCG(BE), July 1945.

18 TNA, FO 1031/69, P.M. Wilson to Chief, FIAT (Br), 9 November 1946.

19 TNA, FO 1031/69, V.B. Bennett to HQ, T-Force, 2 December 1946.

20 TNA, FO 1031/75, 8 October 1946.

21 Dan Lomas, 'The Drugs Don't Work: Intelligence, Torture and the London Cage, 1940–48', *Intelligence and National Security*, 33 (2018), 922–3.

22 J.M. Goodchild, 'Exploitation of Displaced European Refugees and Axis Prisoners of War in Britain, 1939–49', in Sandra Barkhof and Angela K. Smith (eds.), *War and Displacement in the Twentieth Century: Global Conflicts* (Abingdon: Routledge, 2014), 112–14; Neitzel (ed.), *Tapping Hitler's Generals*; Helen Fry, *The M Room: Secret Listeners Who Bugged the Nazis in World War 2* (London: Marranos Press, 2012).

23 Frank, *Operation Epsilon*. See also: Mary A. McPartland, 'The Farm Hall Scientists: The United States, Britain and Germany in the New Atomic Age, 1945–46', PhD dissertation, George Washington University (2013).

24 Frank, *Operation Epsilon*, 33.

25 Walker, *Nazi Science*, 214–5.

26 Ibid., 233–35.

27 Richelson, *Spying on the Bomb*, 62.

28 Walker, *Nazi Science*, 236–8.

29 TNA, CAB 126/333, W.A. Akers to P.M. Blackett, 5 September 1945.

30 David Cassidy, *Uncertainty: The Life and Science of Werner Heisenberg* (New York: W.H. Freeman, 1991), 502.

31 TNA, PREM 8/373, 'Interrogation of German Scientists in United Kingdom', 20 December 1945.

32 TNA, PREM 8/373, Letter from Dr J.W. Jeffery to Mr W.J. Proctor, MP, 1 February 1946.

33 TNA, PREM 8/373, Letter to the Prime Minister, 7 November 1945.

34 TNA, PREM 8/373, 18 February 1946; Walker, *Nazi Science*, 239–41.

35 TNA, FO 1031/75, 'EPES: policy', 10 August 1946.

36 TNA, FO 1031/75, 'EPES: policy', 10 August 1946.

37 TNA, FO 1031/59, 'Attachment of Personnel of Enemy Personnel Exploitation Section to FIAT Forward (British)', 14 August 1946.

38 Aldrich, *Hidden Hand*, 188.
39 TNA, FO 1031/22, 'Employment of German Scientists outside Germany', 11 August 1947.
40 TNA, FO 1031/68, 'List of British and American Agencies', 11 September 1946.
41 TNA, FO 1032/1459, 'Field Information Agency Technical (FIAT)', 10 December 1946.
42 TNA, FO 371/71038, 'Unterlüß Works Centre', 31 May 1948.
43 TNA, FO 1031/75, 1946.
44 IWM, 'German Organisation and Personalities Engaged in Research and Development of Armaments during the Second World War', October 1948.
45 TNA, FO 371/71038, 'Disbandment of the Unterluss Works Centre', 12 May 1948.
46 TNA, FO 1031/20, 'Exploitation of German Scientists and Technicians: Policy', 12 September 1945.
47 TNA, CAB 122/357, Cabinet Offices to BJSM, 19 January 1946.
48 TNA, AVIA 54/1403, 'Denial of German Aeronautical Scientists to the Russians', 19 March 1947.
49 Henke, *Die amerikanische Besetzung Deutschlands*, 766.
50 Longden, *T-Force*, 292.
51 TNA, FO 1032/297, 'Interrogation and Detention Policy', 26 June 1946.
52 TNA, FO 1031/50, 'Minutes of 4th BIOS Meeting', 6 March 1946.
53 TNA, FO 1032/164, Brig. Spedding to G(T)&CW, 8 March 1946.
54 TNA, FO 1032/297, 28 October 1946.
55 TNA, FO 1031/69, V.B. Bennett to Intelligence Division, 3 December 1946.
56 TNA, FO 1031/50, 'Minutes of 4th BIOS Meeting', 6 March 1946.
57 See Chapter Eight.
58 TNA, FO 1031/19, E.A. Bearder to Secretariat, T&I Division, 6 August 1946.
59 TNA, FO 1031/19, W.E.H. Grylls to Secretariat, T&I Division, 21 August 1946.
60 Ian Turner, 'British Policy Towards German Industry, 1945-9: Reconstruction, Restriction or Exploitation?', in Ian Turner (ed.), *Reconstruction in Post-War Germany: British Occupation Policy and the Western Zones, 1945–55* (Oxford: Berg, 1989), 67.
61 TNA, FO 1031/19, Lt-Col. D.G. Edwardes to Maintenance Branch, ZEO, CCG, 23 November 1946.
62 TNA, FO 1065/12, 'The Future of T-Force/FIAT Organisation', 16 August 1947.
63 TNA, FO 1031/9, 'Short Visits to the UK by German and Austrian Scientists and Technicians', 24 April 1947.
64 TNA, FO 1031/19, 'General Instructions for Escorts', 9 September 1946.
65 TNA, FO 1031/19, 23 November 1946.
66 TNA, FO 1031/19, F.J. Broomfield to HQ, T-Force, 4 January 1947.
67 Gimbel, *Science, Technology, and Reparations*, 18; Nahum, 'I believe the Americans have not yet taken them all!', 113.
68 TNA, AIR 40/1178, 'ADI(K) Periodical Progress Report No. 5', 11 December 1945.
69 McGovern, *Crossbow and Overcast*, 201–2.
70 TNA, FO 1031/9. Price adjustments calculated on TNA Currency Converter [accessed online 18 July 2018, http://www.nationalarchives.gov.uk/currency-converter/].
71 Reinisch, *Perils of Peace*, 179.
72 TNA, FO 1031/9, 'Definition of Dependants of German Scientists & Technicians', April 1947.
73 TNA, FO 1031/19, 'Minutes of 6th BIOS Reception Centre Panel Meeting', 22 August 1947.
74 TNA, FO 1031/19, 'Pamphlet for Issue toPersonnel at BIOS Reception Centre', 5 December 1946.
75 TNA, FO 1031/25, 'EPES Special Intelligence Report No. 8', 30 June 1947.

6   The brain drain

The term 'brain drain' is a complex and loaded one and has been applied to various exchanges of expertise and know-how throughout history, yet is often warped by a certain political slant or agenda. In the context of the immediate aftermath of the Second World War, when Germany lay at the mercy of its occupiers, the victorious Allies and many other nations besides aimed to claim lasting 'intellectual reparations' by absorbing the specialist human resources of the former Third Reich. In this, they displayed a voracious appetite which was, to some extent, sated and, as such, the term 'brain drain' seems entirely appropriate. Moreover, the recruitment of German scientists by the victorious Allies after the war has become one of the best-known and most enduring images of the exploitation programme. Indeed, the formerly Nazi scientist working for the West has become a vivid cultural archetype, appearing in films such as *Dr Strangelove* (1964) and *Captain America: The Winter Soldier* (2014), largely inspired by prominent cases such as that of Wernher von Braun. Despite this, in the public consciousness, the real beneficiaries of German expertise after the war were the United States and, to a lesser extent, the Soviet Union. Britain is often left out of the story, with the implicit assumption that it did not try to recruit German scientists and technicians after the war or that, if it did try, it largely failed. That is not true on either count – Britain's personnel exploitation scheme was sizeable, well-resourced and generally successful, in both defence and civil-industrial fields – as this chapter will demonstrate.

Certainly, British officials were not slow to recognise the potential of personnel exploitation. As early as 5 May 1945, R.P. Linstead (then deputy head of the Combined Intelligence Objectives Sub-Committee (CIOS) and soon to become the first head of the British Intelligence Objectives Sub-Committee (BIOS)) wrote a letter to Sir Edward Appleton (secretary of the Department of Scientific and Industrial Research and future Nobel laureate) reflecting on a recent tour of research establishments in Germany, in which he noted that:

> A number of German scientists and technologists are being uncovered in the course of our operations who, if they were transplanted to this country would be a vast benefit to our technology. We might have to go back to the Huguenot or Flemish migration to find a parallel.[1]

Furthermore, in July 1947, a Board of Trade circular, sent to numerous British trade associations, described the employment of German scientists and technicians within British industry as 'the climax of the whole [exploitation] operation' and warned that 'the longer we delay our approach to them, the less our chance of getting the best ones'.[2] These comments both suggest a consistent and enthusiastic commitment to the utilisation of German expertise in a long-term capacity in Britain. What they do not reveal are the convoluted debates and developments of policy which made this initiative possible in reality, nor do they even hint at the complex lived experience of those German experts who relocated to Britain as part of these schemes and who attempted to integrate in workplaces and communities where they were often still viewed as 'enemy aliens'. This chapter will therefore chart the growth and evolution of the British programme to recruit German scientists and technicians across its various manifestations and explore the real-world implications of such an endeavour on those most intimately affected.

Defence recruitment

Movement towards a policy of recruitment began less than a month after the war's end, when Anglo-American co-ordination was still of paramount importance. On 5 June 1945, the British Chiefs of Staff received a telegram from the Joint Staff Mission in Washington, DC, which informed them that the US Chiefs of Staff had informally decided 'to bring German civilian scientists and technicians to the United States for the purpose of exploiting their knowledge by the military in the development of weapons which can be used against the Japanese'.[3] Nine days later, after some American prompting, the British Chiefs of Staff signalled their assent, but with the inclusion of several key provisos. Firstly, that the intelligence gained by such investigations be shared between the Americans and British; secondly, that the security risks of letting exploited German experts return to Germany armed with knowledge of British or American research be addressed; and thirdly, that a system of allocation of such human resources between the two powers be devised. They also added, for British eyes only, that they were 'sceptical if German scientists could really contribute to weapon development in time for the Japanese war'.[4]

During considerations of the American proposals, the Deputy Chiefs of Staff (DCOS) committee suggested 'that some German scientists would be of considerable value to our own research'.[5] This acknowledgement, while couched in loose and speculative terms, shows the original germ of the British plan to exploit specialist personnel. The idea quickly gathered momentum and, six weeks later, the DCOS committee considered a report by the Directors of Scientific Research at the Admiralty, Ministry of Aircraft Production and Ministry of Supply, 'advocating that general agreement should now be sought to the limited employment of a number of German scientists in this country under suitable conditions of security'.[6] This concept was gradually fleshed out and, on 1 August, the DCOS committee expressed their belief that 'there is no doubt that very great advantage to our own defence research and development would be derived from bringing to the UK a small number of high-grade experts to carry on their work in specialised fields', of which they included aerodynamics, hydrodynamics and power plants as preliminary examples.

Extensive discussion ensued and several concerns were raised – first, that in so doing they might allow a small part of German war potential to endure, and even grow by learning British military secrets; second, that it would appear hypocritical when they were so strongly trying to deter their wartime Allies from pursuing a similar policy; and third, that it might arouse public discontent in Britain at paying German scientists who so recently had played a key role in the war effort against the Allies.[7] The Joint Intelligence Sub-Committee registered their particular anxiety on the former issue, utilising a Security Service report which argued that the proposals for personnel exploitation were 'based on dangerous assumptions and that the security risks had been under-estimated'.[8] The Home Office, too, resisted the plan, worrying that it would be too difficult to keep tabs on the German experts brought over and practically impossible to prevent them from learning valuable British defence secrets.[9] Moreover, some worried that Britain might become too reliant on German expertise and that, in an emergency, the government would face 'the dilemma either that a large number of key positions are held by men of doubtful loyalty, or that the war effort is adversely affected by dismissing them'.[10]

Ultimately, however, the fear of being left behind proved the deciding factor – at a meeting on 15 August, Professor Charles Ellis, scientific advisor to the War Office, remarked that he was sure the Americans would press ahead with importing a certain number of German experts and asked 'could we afford not to adopt the same policy?[11] Even greater urgency was conveyed by the fear of Germany's brightest minds ending up in the Soviet Union, which, even by 1942, was being perceived by many of the more astute British analysts as the most likely opponent in a future conflict.[12] The DCOS was convinced that 'the Russians will in any event employ German technicians upon whom they can lay their hands' and the notion of Soviet resources being coupled with German expertise struck terror into the hearts of British policymakers.[13] Thus, while detention in facilities such as Dustbin and Matchbox may have prevented key German specialists from passing into Soviet hands in the short-term, it became necessary to initiate a project to pre-empt and then counteract the USSR's large-scale recruitment of military scientists, technicians and engineers.

While the DCOS considered the risks and benefits of an employment scheme, the key military ministries were already eagerly discussing which fields they would be most interested in exploiting and provisional numbers to be allocated to each. The Admiralty wanted 25 German experts in subjects such as hydrogen peroxide engines and optical crystals; the Ministry of Aircraft Production wanted 40 on topics such as supersonic aircraft and infrared-guided missiles; and the Ministry of Supply estimated it would need 85 covering rockets, ceramics, fuses and internal and external ballistics.[14] These wish-lists, and the DCOS recommendations, were passed up the chain of command to the Chiefs of Staff Committee, at which level the reactions were more equivocal, albeit with a general positive leaning.

Chief of the Imperial General Staff, Sir Alan Brooke, thought that 'the pros and cons were very evenly balanced', while First Sea Lord Andrew Cunningham and Chief of the Air Staff Charles Portal felt that 'on the whole, we stood to gain more than we might lose by bringing these scientists to this country'.[15]

Indeed, this view won out and on 24 August 1945, nine days after the Japanese surrender and the official end of the Second World War, the Chiefs of Staff 'agreed in principle to the employment of German scientists and technicians in this country, notwithstanding the security risks involved', and approved the recruitment of the 150 German specialists sought by the armed forces.[16] One week later, the Chiefs of Staff sent a paper outlining the policy to the highest body of military decision-making in Britain, the Defence Committee of the Cabinet, for their consideration. The Committee, chaired by the recently elected Prime Minister Clement Attlee, discussed this paper thoroughly and then approved it on 31 August.[17] In the official terms of this approval, they agreed 'in principle that German scientists should be brought to this country to be employed on research in the national interest, provided that they are regarded as servants of the State, and subject to certain safeguards'.[18]

This programme became known as the DCOS Scheme, after its origins with the Deputy Chiefs of Staff, and that committee became the co-ordinating body for the recruitment of all German defence specialists to Britain in the post-war period. However, the scheme did not take off at any speed. By October 1946, when the programme had been in operation for just over a year, the Americans announced that they were extending the limit of their German military specialist recruitment to 1,000 individuals; at the same time, the British had only managed to secure under contract 33 German defence scientists, a deficit which they attributed to the 'slowness of procedure for reception in UK'.[19] By April 1947, there were 60 German experts employed under the DCOS Scheme with 28 more contracts pending.[20] In November 1948, Sir Ben Lockspeiser, the Chief Scientist at the Ministry of Supply, reported that, 'broadly speaking', there were about 90 German scientists working at British defence establishments. He also registered his concern that, due to legal restrictions, it would be very difficult to extend any of these contracts beyond 1950.[21] These numbers were pretty uninspiring considering that even the most tentative target expressed during the policy development stage was for Britain to recruit 150 individuals.

One of the main reasons why the British recruitment figures remained so low was because they were reluctant to use coercion (as the Soviets often did) and thus insisted on only bringing over German specialists who were 'fully prepared to work abroad' so as to 'encourage their whole-hearted co-operation'[22] and because they were unable to make offers which could compete with those of the Americans. Britain did, however, have some potential appeal for German scientists to the point where they 'might prefer less favourable terms from the British to apparently more attractive offers from the American and Russian authorities'.[23] In some cases, this was because many German citizens viewed the Western Allies, and particularly the British, as far more palatable employers than the Soviets, due to pre-existing political convictions as well as, in some cases, skewed Nazi-era racial theories.[24] This was certainly true of Hellmuth Walter, the submarine and rocketry expert whose *Walterwerke* facility in Kiel had been one of the most significant early exploitation targets. He proved more than willing to co-operate with the British (he worked for the Admiralty at the Vickers shipyard in Barrow-in-Furness between 1945 and 1948), supposedly because he felt it was imperative that the 'Anglo-Saxons' controlled Europe in the face of the Soviet threat.[25]

One of Walter's employees, Hermann Treutler, a peroxide fuel expert, took this even further – he felt it was 'justifiable to come to the enemy's country because we [the Germans] were the master race and Britain was part of our Anglo-Saxon race'.[26] The larger issue this represented for the British exploitation authorities was that many of the German experts most willing to work for Britain were also those least repentant about their Nazi pasts. Indeed, Hellmuth Walter had joined the NSDAP (National Socialist German Workers Party) in 1932 (before they came to power in Germany), rose to the rank of *Amtsleiter* in the pre-war years and was 'regarded locally as a complete Nazi'.[27] The Admiralty recruiters knew that if they did not employ him, he would either find work elsewhere or they would have to monitor him constantly at great expense to prevent him from doing so – in short, it was judged to be better to have him 'with us than against us', an attitude that would be echoed time and again during the recruitment process and reflected the stark polarisation of the new Cold War world.

Cases such as these proved deeply problematic for the British authorities because, as desperate as they were to secure the best and brightest of Germany's scientific and technical talent, they were also deeply concerned about security in Europe, particularly with regard to a German military resurgence.[28] To this end, official policy made it very clear that 'nobody whose record indicates that he was a convinced Nazi should be brought to the UK to work, however high his scientific qualifications.'[29] Yet, even within this clarification, the term 'convinced Nazi' presented some room for manoeuvre, as highlighted by a Ministry of Aircraft Production representative who commented 'exactly what this means no-one appeared to know'.[30] Furthermore, the perceived risk was not just restricted to Nazis. In August 1945, the British Joint Intelligence Sub-Committee (JIC) expressed the opinion that all Germans, even those who had come to Britain before the war, such as Social Democrats and Jews, maintained 'their fundamental loyalty to the Fatherland' and, further, that 'if there were any possibility of Germany's regeneration they would be likely as any to take advantage of it, so long as it was not Nazi'. The JIC felt that even if an independent Germany hostile to Britain was not very likely, one which was 'absorbed in the Russian orbit' presented a genuine danger.[31]

Several of the agencies involved in exploitation prepared reports on the risk of any form of hostile German military revival, and how it could potentially be aided by German experts who had worked for Britain and gained in-depth knowledge of British military research projects. Even during the initial Anglo-American discussions about short-term utilisation of German expertise to contribute towards the war against Japan, British officials felt that 'we should not be bound to return any such scientists to Germany, either at the end of the Japanese war, or indeed at any time.' The reasoning they gave for this hard line, which was in direct contrast with the initial American proposals, was this:

> In working in the United States or in this country German scientists will necessarily become acquainted in some measure with our techniques and it is obviously undesirable that such men should return to Germany armed not only with the knowledge they now possess of German science, but British or United States knowledge.[32]

One suggestion for how to handle this risk came from the Scientific and Technical Research Board which proposed that all German scientists brought over to work in Britain be made to sign a document which 'renders them liable to prosecution if they disclose to unauthorised persons, details of the work on which they had been engaged'. The Intelligence Division of the Control Council felt this would not be sufficiently effective and suggested in addition that each German expert be given an 'informal talk' on the importance of discretion and instructed to report any attempts to elicit secret information from them to the nearest Intelligence Division office.[33]

In terms of the execution of personnel exploitation itself, Operation Surgeon proved to be one of the more successful vehicles for British recruitment, perhaps because a good understanding of the capabilities of the individual German experts involved had been reached during the intensive work conducted at the 'Surgeon stations' in Germany. The Ministry of Supply was also able to reduce the vast amount of time necessary to obtain clearance to bring a German expert to work in Britain to a much more manageable two months, and by late November 1946, the Air Division of the Ministry had already brought over 16 of their 74 target scientists (selected from a total list of 500), and they had begun work in Britain.[34] By January 1947, the number of German aeronautical experts employed in Britain had increased to 30, and as the process was wound down in 1948, it reached a peak of 87.[35] The process by which these individuals were recruited was far from standardised and, even just within the field of aeronautics, there was considerable diversity in the methods used. Adolf Busemann, one of the world's foremost experts on swept wings and supersonic flow, was contracted to stay in Britain directly from his detention at Inkpot in Wimbledon; the experienced aerodynamicists Dietrich Küchemann and Johanna Weber (one of only a handful of women who were targeted) were offered employment freely while still writing up reports on their wartime work in Germany, and were enticed by the brevity of the initial six-month term; while Karl Doetsch, an accomplished research engineer and test pilot, who had fled Berlin and taken refuge in Bavaria in the American zone after the war, was specifically sought out by British investigators who believed he had the expertise to help solve directional instability problems which were plaguing the Gloster Meteor, Britain's new jet fighter.[36]

An area in which Britain's recruitment efforts were notably less successful was rocketry. Despite taking the lead on the Backfire test launchings, Britain found itself on the back foot when making offers of employment, as they frequently paled in comparison to those made by the Americans, who had the appetite and capacity to absorb as many German rocket experts as they deemed necessary. During their brief stay at Inkpot, Wernher von Braun and five of his colleagues faced entreaties from British officials to renege on their existing contracts with the Americans and relocate to Britain instead. This endeavour failed largely because von Braun had the prescience to see that only the United States and the Soviet Union really had the capacity for a large-scale guided weapons programme in the future.[37] The military head of German rocket development, General Walter Dornberger, did spend almost two years in Britain after the war but, for much of that time, he was incarcerated in the repurposed Island Farm POW camp near Bridgend in Wales, awaiting trial for ordering the V-2 attacks on Britain. The trial never materialised – in part because it would have been absurdly hypocritical as the damage wrought by the RAF and USAAF on German cities, and by the atomic bombs on Hiroshima and Nagasaki, came to light – and Dornberger was released into American custody in July 1947. The main group of 12 rocket scientists and technicians Britain did manage to recruit were largely second-rate. On paper, their biggest catch was Walter 'Papa' Riedel, who had been Head of Design at Peenemünde in the early war years, but who had been exiled to a peripheral project of limited importance in the Austrian Alps from 1943.[38] Many other members of this group, including the aforementioned Hermann Treutler and the relatively talented engineer Johannes Schmidt, had actually spent much of the war working on submarine and aircraft engines under Hellmuth Walter and, in an act of all too common post-war opportunism, reinvented themselves as rocket experts and secured work at the Ministry of Supply's new Rocket Propulsion Establishment at Westcott in Buckinghamshire.[39]

Most of the German specialists who were brought to work for military establishments in Britain after the war, under the DCOS Scheme or one of its smaller parallels, were subject to roughly the same basic contract terms. They were initially 'landed' for a period of six months which could be extended 'for a further limited period if justified in the national interest' and often was. During this time, they were always contracted to a government ministry or department, which meant that they were '*ipso facto* a temporary government employee'.[40] The salary scale offered to the experts covered quite a considerable range and was divided into six sections: Grades I to III for scientists and the same for technicians. A Grade I scientist (the highest rate) could earn between £700 and £800 p.a. (approximately £25,000 today), while a Grade III technician (the lowest) could only earn up to half that (for reference, the average wage for a skilled tradesman in Britain at this time would have been approximately £500 p.a.). The German employees were expected to pay British income tax on this salary, and they were allowed to remit up to 50 per cent of their earnings back to their families in Germany, where it was also subject to German income tax.[41] As in the case of German specialists brought over for short-term interrogation, the families of German experts employed in Britain were also fairly well looked after; alongside this 50 per cent remittance, they were guaranteed protection, 'food value of 2,300 calories for wives and children and a certain amount of heat and light for essential warmth and cooking requirements'.[42] Factors such as these were considered vital as British exploitation agencies had noted, as though it surprised them, that the scientists attached 'great importance to the protection of their homes and families during their absence'.[43] Offering good terms to both the experts and their loved ones was therefore deemed essential to a successful recruitment initiative.

The employment of German scientists and technicians on government contracts to work on defence projects in Britain played a significant role in reshaping the nature of the post-war exploitation programme. Nonetheless, its lifespan was necessarily limited and, in July 1949, it was announced that:

> The DCOS Scheme for the recruitment of German scientists had now been terminated. It is assumed that the exceptional type of men, whom this scheme was intended to cover, have all been considered by now. Furthermore the special recruiting arrangements which formerly existed are no longer required since a routine procedure whereby anyone, government department, firm or individual can apply to employ Germans has now been laid on.

Throughout its four years of operation, the DCOS Scheme had secured the recruitment of 172 German specialists, and many of those stayed on beyond 1950 and became naturalised British citizens.[44] This figure only reveals a comparatively small segment of the British post-war recruitment of German scientists and technicians as it only accounts for those employed on government defence work; a far larger programme was created to exploit the best of German civil industry.

Civil recruitment

As has been seen with the exploitation of facilities, documents and equipment which occurred on the ground in Germany, although the initial stated focus was to acquire only intelligence pertaining to defence technology, with the aim of strengthening Allied arsenals at the expense of Germany's, this was soon augmented and then outstripped by commercial and industrial exploitation. The same was true in the recruitment of personnel. It has been argued that even before the end of the war Britain was laying plans for an ambitious expansion of its export capacity, to be effected largely by transplanting export-relevant technologies, and the associated personnel, from Germany.[45] As early as August 1945, during the initial DCOS Scheme discussions, Dr Charles Goodeve of the Royal Navy's Research and Development Department 'raised the question of bringing German technicians to this country for use in industry not wholly connected with defence', with particular reference to the instruments industry.[46] Goodeve's query was dismissed at the time, as recruitment on defence matters alone would be a much easier policy to push through, but the concept as a whole did not disappear.

In September, the Board of Trade issued a memo commenting on the benefits to British industry, and even the war effort, provided by German craftsmen who had emigrated to Britain before the war, but cautioning that admitting German scientists and technicians so soon after the end of the war might seem to many to be 'objectionable and undesirable'.[47] By December, these qualms carried less weight and the Cabinet Office contacted the British Joint Staff Mission in Washington to inform them that 'Civil Departments are so impressed with successful Combined arrangements made for Germans in the Defence field that they would like to follow similar procedure in the Civil Field.' It was also suggested that any exploitation of German specialists for civil purposes would follow the same technique as that used for defence recruitment.[48] At this stage, it was considered desirable to develop a joint policy with the Americans for any form of civil-industrial recruitment and the Joint Staff Mission tasked Sir John Magowan, the Minister in Charge of the Commercial Department at the British Embassy, and R.D. Fennelly, Head of the British Raw Materials Mission, both in Washington, to discover the extent to which the US was pressing ahead with any such scheme.

The British were impatient to begin this phase of exploitation and gave the Americans one month from the official presentation of the British proposals to signal their participation and prepare to exchange target lists or else, the British warned, 'we shall consider ourselves free to go ahead on a unilateral basis'.[49] In order to be in a position to press ahead with some urgency, whether in an Anglo-American arrangement or alone, the British authorities had already begun developing plans for the execution of such a policy. In November 1945, the Board of Trade convened a meeting to form a panel which would enact the personnel exploitation of German civil industry; roughly a non-military equivalent to the DCOS Committee. The product of this meeting was the formation of the Darwin Panel, named for its chair, Sir Charles Darwin: physicist, director of the National Physical Laboratory and grandson of the illustrious naturalist whose name he shared.[50] The panel was comprised of members from the Department of Scientific and Industrial Research, Board of Trade, Control Commission for Germany, Home Office, Treasury, German Economic Division, Admiralty, Security Services and the Ministries of Supply, Labour, Health, Agriculture and Fisheries, Aircraft Production, and Fuel and Power.[51]

The Darwin Panel did not wait to hear from Washington whether the scheme would be a combined or unilateral one, and instead began surreptitiously co-ordinating policy. It instructed all divisions of the Control Commission for Germany (British Element) (CCG(BE)) that information about the scheme 'should immediately become known to the Germans, by "bush telegraph", while precise conditions of service are being worked out in England'. They also asked all Branches to submit the names of 'really first class men known to them and who they consider would be suitable for employment in England' though they emphasised that 'the type of person required is the scientist or technical expert and not the "business executive" type'.[52] Another important tenet of the policy was the accessibility of the German specialists who were brought over. They were not to be employed 'in the ordinary sense of the term by the firms with whom they work ... but will, in a sense, be loaned by the government, roughly in the capacity of consultants'; this was largely to ensure that their expertise was 'made available to the whole of the scientific field or the whole of industry, and not ... to individual firms who could more or less copyright the results'.[53] Although this was a necessary step, it did, in fact, hamper the early stages of Darwin Panel recruitment, as many Research Associations (who were among the only bodies who were actually allowed to hire German experts) considered it to be too much hassle to employ these men or, alternatively, they were interested in only utilising them for a matter of weeks, which fell under the remit of BIOS interrogations and not Darwin Panel work.[54]

In the meantime, in Washington, the State Department, which had acted as the most consistent obstacle to a joint Anglo-American policy on civil-industrial recruitment, began to relent. On 21 August 1946, the US Chiefs of Staff circulated a memo, outlining their new policy 'to facilitate the entry into the United States, under the immigration laws, of a limited number of outstanding German and Austrian scientists and technicians'.[55] With US concordance, the British policy was able to increase in scale and speed. The Darwin Panel issued its first 'Comprehensive List' of 132 German scientists and technicians approved for employment in Britain, which included: head toolmaker, highly skilled spectrograph mechanic, chief camera designer, specialist in button manufacture, technical supervisor of alarm clock production, leather expert, superintendent of rubber department, authority on sugar beet, consultant on die design for turbine blades, and experts on needles and fishhooks, production of Rayon thread from viscose, and x-ray analysis, to provide just a fairly limited spectrum.[56] The inclusion of a 'principal scientist in manufacture of gyroscopic gunsights' on this list goes to show that, despite the focus on civilian industry, the expertise required for the production of military technology outside the government research establishments was also factored in.

The procedure utilised for Darwin Panel recruitment was very similar to that of the DCOS Scheme and, naturally, there were a number of crossovers. Indeed, in January 1946, Sir William Palmer of the Lord President's Committee (which had oversight of the Darwin Panel) commented that 'it is almost impossible to draw a rigid dividing line between research for industry and for defence purposes'.[57] All of those brought over under the Darwin Panel were salaried on the same scale as those who came to Britain to do defence work, and their families were entitled to the same amenities. Another element common to both schemes was the degree to which bureaucracy and adherence to guidelines could severely hinder the process of recruitment, and this was often exacerbated when defence and civil interests coincided. This is especially clear in the case of Otto Reder, a talented German aeronautical specialist and expert in the field of helicopter technology, who was brought over to Britain in October 1946 to be interrogated under the BIOS Scheme. He was then returned to Germany with only a vague assurance about future employment but asked not to take another job in the interim. In December, Reder wrote to L.R. Allum, the British supervising officer of German scientists at the Royal Aircraft Establishment (RAE), Farnborough, noting that he was 'still awaiting very anxiously any news from your authorities about my eventual immigration' and enquiring optimistically about when he and his family would be able to come to Britain, how much luggage they could bring, and how much notice he would have of his move. He signed off by saying 'I am very sorry to give you so much trouble, after all the trouble you have already had with me, and I thank you for all your help.'[58]

What followed was a run of correspondence between the Fairey Aviation Company, who were extremely keen to secure Reder's employment, and various government offices and agencies who appeared to be perpetually mired in bureaucracy. By May 1947, Major Malet-Warden, of FIAT, contacted BIOS and described the situation in no uncertain terms:

> We are still without definite word of subject's engagement by Fairey Aviation Co Helicopter Dept. We cannot stall Reder off any longer than two more weeks as on financial grounds he will be compelled to seek other employment. Please treat as urgent or this office will not stand in Reder's way if he wants employment elsewhere.[59]

In August, Reder was brought back to Britain under BIOS auspices and housed at Spedan Towers again, in order to prevent him being employed elsewhere, while the details of his recruitment by Fairey were hammered out.

However, just as progress appeared to be being made in this case, a new obstacle presented itself; in September, the German Division at the Board of Trade discovered that Reder's name appeared on the US 16th Defence List and they were therefore unable to move forward at all on the matter of his recruitment.[60] This further issue was, in fact, cleared up by Reder himself who stated that between January 1946 and January 1947 (excepting his time as a BIOS interrogation subject), he had been employed in the Science Department at an unspecified American-run university in Berlin and the head of that Department had informed him that he had been removed from all American recruitment lists. In October 1947, this fact was confirmed by Group Captain J.R. Wilson of the British Joint Staff Mission in Washington and Reder was finally released for employment by Fairey Aviation, one full year since he had initially been brought over by BIOS and promised future employment in Britain.[61] That was not quite the end of Reder's tale, however, as within months he had been returned to Germany on account of trade union resistance to his appointment made by the local branch of the Association of Engineering and Shipbuilding Draughtsmen.[62]

However, Reder's protracted experience seems to have been the exception rather than the rule. In fact, the initial limit placed on the Darwin Panel scheme of 200 specialists was reached easily by the end of 1946,[63] and in May 1947, the scheme expanded significantly and the ceiling was lifted to 500.[64] This expansion was necessary in order to allow for a major change in the British recruitment policy – this was the beginning of 'exclusive exploitation', which allowed individual companies (including those working on government defence contracts) to have unrestricted access to a particular German specialist for their own utilisation as they saw fit (as seen above with Fairey Aviation's pursuit of Reder). Previously, the German scientists and technicians were 'employed by government departments and paid out of public funds' and their service, 'like the specialised plant ... seized as reparations or booty', was considered to be 'national property' so that the products of it could be circulated 'throughout the industries concerned for the benefit of the country as a whole'. However, by December 1946, it was noted that this was considered to be unsatisfactory by many:

> Private firms would, for the most part, prefer to pay the Germans' salaries themselves, and keep the technical advice and information gained as their own property *vis-à-vis* their competitors. The Germans would also prefer to be employed by the private firms since there is a ceiling of £800 per year on the salaries paid to them by government departments.

Initial reaction to this was not positive; the Board of Trade did not feel it could work in tandem with the standard Darwin Panel scheme, and predicted further opposition from the Treasury, Home Office and other interested departments, but did concede that it might be possible as long as the private firms still allowed their German employees to publish their findings more widely, through learned societies for instance.[65]

Discussions on the possibility of this 'exclusive exploitation' scheme continued throughout early 1947, with the ever-important denial policy very firmly borne in mind, and concerns grew that the 'Research Associations and organised trade bodies' had already absorbed all the German experts they could, and 'the widening of the scope of the transfer of German knowledge and ideas will become increasingly dependent on employment by private firms'.[66] This raised a couple of issues which the Inter-Departmental Committee on German Scientists (IDCGS) worked to smooth over. First, it was decided to remove a clause which informed the German experts that they could be ejected from Britain at any time; the Home Office reserved that power over aliens at all times anyway and it was felt that explicitly stating such 'could have a discouraging effect'. Second, it was noted that although 'firms generally appeared to think that their chances of getting government contracts would be improved if their staff was strengthened with German experts', there would also be 'cases where a firm employing a German could not, for that very reason, be given work of particular secrecy or importance'. In short, it was concluded that 'the employment of Germans in firms is generally likely to increase the firms' efficiency, but there may well be cases where the security value is consequently diminished.'[67]

Nonetheless, despite these reservations, the policy of 'exclusive exploitation' was formally approved by the Defence Committee of the Cabinet on 14 May 1947. The Darwin Panel still played an important part in its execution. German specialists could not be recruited directly from Germany by private British firms and so still had to spend an initial six-month period in Britain on government contract. During this time, it was the responsibility of the Darwin Panel to ensure that the specialist got maximum commercial exposure, 'either by posting him to a Research Establishment or by arranging for him to visit a number of firms', as this would help him to build a network of contacts and increase 'the likelihood of his finding ultimate employment in which his knowledge would be fully utilised'. This had the additional benefit of increasing the chances of the expert securing fair financial reward for his services to British industry on the whole.[68] This move also saw an enlargement in the Darwin Panel scheme as it was felt that the employment of 'key men' who had designed a novel process was 'not always sufficient to permit its rapid and economic exploitation' and that acquiring the services of engineers and technicians who had been 'responsible for the application in the industrial sphere' was worthwhile too.[69]

The move to 'exclusive exploitation' proved a sizeable success for the exploitation of German expert personnel in civil industry in Britain. Some 70 per cent of the individuals initially recruited by the Ministry of Supply came to be retained by private firms once their initial Government contract was concluded and over 150 firms made use of the services of German experts across this period. AFN Ltd, which manufactured Frazer-Nash and Bristol sports cars, employed former BMW Technical Director Fritz Fiedler to work on chassis design; the Anglo-Iranian Oil Company (an antecedent of British Petroleum) brought in an IG Farben research chemist to advise on the production of lubricating oils at their flagship Llandarcy refinery in South Wales; various electronics companies (including EMI and Pye) hired German scientists after the war to work on television cameras and cathode ray tubes – these are just a few examples which give a sense of the range and the scope of this recruitment programme. Other individuals spread their knowledge across various firms, such as sugar beet expert Otto Koenig, who raised the sugar yield in 18 different factories, and Karl Hantlemann, who supervised the installation of new brick-making machinery in some 25 firms – invaluable during the period of post-war reconstruction in Britain. Nonetheless, the process was not tenable indefinitely and, on 5 April 1949, the Darwin Panel stopped receiving new nominations and spent the following months dealing with the backlog of existing cases. By this time, a more conventional overseas labour permit system was in place, which allowed German citizens to seek work with greater ease in Britain and elsewhere.[70]

The émigrés

Often, these recruitment schemes, discussed and developed at senior levels of the British government and civil service, were couched in abstract terms – it was considered sufficient just to assert that the military and private industry alike were set to benefit from the influx of German expertise, a distinctly amorphous concept. However, there were, of course, real-world implications of these initiatives, not least for the German scientists and technicians, and their families, who relocated to Britain, as well as for the British workplaces and communities which absorbed them. Much of this was governed by factors beyond official control, but part of it was shaped by deliberate policy (as in the salary and tax arrangements discussed above). One of the primary concerns of the authorities responsible was security. As such, German experts brought to Britain were still subject to the 'usual enemy alien restrictions' including movement only within a five-mile radius, no ownership of a car, motorcycle or camera, only one letter home per week, no contact with the press or attendance at party political meetings and adherence to a curfew of midnight to 6 a.m. They were, however, permitted to visit 'local licensed premises'.[71]

The German specialists were also permitted to take occasional leave (though not until their initial six months of employment was complete) to return to Germany and visit their families. On some rare instances, it was possible for family members to travel to Britain instead, particularly in cases where it was feared that if the expert returned to Germany, he might be at risk of kidnap by the Soviets (this was particularly pertinent if his home was in Berlin).[72] A policy was also considered wherein the wives and children of German scientists could be relocated to Britain on a semi-permanent basis. This initially encountered considerable opposition, largely due to wanting to avoid double standards 'in view of the impossibility of permitting British Officers employed by the Allied Control Commission and British Army Officers to take their families to Germany at the present time'.[73] However, by January 1947, it was agreed that in the case of scientists whose contracts were being extended for a second term, their families could be moved to Britain to join them, as long as the scientist was 'to work there long enough to warrant the trouble and expense of getting his family to England'. In addition, while they were living in Britain, the families were given assurances that 'their homes will be made available to them on their return and no furniture will be moved out during their absence'.[74]

Despite the seemingly generous terms of the contracts offered to the German specialists, this was a period of great change and hardship for the majority of the German population and areas of complaint soon presented themselves, particularly in terms of money and amenities for their families. Many experts expressed 'extreme dissatisfaction' at potentially having to pay both British and German income tax.[75] However, perhaps the biggest issue raised by German scientists was the uncertainty of their fate. In April 1948, Dietrich Küchemann, the eminent aerodynamicist who had been recruited to the RAE at Farnborough, wrote a letter to the Ministry of Supply on behalf of all his German colleagues at the RAE, describing the present state of affairs as 'characterised by uncertainty, a number of special regulations, and individual promises' and requesting a move towards 'normalisation'. The key elements which were desired were parity with British colleagues, 'a civic state of life for us and our families' and greater input into terms of contracts.[76]

Three months later, a similar letter was received by Sir Ben Lockspeiser at the Ministry of Supply from six scientists working for the government at the steam turbine firm of C.A. Parsons in Newcastle, stating that until this time they had 'enjoyed the work which we carried out wholeheartedly' but requesting greater clarity on their future prospects, as this would have 'significant bearing on the settling of our families and the planning of our children's education'.[77] Küchemann received no response and the reply that the entreaty from the C.A. Parsons group received was mostly full of equivocation. The Ministry of Supply was unwilling to commit to anything or to give any 'official promises about the future' and suggested instead that the scientists would be better off seeking employment by a private firm and perhaps naturalisation as British citizens at some point further down the line.[78] Unsurprisingly, this was cold comfort for the men involved.

The government were not only concerned with policing the behaviour, and fielding the complaints, of the German scientists and technicians but were also very aware that bringing these individuals to work in Britain could very easily cause uproar among their new neighbours and colleagues. One issue was raised by the dire shortage of housing throughout Britain and the Ministry of Health insisted that the German experts receive no preferential treatment and compete on the same equal terms as British citizens in their search for accommodation. Further problems which were foreseen included difficulties of assimilating the new German employees into British workplaces and opposition from British professional staff organisations to the recruitment of foreigners.[79] In particular, it was felt that if a German technician was 'to hold a position in which he will be giving directions to British workers ... there is bound to be trouble'.[80] The British authorities were not just concerned with keeping the peace in the workplaces which took on German staff. They also acknowledged that, with the German scientists receiving increasingly attractive offers from the Soviet Union, among others, it was important to make sure that those experts working in Britain were satisfied and felt like valued colleagues.[81]

As a result, the government were somewhat proactive in their attempts to ameliorate these potentially hostile conditions. At a Board of Trade meeting in November 1945, attention was drawn to 'the necessity of preparing the ground for the arrival of German technicians and scientists by explaining ... the reasons for their arrival, so that unrest amongst workpeople employed in particular organisations to which they might be attached could be avoided as much as possible'.[82] One tactic which was devised to manage public reaction was the use of the Ministry of Labour's Regional Industrial Relations Officers (RIROs), who would be well-informed about the principles of the scheme and would be deployed to visit British firms and issue reassurances, such as that none of the German recruits were pro-Nazi and that no British jobs would be lost. The RIROs were also instructed to make a report if they became aware of 'difficulties arising from the attitude of workpeople to the introduction of a German scientist'.[83]

A similar procedure was enacted through co-ordination with the Trade Unions Congress (TUC) which, by the end of the war and following the 1945 election of a Labour government, had become an indispensable forum of debate on all industrial matters.[84] Union consultation took place at a high level – the Board of Trade met directly with Sir Walter Citrine, the General Secretary of the TUC, for instance – and then information was filtered down to local branches, to ensure that all unions concerned were 'fully acquainted with the reasons for the adoption of the scheme, so that they would be able to inform employees of the factories to which Germans would be attached and also meet uninformed criticism'.[85] Nonetheless, trade unions had plenty of reasons to be hostile towards an influx of migrants (especially from a former enemy nation), fearing that it could keep wages low and harm their efforts to obtain better working conditions.[86] In fact, trade unions (especially at the local branch level) actually remained among the most vocal and consistent critics of the recruitment of German specialists and, thanks to their links with the Labour government, were able to genuinely limit the absorption of German personnel into the British workforce.[87]

Ultimately, the experts arrived at a time when British perceptions of the German people were decidedly mixed, a combination of horror at the evil which had brought about the Holocaust, pity for the suffering now endured by ordinary German citizens in the aftermath of war and a sense of moral superiority as a victorious occupier in the position to re-educate and reform.[88] In August 1945, a Gallup poll asked members of the British public about their feelings towards the German people – 25 per cent said sympathy, 21 per cent hatred and 15 per cent dislike; 6 per cent felt it was necessary to keep them under strict control, 4 per cent thought that treatment should be harsher, while a further 3 per cent simply asserted that their present suffering 'served them right'. Eighteen months later, in January 1947, another Gallup poll found that 42 per cent of respondents felt friendly towards the German people while 36 per cent felt unfriendly.[89] Winston Churchill, who took quite a pragmatic approach to Germany after the war (inspired in part by his concerns over Soviet aggression in Europe), was disappointed in his countrymen's lack of generosity and unwillingness to 'close the German account' – the Allies might aim to re-educate the Germans, but he wondered who would re-educate the British.[90] On the whole, though, British perceptions of Germany throughout the first half of the twentieth century (aside from times of war) were highly changeable and non-ideological and this state of flux persisted into the post-war period, though negative attitudes were usually dominant.[91]

The years immediately after the war actually saw a relatively large number of German migrants come to Britain – approximately 60,000 between 1945 and 1951 – either to find work or for personal reasons. As suggested by the polling results above, they encountered mixed attitudes from the people they encountered. In the workplace, trade unions served as a hub of particular resentment and the strong work ethic on which the German employees prided themselves was often seen by their colleagues as representing a lack of solidarity and being indicative of their 'foreignness'. Those migrants who lived in close-knit communities, especially rural villages, found themselves generally accepted and even welcomed, while those who lived in larger towns and cities faced greater hostility. There were accounts of being refused service in pubs, having swastikas daubed on property and receiving anonymous threatening phone calls. Even aside from specific incidents such as these, the general environment could often be cold and bitter – one post-war *émigré* recalled her fellow church-goers refusing to share her pew on a Sunday, instead preferring to squeeze together where there was less room – she described this underlying rejection by her new neighbours as 'a wall'.[92]

Many of these issues were heightened for the German scientists, in part because their arrival was often reported widely in the local and national press and also because, if they were specialists in a military field, they were directly associated with attacks on British soldiers and civilians during the war. This potential for friction was especially clear when the first group of German scientists arrived in Britain in January 1946.[93] These 23 submarine experts (counting Hellmuth Walter among their number) arrived by ship in Barrow-in-Furness on the Cumbrian coast and were to work in the shipyards of Vickers-Armstrongs. The *Daily Mail* announced their arrival with typical aplomb, running a front-page splash with the headline 'Hush-Hush Germans Land in Britain' and describing their entrance as being 'behind a security curtain of wartime rigidity'. It also reported that a 'wave of anger' was sweeping the town of Barrow and that the shipyard workers resented 'the arrival of former enemies, who are said to be still pro-Nazi'. The intrepid reporter recounted how that very night he had proceeded to Rock Lea, the 'double-fronted, three-storeyed, red-bricked building' where the scientists were being housed, discovering, to his evident shock, that 'there was no guard on duty and the wrought iron gates were wide open'. Furthermore, he continued in his tone of thinly veiled suspicion, when he rang the doorbell, the door was opened by a senior Women's Royal Naval Service (WRNS) officer, who 'would not allow me near the glass-panelled door, through which peals of laughter could be heard'.[94]

Naturally, the sense of outrage simmering below the surface of this article has to be attributed, to at least some degree, to journalistic flair and the quest for a good story. However, as it turned out, this *Daily Mail* reporter was not the only one scandalised by the arrival of these German experts and their accommodation at Rock Lea. On 11 January, the *Manchester Guardian* reported that there had been protests by both the Co-operative Youth Club and the Townswomen's Guild in Barrow against the accommodation of the scientists in Rock Lea (which had been requisitioned by the Admiralty) and that the latter had written to the local Member of Parliament (MP) to register their disapproval.[95] A former mayoress of Barrow and leading member of the Townswomen's Guild, Mrs F.M. Poyntz gave a statement to the press in which she said, 'we feel that these scientists ought not to be living on the fat of the land. We hear on the wireless all about the atrocities [committed] by the Germans during the war ... these men are objectionable and will not give of their best [*sic*].'[96]

Four days later, the *Daily Mail* printed a letter from an anonymous correspondent in Glasgow, writing under the *nom de plume* 'Veritas', who congratulated the newspaper for its exposé on the 'unnecessary pampering of the German scientists', attributing it to Britain's 'age-old custom' of 'forgive and forget' and lamenting that this very attitude was responsible for the failed policy of appeasement before 1939. He went on:

> Less than a year ago these same German scientists were racking their brains to invent means of exterminating us and now we bring them safely to our island and give them the finest housing accommodation, while any old thing will do for Britain's workers, soldiers, sailors and airmen.[97]

In fact, it rapidly became clear that the major grievance the British public, both local and national, had about the scientists in Barrow, was the apparently luxurious lodgings which they had been given, during a period of nationwide housing shortage. As a result, following the protests from local groups, Walter Monslow, Barrow's MP, visited Rock Lea and reported back that the scientists 'were not living in luxury', even deigning to describe the 'coconut matting on the floor' and pointing out that the men 'did all their own work except making beds and cooking'. He concluded that he was 'satisfied that their work here ... must obviously be to the best interests of Barrow in the ultimate'.[98]

The Barrow brouhaha was really little more than a storm in a teacup, beginning as the bitter griping of a few local vested interests and whipped up into an almost xenophobic frenzy, albeit a short-lived one, by some national media outlets. In reality, there were even some who were quite positive about the arrival of these experts – one British submarine engineer from the Vickers yard observed: 'If the Germans can show us something new, and something which is going to advance engineering science, then we should be grateful for having them with us.'[99] The German rocket scientists who came to the Guided Weapons Establishment at Westcott, near Aylesbury in Buckinghamshire, in early November 1946 met a similarly mixed reception. Two weeks after their arrival, three of them were invited to the local church by the parish priest to hear a series of organ recitals and a local newspaper proudly reported their attendance, alongside several German POWs from the area.[100] On the other hand, when it was suggested, in February 1947, that several of these German experts be housed at The Firs, a country house in the nearby village of Whitchurch, which had hosted 'M.D.1' – a small arms design facility known as 'Churchill's Toyshop' – during the war, there was considerable opposition from local residents. This was most clearly voiced in a petition sent to the local MP, which stated:

> As residents in the area, we very strongly object to this move, which would almost certainly result in many Germans living in the district and being free to mix with our people. ... It seems to us that they could be accommodated elsewhere, in some place where they would not be able to exercise what might be a harmful influence on village life.

The former head of 'M.D.1', Lieutenant-Colonel R. Stuart Macrae, also jibed that, during the war, he had feared the Germans might take over, but he had never thought that when the war was won, the Germans would take over just the same.[101]

Over time, attitudes in the area surrounding Westcott softened and the German scientists became welcomed, and often well-respected, members of the local community. This became very apparent when an explosion at the Guided Weapons Establishment claimed the lives of three men and injured eleven others in November 1947. Among the dead was the *de facto* leader of the German group, Johannes Schmidt, while his colleague Heinz Walder was one of the injured. The local newspaper reported that German scientists attended the funerals of the deceased British specialists, and vice versa, and also noted that the accident had given 'local people a deep sense of sympathy and distress, and its poignancy was heightened by the fact that a great German scientist whose knowledge and skill were being applied to British research work was among the three who lost their lives.'[102] This public display of grief, and the fact that it was supposedly directed especially strongly towards the German victim, suggests a greater degree of integration than earlier reports. A brief epilogue to the Westcott explosion story was the attempt by Dr Schmidt's widow to secure a larger compensation payout than the £2,920 (over £100,000 today) which she had initially been offered. However, the Treasury decided that it was not possible to 'treat Frau Schmidt more favourably than a British national – a consideration which is emphasised by the fact that two established [British] civil servants were killed in the same explosion which resulted in Dr Schmidt's death'.[103]

One interesting point to consider is the way in which the experience of German rocket scientists brought to Britain differed from those who went to the United States. Numbering around 100 individuals and based at Redstone Arsenal near Huntsville, Alabama, they so wholeheartedly recreated their original community there that even local residents began referring to it as Peenemünde-South.[104] They were generally perceived as having a unique and positive impact on the local area and were the subject of great admiration and pride. This was especially true of their most famous member, Wernher von Braun, who was integral in the US space programme and who designed the *Saturn V* rocket which first took man to the moon in 1969. He became both a local and national celebrity and quite central to Huntsville's identity and is now commemorated in the names of the local civic centre and a research hall at the University of Alabama campus, a mural at the airport, as well as displays at the nearby US Space and Rocket Center.[105]

While nothing to this scale or extent can be seen among the German expert *émigrés* to Britain, many did still have a lasting impact on the facilities and communities which they joined. This is perhaps most evident at the Royal Aircraft Establishment in Farnborough, where a number of German scientists forged their post-war careers. Most notable among these are Dietrich Küchemann and Johanna Weber. Both had come to Britain on short-term contracts arranged via the DCOS Scheme and emerging directly from Operation Surgeon (Küchemann first in October 1946 and then, at his urging, Weber in August 1947). Their contracts were continually renewed and then made permanent in 1953, when they both became naturalised British citizens. Both Küchemann and Weber spent the rest of their careers at the RAE – the latter reaching the rank of Senior Principal Science Officer, and the former becoming Head of the Aerodynamics Department in 1966 – and both made lives in Britain. Küchemann brought his wife and three children over in 1948 and they initially shared a rented house with a British colleague and his family; later, they bought a place of their own, where Weber also lived until the bungalow next door became available in 1961. Küchemann died unexpectedly in 1976 but Weber remained in that bungalow until 2010, by which point she was the oldest resident on the street. It is worth noting that, in an interview in 2000, Weber described the British people she met when she first arrived as 'extremely friendly. I have never heard a bad word in all my life.'[106] Not only do the stories of Dietrich Küchemann and Johanna Weber show us that successful integration was possible for the German experts who came to Britain after the Second World War, they also reveal the benefits the programme could have for Britain as a whole – both Küchemann and Weber were integral to the final design of Concorde, as well as a number of other important and high-profile British aeronautical projects.[107]

Overall, it is clear why many of those involved saw the recruitment of German scientists and technicians as the most valuable phase of the post-war exploitation of science and technology. While visits to facilities in Germany and the examination of confiscated documents and equipment had loose antecedents in the conventional looting and plundering which traditionally accompanied the occupation of enemy territory at the end of a conflict, the interrogation and recruitment of individuals with specialist knowledge were all but unprecedented. This new development was the product of many factors: an intelligence network which had expanded considerably in wartime and which understood the benefits of human intelligence, domestic British industry which had built closer links to the state during the war and was now crying out for support and investment in return for the contributions it had made to the national war effort and, perhaps most importantly, the looming spectre of some future war, in which it was assumed that a nation's scientists and technicians would be as significant as its soldiers, sailors and airmen.

While these intentions were clear, the general success of British recruitment strategy is harder to establish. For a start, on account of the myriad and often unofficial ways in which German specialists could be brought over to Britain, estimates of the total number are notoriously hard to pin down. Perhaps the best attempt in this direction has been made by Carl Glatt, in the course of his exhaustive PhD research. He estimates that between 807 and 1,052 German scientists, technicians, engineers and other specialists took up employment in Britain between 1945 and 1949, across both defence and civil-industrial recruitment schemes.[108] This is consistent with the patchwork of other available data and reveals Britain to have finished a close fourth in the race for these expert personnel, behind the United States, the Soviet Union and France.[109] Glatt goes on to argue that this comparatively smaller number of German personnel employed in Britain was not purely a result of German reluctance to take up employment there, but rather that the capacity of the British industrial and social system to absorb them was too low, in part due to issues such as trade union resistance.[110] In some areas, Britain did not pursue German expertise at all, which can be seen as an indication of the British approach to personnel exploitation being purposeful and highly selective.[111] Of course, it is also true that, in some cases, Britain simply could not match the employment offers or conditions of service made by rival nations, especially the USA and the Soviet Union. The nature of this contest with other countries, and the broader diaspora of German specialists after the war, forms the focus of the next two chapters.

Notes

1 TNA, AVIA 15/2209, R.P. Linstead to Edward Appleton, 5 May 1945.
2 TNA, AVIA 54/1403, 'The Exploitation by British Industry of German Scientific and Industrial Knowledge', 29 July 1947.

3 TNA, CAB 122/343, British Joint Staff Mission to Chiefs of Staff, 5 June 1945.
4 TNA, CAB 122/343, Chiefs of Staff to British Joint Staff Mission, 14 June 1945.
5 TNA, CAB 82/3, 'Minutes of 4th Meeting of DCOS', 6 June 1945.
6 TNA, CAB 82/3, 'Minutes of 7th Meeting of DCOS', 18 July 1945.
7 TNA, CAB 69/7, 'Minutes of 4th Meeting of Defence Committee', 31 August 1945.
8 TNA, CAB 81/93, 'Minutes of 53rd Meeting of JIC', 8 August 1945.
9 TNA, CAB 81/93, 'Minutes of 55th Meeting of JIC', 14 August 1945.
10 TNA, CAB 121/429, 'Employment of German Scientists in Defence Establishments', 5 February 1947.
11 TNA, CAB 82/3, 'Minutes of 10th Meeting of DCOS', 15 August 1945.
12 Julian Lewis, *Changing Direction: British Military Planning for Post-War Strategic Defence, 1942-47* (London: Sherwood, 1988), xcvii.
13 TNA, CAB 122/343, 'Policy for the Exploitation of German Science and Technology', 1 August 1945.
14 TNA, CAB 122/343, 'Policy for the Exploitation of German Science and Technology', 17 August 1945.
15 TNA, CAB 79/37, 'Minutes of 202nd Meeting of COS', 21 August 1945.
16 TNA, CAB 122/343, 'Minutes of 206th CoS Meeting', 24 August 1945.
17 TNA, CAB 69/7, 'Minutes of 4th Meeting of Defence Committee', 31 August 1945.
18 TNA, FO 1032/300, 'Employment of German Scientists and Technicians', September 1945.
19 TNA, CAB 122/349, FIAT Main (Br) to Intelligence Division, 4 October 1946.
20 TNA, AVIA 54/1403, 'Employment of German scientists and technicians', 21 April 1947.
21 TNA, AVIA 54/1403, Sir Ben Lockspeiser to Ministry of Supply, 12 November 1948.
22 TNA, AVIA 54/1403, 'Policy for Exploitation of German Science and Technology', 2 May 1947.
23 TNA, CAB 122/342, 'Minutes of DCOS Meeting', 12 December 1945.
24 Hans-Peter Schwarz, 'The Division of Germany, 1945–1949', in Mervyn P. Leffler and Odd Arne Westad (eds.), *The Cambridge History of the Cold War*, vol. I (Cambridge: Cambridge University Press, 2012), 149–50.
25 Longden, *T-Force*, 260.
26 Hermann Treutler, cited in Bower, *Paperclip Conspiracy*, 201.
27 TNA, ADM 178/392, 'Takeover of German Naval War Factories', 1945.
28 Deighton, *Impossible Peace*, 5.
29 TNA, AVIA 54/1403, 'Policy for Exploitation of German Science and Technology', 2 May 1947.
30 TNA, LAB 8/1198, MAP to T. Brind, 17 November 1945.
31 TNA, CAB 82/6, 'DCOS papers: 1–66', 14 August 1945.
32 TNA, CAB 122/343, Chiefs of Staff to BJSM, 14 June 1945.
33 TNA, FO 1031/19, Intelligence Division to Research Branch, 29 July 1946.
34 TNA, AVIA 12/82, 'Operation Surgeon: Memorandum', 23 November 1946.
35 Uttley, 'Operation Surgeon', 9.
36 Nahum, 'I believe the Americans have not yet taken them all!', 113–4.
37 McGovern, *Crossbow and Overcast*, 201–2.
38 Neufeld, *Rocket and the Reich*, 269.
39 Bower, *Paperclip Conspiracy*, 199–202.
40 TNA, FO 1031/19, Lt-Col. D.G. Edwardes to Maintenance Branch, ZEO, CCG, 23 November 1946.
41 TNA, CAB 122/357, H.D.B. Wood to Sir John Magowan, 26 January 1946. Price adjustments calculated on TNA Currency Converter [accessed online 18 July 2018, http://www.nationalarchives.gov.uk/currency-converter/].
42 TNA, CAB 122/357, 'Employment of German Scientists in a Civil Capacity', 29 April 1946.

43 TNA, CAB 122/357, 'Employment of German Scientists in a Civil Capacity', 29 April 1946.
44 Uttley, 'Operation Surgeon', 19.
45 Werner Abelshauser, 'Immaterial Reparations and the Reintegration of West Germany into the World Market', in Judt and Ciesla (eds.), *Technology Transfer*, 110.
46 TNA, CAB 122/343, 'Minutes of 10th DCOS Meeting', 15 August 1945.
47 TNA, FO 1032/164, 'Employment of German Scientists and Technicians in Civil Industry in the United Kingdom', 28 September 1945.
48 TNA, CAB 122/357, Cabinet Office to BJSM, 22 December 1945.
49 TNA, CAB 122/357, 19 January 1946.
50 A.G.M. Barrett and D.H.R. Barton, 'Darwin, Sir Charles Galton (1887–1962)', *Oxford Dictionary of National Biography* (Oxford: Oxford University Press, 2004) [accessed online 6 August 2015, http://www.oxforddnb.com/view/article/32716].
51 TNA, FO 1032/164, 'Minutes of 1st Darwin Panel Meeting', 21 November 1945.
52 TNA, FO 1032/164, 'Memo from Econ 4', 29 November 1945.
53 TNA, LAB 8/1198, 'Minutes of Board of Trade Meeting', 22 November 1945.
54 TNA, LAB 8/1198, 'Minutes of 2nd Darwin Panel Meeting', 10 December 1945.
55 TNA, CAB 122/357, 'US Chiefs of Staff Memo', 21 August 1946.
56 TNA, CAB 122/357, 'Comprehensive List of Names', 31 July 1946.
57 TNA, CAB 122/357, 'Memo by Sir William Palmer', 11 January 1946.
58 TNA, AVIA 54/1403, Otto Reder to L.R. Allum, 14 December 1946.
59 TNA, AVIA 54/1403, Malet-Warden to BIOS Group II, 9 May 1947.
60 TNA, AVIA 54/1403, G.M. Judges to J. Protheroe, 5 September 1947.
61 TNA, CAB 122/360, J.R. Wilson to A.H. Thorold, 2 October 1947.
62 Glatt, 'Reparations', 963.
63 TNA, AVIA 54/1403, 'Minutes of Defence Committee Meeting', 11 December 1946.
64 TNA, FO 1032/164, 'Minutes of 17th Darwin Panel Meeting', 16 May 1947.
65 TNA, AVIA 54/1403, 'Employment of German Scientists and Technicians', 9 December 1946.
66 TNA, AVIA 54/1403, 'Russian Enticement of German Scientists', 18 April 1947.
67 TNA, AVIA 54/1403, 'Minutes of IDCGS Meeting', 11 February 1947.
68 TNA, FO 1032/164, 'Minutes of 18th Darwin Panel Meeting', 30 May 1947.
69 TNA, LAB 8/1450, 'Proposals to Widen Scope of Darwin Scheme', April/May 1947.
70 Glatt, 'Reparations', 919–53.
71 TNA, LAB 8/1198, 'Minutes of Inter-departmental Meeting of Security Officers', 17 December 1945.
72 TNA, AVIA 54/1295, J.H. Keane to B. Barrett, 25 June 1948.
73 TNA, LAB 8/1198, 'Minutes of Board of Trade Meeting', 22 November 1945.
74 TNA, FO 1032/302, 'Policy for Families', 4 January 1947.
75 TNA, AVIA 54/1294, 'Minutes of DCOS Meeting', 17 April 1946.
76 TNA, AVIA 54/1294, D. Küchemann to The Secretary, Ministry of Supply, 22 April 1948.
77 TNA, AVIA 54/1294, Letter to Sir Ben Lockspeiser, 31 July 1948.
78 TNA, AVIA 54/1294, Lockspeiser to C.A. Parsons group, 6 August 1948.
79 TNA, AVIA 54/1403, 'Memo from Director of TPA', April 1947.
80 TNA, LAB 8/1198, 7 November 1945.
81 Patricia Meehan, *A Strange Enemy People: Germans Under the British, 1945–50* (London: Peter Owen, 2001), 211.
82 TNA, LAB 8/1198, 'Minutes of Board of Trade meeting', 22 November 1945.
83 TNA, LAB 8/1198, 'Memo to RIROs', 27 February 1946.
84 Chris Wrigley, 'Introduction', in Chris Wrigley (ed.), *British Trade Unions, 1945–1995* (Manchester: Manchester University Press, 1997), 2.
85 TNA, FO 1032/164, 'Minutes of 6th Darwin Panel Meeting', 30 January 1946.

86 Inge Weber-Newth and Johannes-Dieter Steinert, *German Migrants in Post-War Britain: An Enemy Embrace* (Abingdon: Routledge, 2006), 33–4.
87 Glatt, 'Reparations', 962–4.
88 Evgenios Michail, 'After the War and After the Wall: British Perceptions of Germany following 1945 and 1989', *University of Sussex Journal of Contemporary History*, 3 (2001), 3.
89 George H. Gallup (ed.), *The Gallup International Public Opinion Polls: Great Britain, 1937–1975* (New York: Random House, 1976), 117, 148.
90 John Ramsden, *Don't Mention the War: The British and the Germans since 1890* (London: Little, Brown, 2006), 213.
91 Jan Rüger, 'Revisiting the Anglo-German Antagonism', *The Journal of Modern History*, 83 (2011), 588.
92 Weber-Newth and Steinert, *German Migrants in Post-War Britain*, 1, 105, 152–3.
93 Lasby, *Project Paperclip*, 169–70.
94 'Hush-Hush Germans Land in Britain', *Daily Mail*, 5 January 1946, 1.
95 'German Scientists at Barrow', *Manchester Guardian*, 11 January 1946, 8.
96 'Women Don't Want German Experts Here', *Dundee Courier*, 9 January 1946, 2.
97 'Letters to the Editor', *Daily Mail*, 15 January 1946, 5.
98 'German Scientists at Barrow', *Manchester Guardian*, 15 January 1946, 6.
99 'Arrival of German Scientists: Barrow More Interested than Disturbed', *Lancashire Evening Post*, 5 January 1946, 1.
100 'Headmaster's Organ Recital', *Bucks Herald*, 22 November 1946, 7.
101 'Whitchurch Could Help World Trade', *Bucks Herald*, 7 February 1947, 7.
102 'A Local Tragedy', *Bucks Herald*, 21 November 1947, 5.
103 TNA, AVIA 54/1295, W.S. Polley to J.A.H. Smith, 24 May 1948.
104 Burghard Ciesla and Helmuth Trischler, 'Legitimation through Use: Rocket and Aeronautical Research in the Third Reich and the USA', in Mark Walker (ed.), *Science and Ideology: A Comparative History* (Abingdon: Routledge, 2003), 171.
105 Monique Laney, *German Rocketeers in the Heart of Dixie: Making Sense of the Nazi Past in the Civil Rights Era* (New Haven, CT: Yale University Press, 2015), 71–2.
106 John Green, 'Obituary – Dr Johanna Weber', Royal Aeronautical Society [accessed online 20 July 2018, https://www.aerosociety.com/news/obituary-dr-johanna-weber/].
107 Nahum, 'I believe the Americans have not yet taken them all!', 126.
108 Glatt, 'Reparations', 937.
109 Neufeld, 'Nazi Aerospace Exodus', 53.
110 Glatt, 'Reparations', 964.
111 Nahum, 'I believe the Americans have not yet taken them all!', 124.

7   Allies and rivals

The period immediately following the end of the Second World War saw a considerable realignment of geopolitical allegiances and saw many nations re-evaluate their priorities, often resulting in major policy shifts. For example, the United States completed its transition to primary global superpower, eclipsing Britain's earlier hegemony, while processes such as European integration and decolonisation took their first tentative steps.[1] The line separating allies and rivals quickly became blurred. From a British perspective, the United States remained a close, if not wholly trustworthy, friend; France and other European nations occupied a middle ground of ambiguous allegiance; while Britain's relationship with its Empire, Commonwealth and Dominions was subject to much soul-searching and reassessment. The growing threat of the Soviet Union was also of enormous significance and will be discussed in the next chapter. Beyond this, independent nations around the world sought to take advantage of the new world order and, in looking to advance and progress, often saw German science and technology as a potential catalyst for their development.

Ultimately, the resources to be exploited in Germany were finite so an element of international competition over the most valuable elements was to be expected. Tom Bower has written that it was the 'plunder' of Germany which split the tight wartime alliance, while Alec Cairncross and others have argued that reparations (of which exploitation was one component) proved to be the biggest bone of contention between the former Allies.[2] Conclusions such as this should be treated with caution, however; there were bigger, ideological issues involved while long-term strategic concerns often played a decisive role. The competitive process of exploitation was, in itself, potentially divisive but it was also a product of the rapidly polarising international relations of the immediate post-war period.

The main thrust of this chapter, therefore, is to contextualise British exploitation on the international stage. Britain occupied a unique position at this time as it sat at the centre of three interlocking circles – the Transatlantic Alliance, Europe, and the Empire and Commonwealth.[3] This chapter will explore how British exploitation was affected by each of these three spheres of international engagement in turn, and will reveal the way in which close wartime allies (the United States and France) became post-war rivals, as well as how exploitation, especially of personnel, became a global phenomenon in the years after 1945. It is important to note that there were often differences between official policy (decided by those who were perpetually mindful of geopolitical considerations) and the actions taken by those on the ground (for whom pragmatism was essential to overcome various practical obstacles) – this chapter will address both elements in tandem. On the whole, this chapter will show how scientific and technical exploitation fitted into the wider diplomatic framework of the new post-war world, and how Britain navigated this complex web of alliances and rivalries.

The United States

In wartime, the relationship between Britain and the United States was the closest of any of the Allies, leading John Baylis to argue that it was so intimate and informal that traditional state sovereignty was eroded and replaced by a common Anglo-Saxon identity and purpose.[4] Although the speed with which this 'special relationship' was set aside immediately after the war challenges this assumption, there is no doubt that, in wartime, good communication and frequent consultations meant that many Anglo-American actions were taken essentially bilaterally. Scientific and technological exchange played a significant role in both forging and maintaining this bond throughout the war. As discussed in Chapter One, the Tizard Mission was an unprecedented example of military collaboration by two unallied countries, but even this was not unequivocal. For example, British officials remained concerned that American security was not sufficient to safeguard British secrets, while the Americans deliberately withheld details of their advanced Norden bombsight, despite ardent British interest.[5]

Another, later example was the atomic bomb project. Although the Manhattan Engineer District (the codename given to the project) was an American affair, many of the experts from the British 'Tube Alloys' programme were involved closely too, and several even relocated to the US to contribute directly. As a result, when Manhattan head Brigadier Groves put together the Alsos initiative, though it was US-led, it operated in close alignment with the British too.[6] Unsurprisingly, this attitude of Anglo-American collaboration carried over into the origins of the scientific and technical exploitation programme proper. When the Combined Intelligence Priorities Committee (CIPC) was first proposed, it rested firmly on the assumption that it would be responsible for gathering intelligence for the benefit of both nations. Despite this, there was a simmering sense of competition beneath the surface, and the British were keen to have the committee chaired by one of their own, partly because it was to be based in London and would utilise intelligence primarily from British sources, and also because they imagined that the US would institute and lead a parallel committee in Washington to handle similar matters in the Far East, and they wished to have balance.[7]

When the CIPC came into existence (quickly becoming CIOS, or the Combined Intelligence Objectives Sub-Committee), balance was actually achieved by having an American military chairman, Brigadier-General T.J. Betts, and a British civilian deputy, R.P. Linstead. This was palatable to the British as, for the time being, there was no parallel organisation in the US tasked with Japanese investigations. The way in which exploitation was to be conducted was also decided upon collaboratively. In April 1944, Alastair Balfour of the War Office's Civil Affairs department wrote to Lieutenant-Colonel Alexander Geddes, at the Deputy Chief of the Imperial General Staff's office, cautioning him that it would be premature to submit any outline blueprint for exploitation to SHAEF (Supreme Headquarters Allied Expeditionary Force) before the Americans had a chance to do so as 'SHAEF is entirely a joint affair and I feel we should be treading on dangerous ground if we were to submit our plans unilaterally'.[8] In this spirit, CIOS operated fairly effectively in its Anglo-American format under the jurisdiction of SHAEF during the last year of the war and the first two months of peace, but a major change was visible on the horizon.

The extensive wartime collaboration between Britain and the United States came to an abrupt end once the war was over. This was perhaps signalled most clearly by the sudden termination of Lend-Lease aid, on which Britain had become increasingly reliant, on 2 September 1945, but it was also echoed in scientific and technological matters too.[9] The passage of the McMahon Act in August 1946 meant that, despite British contributions to the Manhattan Project, the USA had no intention of sharing any of its atomic secrets with Britain and wished to remain the world's only nuclear power for as long as possible.[10] The era of Anglo-American scientific co-operation was seemingly over and, even though American scientists felt that British contributions such as the Tizard Mission had been of 'inestimable value', no efforts were made to reciprocate or repay this generosity after 1945.[11] This was indicative of the new age of American supremacy, with the United States emerging from the war with full superpower status and the intention of pursuing its global agenda, with little regard for the opinions or wishes of former allies. This attitude even extended to Britain, the other half of the so-called 'special relationship', and the nation whose worldwide hegemony the United States had essentially usurped.[12] Well-worn narratives of post-war British decline may often be exaggerated, but there is no denying that the Anglo-American power dynamic irrevocably changed between 1939 and 1945.

The impacts of this realignment were evident in terms of exploitation, too. As the dissolution of SHAEF and CIOS drew closer, officials on both sides of the Atlantic began to consider how to proceed on a more unilateral basis. For the Americans, this meant pushing forward with a scheme to bring German experts over to the United States for work on a number of military projects to aid in the Pacific War – a theatre in which American efforts dwarfed those of its allies.[13] Still constrained by their close relationship with Britain, they pestered their ally for their assent to such a scheme, which the British gave on 14 June 1945, with the key proviso that 'the knowledge so obtained will be available not only to interested United States agencies but our own as well'.[14] The common view among British officials was that 'it seemed most unlikely that any research and development work could fructify before the end of the Japanese war' but they kept this opinion to themselves. A bigger concern was that of allocation – that is, a system to ensure that the best spoils of war were fairly distributed between the two nations. At this juncture, CIOS was considered to be a suitable body to adjudicate on these matters, but it was an issue which would bedevil Anglo-American relations throughout this period.[15]

The British began considering this issue with even greater concern in the immediate run-up to the demise of SHAEF. On 4 July 1945, nine days before the Supreme Headquarters ceased to exist, the British Chiefs of Staff sent a note to the Joint Staff Mission in Washington, acknowledging that, on account of their greater commitment to the Pacific War, the Americans should be given preference on any exploitation material in cases where the quantity was insufficient to meet both US and British demands, but mitigated this by insisting that the Americans must 'not be given *carte blanche* to remove equipment, scientific personnel and documents without consultation' and that they should keep and share records of all that they did evacuate.[16] The Deputy Chiefs of Staff committee shared this sentiment too. In a meeting held the next day, they warned of the Americans conducting a 'somewhat piratical policy ... which was likely to prove extremely effective' in the absence of any relative British procedure, and also expressed concern that German scientists taken to the US may well be reluctant to subsequently come and work in Britain.[17] That is not to say that the US pursued exploitation more aggressively than the British. Indeed, this is a subject of some debate – on the one hand, it was argued that as the USA was so industrially advanced, they would be able to more easily make use of the captured science and technology while, on the other, the fact that they were so advanced could mean they had far less need of it.[18]

In any case, it was not always the Americans taking the lead – for example, when the British wanted to open personnel exploitation up to include German specialists who had no connection with defence, they were met with considerable reluctance on the American side, especially from the State Department. This disinclination was the result of a number of sticking points, including concerns about immigration (defence scientists had been brought over as prisonersof war, a categorisation which could not be extended to civilian experts) and potential accusations of hypocrisy, regarding US efforts to deter Latin American countries from recruiting Germans while doing it themselves.[19] Nonetheless, the British considered the State Department's obstacles to be delaying, rather than prohibitive, factors and turned their mind to concerns about a fair allocation policy for these civil-industrial specialists. Officials at the Board of Trade, who were masterminding this element of personnel exploitation, worried about the 'influence wielded by private American interests' and sought, therefore, a 'UK/US agreement which would allocate demands for scientists on a non-competitive basis. Otherwise we might find ourselves outbid both in respect of salaries and conditions of service.'[20]

The nature of these negotiations did not necessarily mean a lack of co-operative spirit between the two nations. Investigations in Japan, for example, were primarily conducted by the Americans, as had been predicted, but they happily shared the fruits of their labours with the British, through BIOS channels.[21] BIOS also received reports, though, usually, only single copies, of any interrogations of German scientists carried out in the USA.[22] Policy on documents and archives also revealed a positive Anglo-American relationship – for maximum convenience, the records of the German government were held at the Ministerial Control Centre at Kassel which was situated practically on the border between the British and US zones of occupation.[23] Generally speaking, it was the agents and organisations in the field which showed the greatest propensity for co-operation between Britain and the US, driven by pragmatism rather than geopolitical posturing. Paul Maddrell has highlighted general collaboration in scientific intelligence-gathering on the ground during the period of occupation, much of which was based on the sharing of information and the mutual imitation of technique.[24] Furthermore, in the conclusion to an official history of T-Force, it was noted that the unit had 'provided the most striking and happy occasions for Anglo-American co-operation'.[25] FIAT (Br), meanwhile, being the only British unit based in the American zone, 'had built up a goodwill with the Americans, which is of immense value to the Control Commission, and to Ministries and Departments in the UK'.[26] This was important as FIAT was handling roughly 80 individual British visits to the US zone every week, most of which lasted for an average of three weeks.[27]

These British visitors to the American zone did not always experience this goodwill and instead encountered strict rules and regulations, many of which were in place to avoid incidents which could jeopardise friendly and collaborative relations between the two countries. According to orders issued in July 1945, the visiting investigators were told that, at all times, 'military courtesy, discipline and proper wearing of uniform will be observed'.[28] Furthermore, in terms which some would have undoubtedly found condescending, they were told: 'you will find patience, good temper and tolerance the most important assets.'[29] Writing in his memoirs after the war, R.V. Jones, the Air Ministry's head of scientific intelligence, who briefly travelled to Europe with the Assistant Directorate of Intelligence (Science) mission, recalled that 'wherever one wanted to go in the American zone ... a piece of paper was essential because no American officer would act without written authority', jokingly surmising that this predisposition 'stemmed from having a written Constitution'.[30] When these American guidelines were not met, it could have serious repercussions. At the beginning of 1946, the Americans became increasingly reluctant to allow British reparations teams to enter the US zone, and the reason given was that the US officials 'were aggrieved by the conduct of certain of our reparations assessment teams in the past when visiting factories'. One aspect of this poor conduct was British assessors informing German workers that their factory was due to be dismantled which supposedly reduced worker efficiency and output and risked sabotage.[31]

These differences and disputes did not prevent Britain from learning from, and even imitating, the American strategy in exploitation. As a result of the greater resources at the United States' disposal, they were often able to take bold new steps first, and once their success had been proven, the more tentative British were able to follow suit. For example, the US created the first major detention and denial centre for German scientists and technicians, and their families, at Landshut in Bavaria towards the end of 1945. Conditions at the Landshut centre were reportedly dire, which concurs with accounts of many prisoner of war camps in the American zone, which suffered from woeful shortages of shelter, food and sanitation.[32] One German scientist was recorded as flippantly remarking that 'even the Nazis fed the inmates of concentration camps'.[33] Nonetheless, Landshut allowed the US Army to keep an eye on their prizes, arrange for convenient travel to America and prevent any of the experts falling into Soviet hands.[34] This arrangement certainly impressed the British and in January 1946, FIAT (Br) strongly recommended that 'arrangements be made to establish [a] concentration area in [the] British zone similar to that operated by US authorities at Landshut'.[35] As discussed elsewhere, the Americans also led the way in reactivating the German hotel industry to facilitate easier travel to Germany for investigators and industrialists – an initiative which was far less successful when the British attempted to emulate it some months later.[36]

Not all interactions between the US and British were as positive or as productive as those described here. There were occasions when a lack of co-ordination, especially over valuable and sought-after targets like chemical warfare installations, led to confusion and redundancy.[37] Beyond this, though, there was also a genuine rivalry; Clarence Lasby has defined the two countries as 'resolute adversaries' in the race for the spoils, describing a relationship characterised by 'strong feelings of suspicion and competition'.[38] Indeed, in a Gallup poll of July 1948, 14 per cent of British respondents stated their belief that the USA was trying to dominate the world (though, to qualify, a staggering 70 per cent felt that world domination was the goal of the USSR).[39] In order to defend themselves against America's more directly acquisitive approach to exploitation, the British expended considerable effort trying to wrangle a fair allocation policy on all samples of 'secret weapons'. This was built around the principle of 40 per cent each to Britain and the US, and 10 per cent each to France and the Soviet Union, though preference was granted to the Americans 'in all cases where there are insufficient samples, personnel or equipment available' for both parties.[40]

Despite these measures, Britain still often lost out to the US, especially when they chose to flex some military or diplomatic muscle, or resorted to underhand tactics. For example, during his Ministry of Aircraft Production mission to Germany in summer 1945, Roy Fedden alleged that he had two trucks loaded with aeronautical equipment taken away from him by US forces at gunpoint, and that American investigators with whom he had examined wind-tunnel models of swept-wing aircraft at Völkenrode (one of the sites utilised as part of Operation Surgeon) went back secretly by night and took them away. Similarly, in May, sixteen ships carrying 100 V-2 rockets from Antwerp to New Orleans were intercepted and forced to halt by the Royal Navy in the North Sea. The captains of the Navy ships demanded that, in accordance with the allocation agreements, fifty of the rockets be handed over to them. The Americans refused, and continued to refuse, even when the Foreign Office sent a direct request to the State Department, and eventually the British relented and the ships continued on their way, their precious cargo intact.[41]

Examples of Anglo-American competition in the exploitation of documents and material were, therefore, not uncommon, but the exploitation of personnel provided even more potential for friction. The American equivalent of the

British DCOS and Darwin Panel recruitment schemes was Project Paperclip which targeted the employment of around 1,000 German scientists and technicians in a variety of fields, most of which with some direct military utility. Project Paperclip was a child of many parents, but it was primarily co-ordinated by the Joint Intelligence Objectives Agency (JIOA), an organisation which, like its counterpart BIOS, had come about following the dissolution of CIOS.[42] Once again, America's economic supremacy gave them a considerable edge in recruitment, allowing them to offer much better terms than the British could. In addition, many German people viewed Britain as a country of the past and imagined that the future belonged to the United States (and the Soviet Union).[43] As early as July 1945, an exasperated civil servant at the Ministry of Aircraft Production, tasked with planning for British recruitment through Operation Surgeon, had cause to sarcastically exclaim: 'I believe the Americans have not yet taken them all'![44] A year later, Piers Synott, Under-Secretary of State for the Admiralty, voiced concerns about this in a letter to G.H. Curtis at the Treasury, worrying that dissatisfied German scientists were 'passing out of our hands' and into those of the Americans, as well as the Soviets and French.[45]

The Americans also did much to frustrate British recruitment efforts, sometimes deliberately and sometimes through generally obtuse behaviour. For instance, when BIOS wanted to bring a group of IG Farben employees who were in US custody in Germany to Britain for two months of interrogation, the Americans refused, stating that they would only release three of the men, and for a seemingly arbitrary three-week period.[46] In addition, the British were left in a weak position when the US decided to shift the Dustbin detention centre from an Anglo-American operation to a unilateral one – a process they conducted gradually by steadily increasing the restrictions on, and obstacles to, British access to the detainees until, despite no change to official policy, the British were left with 'virtually no control in any matters appertaining to this Detention Camp'.[47] It should be noted, however, that this is the story as told from the British perspective; many in the United States accused the British of being the more difficult partner. Lasby quotes US Major-General Hugh J. Knerr as saying that the British were uninterested in co-operation and that, for them, 'what is best for the British Empire is the compelling motive', reflecting a widespread American mistrust of Britain as a nation of avowed imperialists.[48] The truth is perhaps somewhere in between – exploitation officials of both countries on the ground in Germany reported to their home governments that the other ally was being less scrupulous than themselves in order to request greater urgency and resources.[49]

As with the competition over physical material, one of the biggest bones of contention between Britain and America in personnel exploitation was rocketry, in this case the V-2 specialists. The post-war discovery of German plans to launch a long-range guided missile at New York by 1946 spurred on fears of such an attack on US soil, possibly with an atomic warhead, and prompted key changes in American strategic defence policy.[50] When the British first tried to move German rocketry experts to Cuxhaven for Operation Backfire, they found that the Americans were reluctant to let them go, with the US Forces headquarters even issuing an order to the Third and Seventh Armies, which stated that 'no V-weapon engineers or scientists will be allowed to leave the US Occupied area without authority from this headquarters.'[51] The Americans, who had no real interest in the supposedly joint-led Backfire, were pressing ahead with their own investigations of rocketry and did not acknowledge the special importance which the British had granted to the operation. They were also concerned that Britain might try to spirit these men away while engaged on Backfire or at least try to get them to renege on their commitments to come and work in the United States.[52]

Of the 500 German personnel involved in Backfire, only 79 were scientists, and of these, the US requested that 26 be transferred to them immediately. Major-General A.M. Cameron, the officer in charge of Backfire, felt he had to make every effort to meet the US requirements, eventually handing over 14 – a decision made 'after consideration of all available substitutes and acceptance of a lower standard of technical skill being available for the operation'.[53] Even after this concession, the Americans were not sated and the British had to be dogged in their efforts to keep hold of the remaining twelve, who they described as 'key men'. The DCOS Committee discussed the US request and surmised that, as the Japanese war was over, the only use the Americans might make of these scientists was in a long-term research project. Therefore, they concluded that:

> Under these circumstances we cannot believe that their retention at Cuxhaven, for what would probably be a maximum of 2 months, could seriously inconvenience the United States Chiefs of Staff, whereas their withdrawal at this juncture would prejudice the success of Backfire into which much hard work and valuable effort has been put.[54]

In this instance, the British were able to retain the men which they required and the operation was able to proceed, but it is indicative of the fierce competition emerging between Britain and America for the best technical experts in Germany.

There were other examples of America's unscrupulous pursuit of German rocketry and aeronautics experts besides this interference with Backfire. In late August 1945, it was noted that a USAAF officer had simply turned up at Völkenrode in order to 'collect' 13 German specialists, '8 of which we had previously agreed to turn over to the Americans for work on Japanese war objectives, and 4 of which they agreed we should have', forcing the RAF officials to take 'delaying action'.[55] This action is even less justifiable as by this point the Japanese war had been over for almost two weeks. Elsewhere, another USAAF representative, Lieutenant Rosenbauer, was reported as having 'already surreptitiously removed one German scientist who was not on the original list of personnel requested by the Americans' and was interested in two more, so the British were forced to place these men 'in safe custody to prevent removal'.[56] This skulduggery does not suggest a harmonious relationship between two of the closest wartime Allies, and the British attributed the friction largely to 'the mysterious and rather parochial way in which it pleased [the US authorities] to work'.[57]

In reality, much of the difficulty which plagued the Anglo-American relationship throughout exploitation had its roots in shifting world power dynamics. Britain emerged as the junior partner after 1945, but its actions did not always reflect this changed status, while the US remained suspicious of the motives of their wartime ally, particularly with regards to imperial concerns. However, British resentment of this new American preponderance was tempered by the growing fear and suspicion of the Soviet Union, and the realisation that Britain, and even a Western European security bloc, could not realistically prevent Soviet encroachment without American support.[58] As such, efforts were made to commit the United States to the defence of Europe, which often meant glossing over lesser grievances. To some extent, this can be seen in the fact that whenever Britain and the US competed for the services of a German scientist, the Americans emerged victorious – as explored in the previous chapter, the British were only able to recruit the German rocket specialists that the US did not want.[59]

The effects of this, combined with some instances of American deception or brute force, and the fact that most German experts viewed the United States as a far more desirable destination, can be seen in the recruitment figures. By October 1946, the USA had contracted approximately 240 German scientists and technicians, the British only 33.[60] Of some 2,500 aeronautical specialists in Germany in 1945, within two years 12 per cent were in American hands while Britain barely had 1 per cent.[61] Final recruitment figures, while notoriously unreliable, suggest that the Americans had relocated around 1,000 German specialists to the USA by the mid-1950s, along with some 2,000 dependants.[62] While this is not drastically more than estimates of overall British recruitment, quantity alone does not tell the whole story. For the most part, the really significant specialists – individuals like Wernher von Braun, for instance – went to the United States and not Britain. The USA focused more on defence experts, whereas the British focus did tend to skew in favour of the civil-industrial sphere.[63] Scientists who went to America were also far more likely to remain, becoming citizens and building a life there, while many of those recruited by Britain returned to Germany in later life or, indeed, immigrated onwards to the USA. Nonetheless, while Britain may have struggled to compete with the United States in terms of recruitment, the notion of the British scheme being dwarfed by its US equivalent, and of it therefore being essentially irrelevant, is a gross oversimplification. Furthermore, while the story of Anglo-American relations on exploitation can be seen to be dominated by themes of conflict and competition, Britain's interactions with America were still more collaborative and successful than with any other power in the post-war world. Writing in April 1945, R.P. Linstead, the deputy chair of CIOS, worried that this co-operation with the USA would not be replicated with other nations, lamenting that 'it would be very much simpler' if it was.[64]

Europe

The continent of Europe endured widespread devastation during the Second World War, much of it at the hands of Nazi Germany. At war's end then, it is hardly surprising that many countries sought recompense for their suffering through the utilisation of German resources, including science and technology. France, as we shall see, was foremost among these, primarily because it had access to its own zone of occupation, and thus pursued the largest and most comprehensive exploitation scheme of any continental European country, and the one which most directly influenced, and was influenced by, Britain's. It was not alone though and many other members of the wartime United Nations, such as Norway, Belgium and the Netherlands, looked to conduct exploitation too. Furthermore, European nations which had remained neutral during the war, such as Switzerland and Spain, also sought to profit through this process, even without a retributive justification, and will also be discussed here.

In the immediate post-war years, France occupied a curious position on the world stage – having been defeated then split and partially occupied by Nazi Germany in 1940, it could not be considered a major member of the wartime alliance dominated by Britain, the United States and the Soviet Union. Once the liberation of Europe began, France half negotiated, half forced its way back to the top table of international politics, thanks in part to Charles de Gaulle's bullish nature.[65] In fact, the other three victorious powers all saw some advantage in rehabilitating France – especially the British, who believed France would take over some of their responsibilities on the continent.[66] As such, Britain was keen for France to be given an occupation zone in Germany – this was because the US seemed unwilling to maintain a large army in Europe once the war was over and Britain did not want to be solely responsible for occupying the western half of Germany.[67] In Winston Churchill's words, 'Germany would surely rise again, and while the Americans could always go home, the French had to live next door to her. A strong France was vital not only to Europe but to Great Britain.'[68] France was indeed granted their own zone in 1945, albeit one notably smaller than that of the other Allies, which in turn entitled France to a seat on the quadripartite Control Council and granted it a say in all matters pertaining to the future of Germany.[69]

The French approach to occupation, or the so-called 'French thesis', was primarily intended to enhance France's national security and speed its economic recovery – it also reflected French public opinion which was very hostile to any attempt to be 'soft' on Germany. Goals included the disarmament of Germany, a permanent French military presence along the Rhine, extensive reparations, and French or international control of German mines.[70] However, France soon found that these aims were largely untenable for several reasons, most notably that they did not marry up with the overall plans of their Anglo-American allies, on whom France was utterly dependent not only for its security but for its very livelihood.[71] A key element of this was French slowness to recognise the Soviet Union, rather than Germany, as the biggest threat to peace and stability in Europe.[72] Even when senior French leaders did acknowledge this new threat, the presence of powerful communist elements in the post-war French coalition government meant that, in public, these figures had to display a willingness to cultivate cordial relationships with both sides.[73] This outward impression caused both British and US authorities to view France with caution and mistrust, fearing that it might soon align itself outright with the USSR.[74] These sentiments, though tempered with the desire to build a Western European defence grouping, did feed into Anglo-French relations in this period, including with regard to exploitation.[75]

The French attitude to exploitation, meanwhile, was driven by the fact that, after almost five years of Nazi occupation, they were more in need of a boost to their military-scientific research complex than any other major power.[76] They saw the extraction of German resources as a way both to relieve the French taxpayer of the costs of occupying their old enemy and to overcome the economic stagnation which France had suffered in the interwar years. In a French National Assembly debate in March 1946, Philippe Livry-Level, a Resistance hero and post-war centrist politician, described Germany as 'a land for us to exploit'.[77] Their position as an occupying power allowed them to conduct exploitation of their own, and not just rely on the unwanted scraps discarded by the three larger powers, as was the case for some other European nations. However, they did not begin as participants but rather as subjects, with facilities across France which had been maintained and utilised under the Nazi occupation investigated by CIOS teams sent over in the immediate aftermath of their liberation. French protests about this fell on deaf ears in Britain and America – as William Hitchcock has put it, at this stage, France remained a dim and distant star in the international constellation.[78]

Perhaps as a result of this, and perhaps from a desire to not let their wartime occupation translate into being left behind in the race for the spoils of Germany, the French were quick off the mark in conducting exploitation. One of the main bodies involved in this was the *Centre National de la Recherche Scientifique* (CNRS), France's main governmental research organisation, which sent several missions into Germany between 1945 and 1950.[79] Generally speaking, the French exploitation policy was resourceful and unscrupulous in equal measure.[80] For a start, they turned their struggle for recognition as a major post-war power into an advantage – having not been represented at the Potsdam Conference in July-August 1945, they concluded that they were not bound by any decisions made there, allowing them to proceed practically unilaterally in securing advance deliveries of reparations from their zone.[81] They also used their shared border with Germany to their benefit. When a group of Luftwaffe ballisticians refused to relocate to France for fear of the hostile attitudes of the native population, the French authorities allowed them to live on the German side of the Rhine but ferried them across by bus every day to work at the Institut Saint-Louis in France.[82] The French also saw no need to limit themselves by working too consistently alongside any one power, often preferring to operate independently and, as a result, they have been described as 'enjoying reasonable relations with all parties, but [being] trusted by none'.[83]

That said, they did show a willingness to collaborate and pool resources with their allies, including Britain. The French felt they had much to offer: troops to guard targets, skilled personnel to help exploit them and intelligence on 'known and unknown German objectives ... through French prisoners and deported workers'; in return, they hoped for the 'complete exchange of information, co-ordination of research plans and association of French and Allied field teams'.[84]

Although collaboration on this scale was never achieved, with the British and Americans unable to fully divest themselves of their misgivings about the French, the core principle of mutual access to each other's zones of occupation was held to by both sides, as confirmed at a BIOS meeting held at the end of August 1945.[85] As with trips to the US zone, British exploitation teams travelling to the French zone were expected to abide by a set of specific bureaucratic rules. Crucially, they were forbidden from removing any documents, material or equipment without 'prior consent of T-Section, French Army of Occupation'.[86] These restrictions were significant as investigations into foreign zones made up a sizeable component of British exploitation activity in the eighteen months after the end of the war – between June 1945 and October 1946, 1,601 BIOS and British Reparations teams visited targets in the French and American zones.[87] Nonetheless, Monica Maurice, who travelled through the French zone *en route* from Frankfurt to Cologne on a BIOS trip in May 1947, felt that the implementation of these regulations was ineffectual, as the control points on the borders of the French zone seemed to be there only to 'make quite sure we came out at the other end'.[88]

French exploitation teams were subject to similar regulations when visiting the British zone and their frequent failure to comply caused much consternation in the British element of FIAT. They protested particularly strongly about scheduled French teams never arriving or reporting to HQ, which was 'a source of extra work and worry for all concerned and in addition it often means that other teams are unnecessarily delayed in getting clearance'.[89] In some cases, though, the British authorities made considerable effort to ensure these visits went well. When a group of French industrialists arrived in the British zone in December 1945 wishing to examine the Volkswagen factory and subsidiaries in Wolfsburg for reparations purposes, S.G. Galpin of the Economic Division suggested that 'they be given special treatment', including the use of more ostentatious accommodation than was normally used. He took particular interest in this case as he felt it would 'improve relations between [Economic Division] and the French Mechanical Engineering Branch'.[90]

On the whole, then, a workable and largely reciprocal relationship was established between the British and the French, albeit one tinged with a persistent edge of suspicion.[91] When the Intelligence Group of the British Element of the Control Council divided foreign countries into three groups on account of security, France was placed in the second group, which entitled them to receive British material bearing classifications up to and including 'Confidential'. This gave them greater access than the Soviets but less than the USA, India and the Dominions – a categorisation indicative of Anglo-French relations on exploitation.[92] For instance, in June 1945, it was agreed that France would be provided with a copy of the CIOS Black List – however, they were only to be given the 'geographical' list which contained the locations of targets, and not the 'technical' list which featured specifications about the items to be investigated. Not only was this second list not directly issued to the French but SHAEF insisted that 'care should be taken to insure [sic] that the French do not become aware of its existence'.[93] This lack of trust was also evident when the War Office decreed that 'no information on German Chemical Warfare developments during the war should be passed on to the French.'[94]

On the other hand, the French were given copies of the majority of CIOS and BIOS Final Reports which were filed by investigating teams;[95] a courtesy which was not returned, though the reason behind this was assumed to be 'not any lack of goodwill on the part of the French, but the absence of any adequate organisation for dissemination'.[96] The semblance of open co-operation which was maintained throughout concealed a more uncertain relationship between Britain and France.[97] Even when Britain did appear willing to co-operate with France, this was sometimes motivated more by the desire to 'keep up appearances' on the world stage, rather than any genuine collaborative spirit. When the initial arrangements were still being developed, Brigadier Maunsell insisted that Anglo-French relations were 'extremely cordial', but worried that 'if the French apply for, and receive, permission to visit the American zone and have access to records but are refused the same facilities by FIAT (Br), it will inevitably place the British in an unfavourable light.' Maunsell warned that this would create an 'obviously anomalous' situation which would 'soon give rise to bad political relations'.[98]

Behind this façade of friendly interactions and shared information, both the British and Americans took action to thwart or hinder French exploitation attempts. In May 1945, the Deputy Chiefs of Staff informed SHAEF that 'we do not wish other secret weapons, such as new rockets, rocket assisted shells, controlled glider bombs ... and successive types to be given to the French, [unless] samples of these weapons are captured by the French, or are already known to them' and instructed that 'any quantities captured in excess of [US and British] requirements should be destroyed.'[99] The same exploitation allocation machinery upon which the British had insisted in order to prevent the Americans from outdoing them was thus also used to help them outdo the French. Sometimes, more direct action was judged necessary. In April 1945, the US-led Alsos mission launched Operation Harborage which aimed to sweep into the German towns of Hechingen, Bisingen and Haigerloch, forty miles south of Stuttgart, and seize or destroy any atomic research equipment, and remove any specialists, before the frontline troops arrived. This was important because these towns were 'in the line of advance of the French Army'[100] and the head of Alsos, Brigadier-General Leslie Groves, was convinced that 'nothing that might be of interest to the Russians should ever be allowed to fall into French hands'.[101]

This attitude of equating the French with the much more serious Soviet threat had some legitimate grounds – for instance, in late 1944, de Gaulle had signed a mutual assistance pact with Stalin – but as the post-war period had progressed, the Soviets had come to see France as little more than a British and American pawn.[102] Nonetheless, suspicion of closeness to the Soviet Union had a direct impact on the approach which Britain and the US took towards France during exploitation, particularly with regard to recruitment of German personnel – an area in which more conventional rivalries also played a part. The French recruitment programme started slowly and in the face of numerous misgivings. When it was first proposed, Professor Louis Cagniard, who headed the CNRS mission to

Germany, argued that German scientists were not as marvellous as they professed to be and urged that France 'must not fall prey to the same fad [of recruitment] as have our Allies'. Instead, he recommended greater investment in domestic French science and only short-term intensive interrogation of German experts – 'we must treat them like milking cows that we abandon once we have finished with them.'[103] Cagniard's reservations were not heeded, however, and France embarked on a fairly substantial recruitment programme.

It would be fair to say that French efforts in this regard made the British and Americans a little nervous, not least because of the potential Soviet connection. In a JIC report from May 1946, it was noted that French attempts to entice German scientists should be monitored carefully, not only on account of 'the general anxiety felt as to French lack of security' but more importantly, 'the possibility of French co-operation with the Russians'.[104] This bred an atmosphere of suspicion and gave rise to clandestine behaviour which would become all too common under the hostile conditions of the Cold War. In Berlin, Enemy Personnel Exploitation Section (EPES) officers were given a list of intelligence desired on Soviet exploitation efforts and were also told to 'pay attention to the obtaining of information on similar activities carried out by the French'.[105] EPES officials also felt that having to share their office in Berlin with the French element of FIAT compromised its security and suggested 'transferring all Top Secret activities at present carried out by this Section to a special office to which only British and American personnel will have access'.[106]

As this suggests, matters were complicated by the fact that the general nature of the Anglo-French relationship often compelled the exploitation agencies of the two countries to work together, or at least in close proximity to one another. The head of EPES, Lieutenant-Colonel P.M. Wilson, summed up the crux of the issue when he noted that we 'require the French to do a great deal for us in regard to finding Germans and in giving clearances for their evacuation from the French zone to UK' and therefore 'any restrictions we place on the French … would have unfavourable effects on our relations with them, and their co-operation in locating for us in the French zone.'[107] That is not to say that this co-operation was always forthcoming – even by June 1947, 'the French authorities took many months to give clearance' on evacuating German specialists from their zone.[108] In short, the British often had to rely on the French but rarely felt they could trust them.

Not only were French agencies often truculent in releasing German specialists in their custody to their allies, they could also be quite devious when it came to trying to poach targeted experts who were held by Britain or America. In one example, French intelligence officers infiltrated a guarded American transit hotel in Bad Kissingen, Bavaria and went from room to room, talking to the scientists accommodated there, casting doubts on their prospects in the United States and offering them a much better future in France. By the time they were discovered and escorted away, they had successfully managed to convince some of the specialists to go with them.[109] In another instance, the French displayed a remarkable lack of scruples when they seized Otto Ambros while he was on his way to trial for war crimes. Ambros, the senior IG Farben chemist who had considerable responsibility for the Nazis' development of the new nerve agents and who had links to unethical human experiments and slave labour, was being interrogated by an Anglo-American team at Gendorf in Bavaria when a warrant for his arrest arrived from SHAEF. According to this warrant, he was to be immediately transported to the Ashcan detention centre at Mondorf-les-Bains in Luxembourg. The route took him through the French zone, where he was held by the French and set to work for them at Ludwigshafen.[110] It took considerable diplomatic protest on the part of the Americans to finally secure his release back into their custody.[111]

As these various examples show, both Britain (alongside America) and France engaged in various tactics to frustrate each other's exploitation schemes. These could range from refusing to grant, or deliberately delaying, necessary clearances, to essentially kidnapping certain German experts so that they could not fall into the hands of a rival. Despite Britain's best efforts to impede it, France still conducted a pretty sizeable exploitation programme, especially in terms of recruitment. They employed around 90 former Peenemünde specialists, including prominent figures such as Eugen Sänger and Wolfgang Pilz, to work on rockets (leading to France becoming the third nation to launch a satellite, in 1965); the state-owned firm SNECMA hired a large number of BMW aeronautical experts who helped develop a line of jet engines; while another firm, SNCASE, brought in a team of 20 helicopter designers, led by Henrich Focke (one of the namesakes of the Focke-Wulf firm). All told, final estimates suggest France recruited around 800 German individuals, over half of whom were connected with the highly significant field of aerospace.[112]

The Anglo-French relationship in the immediate post-war period was an uncertain and often fractious one, and this was reflected in the vigorous competition and reliance on underhand tactics which characterised their rival exploitation efforts. France was not, however, the only European nation with which Britain had to contend during the quest for the scientific spoils of war. First, it should be noted that, like France, many other European countries began the post-war era as targets for exploitation, rather than participants. In an early iteration of the CIOS Black List, prepared in August 1944, potential sites of interest were identified in Belgium, the Netherlands, Denmark and Norway.[113] In addition, both Czechoslovakia and Luxembourg were mentioned as possibly having sources of 'useful technical information related to the German war effort'.[114] Indeed, CIOS and BIOS both dispatched teams to many of these countries as they were liberated from German occupation – in Czechoslovakia, for instance, particular attention was paid to the Škoda works in Plzeň, which had produced tanks, small arms and all manner of other war material for the Wehrmacht.[115]

Once the war in Europe was over, and the exploitation of Germany got underway to a larger extent, many of these other European countries became interested in conducting their own investigations. In the summer of 1946, T-Force received requests for technical specialists from various nations to visit the British zone of Germany to exploit specific topics – a Belgian team were interested in gas plants, a Danish team to investigate pencil manufacture, and a Yugoslav team to study iron and steel works.[116] Other requests were received from Norwegian,

Dutch and Czech units – in the Czech case, it was approved 'in principle but [only if] restricted to a limited number of teams with specified targets'.[117] In addition, permission was only granted if the sites in question had already been sufficiently exploited by British assessors.[118] At one point, Norway even applied to establish its own FIAT-affiliated operation in Germany, but was met with British reluctance – 'if the Norwegians were permitted to set up an office, a precedent would be created.'[119] Where European nations were not able to mount their own exploitation missions, they often sought to secure the benefits by utilising the reports produced by British (and other Allied) agencies. Unclassified reports were generally made available fairly easily, but a more complex and restrictive application process was put in place for the classified ones.[120] This reflected the fact that, according to the Control Commission for Germany (British Element) (CCG(BE)) Intelligence Group security grading system, most countries were 'Class C' and only eligible to receive unclassified material. The exceptions, in 'Class B' and therefore granted access to material up to and including 'Confidential', were Norway, Denmark, Belgium, the Netherlands, Greece, Turkey and Eire, along with France.[121]

In what has been demonstrated as a recurring trend, the issue became more complicated when it concerned the exploitation of personnel. Up until mid-1947, official British policy was not to allow German scientists to immigrate to any countries other than Britain, the Dominions, or the USA. However, the dual risk of Soviet recruitment and prolonged unemployment in Germany leading to subversive activity prompted the authorities to rethink this stance. In May, the Defence Committee of the Cabinet considered asking Belgium and the Netherlands to take some German scientists during a discussion of tactics to counter Soviet recruitment efforts.[122] Moves such as this reflected a broader policy shift which was codified in July 1947, when ministers concluded that 'on balance it would be of advantage to us to permit German scientists and technicians to emigrate to various countries other than the USSR', though they acknowledged that, as a result, 'German scientists might contribute to increasing the war potential and degree of industrialisation of countries whose strength we would prefer not to be increased'.[123]

Of course, not all countries in Europe were viewed as true allies by Britain in the period immediately after 1945, particularly those which had remained neutral during the Second World War or had acted in a friendly way towards Nazi Germany. Some were also seen as dangerously close to the Soviet Union in the tense post-war period. Traces of this more cagey attitude can be seen in the case of Kurt Tank, the eminent aeronautical engineer who had been head of design at Focke-Wulf from 1931 to 1945 and whom the British had decided not to employ as he was too 'big' and 'important' to insert into British industry without major disruption.[124] In May 1947, the Swedish Air Force showed an interest in him and the British authorities were reluctant, fearing that 'any information which he provided would eventually become available to the Russians'. Against their better judgement, they did eventually decide to approve the move, though it never went ahead and Tank's post-war career went in rather a different direction (see below).[125]

In some cases, formerly neutral European countries could end up as competitors for the services of the more desirable specialists. One example of this concerns a group of 52 German hosiery machine needle technicians, who had left Germany illegally and, as of January 1947, were employed in Switzerland. Britain wished to employ these men, and thus capture the world market for hosiery machine needles, but 'it was stressed that to show too much eagerness to get the men ... might defeat the project by causing the Swiss government to put difficulties in the way of their leaving Switzerland.'[126] Britain sought to use their diplomatic power to secure these men, but soon conceded that their only option was to 'rely on offering greater inducements in order to obtain their services in this country'.[127] Another wartime neutral nation which took an interest in German expertise was Spain, which embarked on what has been described as a 'peripheral Paperclip project'. Spanish officials used their pre-war and wartime connections within the Nazi hierarchy to facilitate recruitment after the war, especially of naval and aeronautical experts, and Francisco Franco's fascist regime in Spain no doubt appeared more amenable to certain German specialists.[128] One of the biggest successes of this Spanish programme was securing the services of the aircraft designer Willy Messerschmitt (whom the main Allies had considered too politically toxic to employ) in the early 1950s, which did lead to the production of a Spanish jet aircraft.[129] As these case studies show, even European nations which had not been directly involved in the war, on either side, sought to benefit from Germany's human resources after the Third Reich collapsed.

The global diaspora

The movement of German scientists after the Second World War was not only limited to the four occupying powers and a handful of other European nations. Rather, it soon became a truly global phenomenon and one which lasted well into the 1950s and 1960s, though the roots of this expansive transnational knowledge exchange lie firmly within the period of immediate post-war occupation. In some ways, Britain was an instigator of this movement as it looked to its imperial Dominions to absorb the German expertise which was surplus to British domestic requirements. During the war, as Roy MacLeod has noted, the relationship of scientific exchange between Britain and the Commonwealth became more democratic and equitable, due in part to the mediating influence of the United States.[130] This attitude persisted into the post-war period – at the opening of a 1946 conference on Commonwealth scientific collaboration at Senate House, King George VI described the British Empire as 'a laboratory richly stored with materials' and stated his view that, through co-operation, it would be possible to 'develop a greater and wider field of scientific investigation than any other community'.[131]

Unlike the more tentative and restrictive attitude taken towards European countries such as Belgium and the Netherlands, and even the slight reticence displayed towards the United States and France, the involvement of the Dominions in British exploitation was more intimate and encouraged from the start. In this, Canada and Australia were perhaps the earliest and most eager participants. A representative from the Canadian government was included on the initial membership of the BIOS committee when it was first formed in July 1945.[132] The Australian government followed suit in May 1946, when they were also granted full membership of BIOS.[133] Other parties did not always have the same success, however – in September, the Palestine Board of Scientific and Industrial Research requested to be added to the distribution list for all BIOS reports. The BIOS response was that, 'in view of the present uncertain political situation in Palestine and the possibility of a change of status from the existing British Mandate, the question of the provision of classified reports [to the Board] should be given careful consideration.'[134] This was bureaucratic obfuscation tantamount to an outright refusal.

The larger, less politically problematic Dominions were, with full membership of BIOS, then able to push on to exploitation proper. In October 1945, Canadian officials reported that 'Canada is anxious to share in the allocation of German war material, particularly that type of material that falls under the general heading "research and development."'[135] Shortly thereafter, Australia established a group of three engineers in London who worked full-time assessing German spoils.[136] However, this eagerness did not always translate into success. In July of the following year, Frederic Hudd, acting Canadian High Commissioner in London, wrote to Viscount Addison, Secretary of State for Dominion Affairs, acknowledging that 'a considerable amount of war material has been acquired' but complaining that 'the work of allocation of war material … has been, in a number of cases, unduly slow.'[137] There were also accounts of interference on the ground in Germany by both British and American officials, including one instance where British guards refused to recognise the credentials of a visiting Canadian team and barred them from examining a former Luftwaffe installation.[138] Incidents like these were isolated however and, for the most part, the Dominions were active partners in the British exploitation programme.

The move towards personnel recruitment led to an even closer working relationship between Britain and its Empire and Commonwealth. During the London conference mentioned above, the Imperial College physicist W.B. Mann argued that the movement of a British scientist to, say, Canada would no longer 'connote a weaker or poorer Britain; it may imply a better-balanced use of our scientists and technologists throughout the Commonwealth, and, therefore, a stronger Commonwealth.'[139] This attitude filtered into personnel exploitation policy too. At the first meeting of the Inter-Departmental Committee on German Scientists, held on 16 January 1947, there were representatives from the Colonial Office, the Dominions Office, and the India Office in attendance.[140] In the early months of 1947, as the relevant British agencies considered various strategies to keep valuable German experts out of the hands of the Soviet Union, it was frequently suggested that 'the source which will probably provide the greatest assistance in this respect is the Dominions.'[141] In April, it was noted that 'the Australian Scientific Mission has expressed great interest in the possibility of finding employment for some of them in Australia.'[142] The momentum behind this policy increased consistently throughout this period, especially as the British cast around for any option which would help to deny these individuals to the Soviets and, in late July, the Board of Trade reported that, in regard to 'the extension of the recruitment to the Dominions and India ... it must be remembered that in the implementation of the Government's policy every encouragement has been given to recruit these Germans.'[143] From this point onwards, concerted efforts were made to try and find jobs for German scientists in places such as Australia, Canada, New Zealand, and South Africa.[144]

Of these, the two Dominions which were most actively involved in this process were, as before, Australia and Canada. In Australia, the programme for recruitment was known as the 'Employment of Scientific and Technical Enemy Aliens' (ESTEA), a name which caused issues when it came to appealing to the brightest and best of German science. The esteemed chemist Otto Bayer, who had helped develop polyurethane and had risen to a senior position in IG Farben during the war, contemplated moving to Australia but wished to do so only as a 'free man, not as an alien'. Another scientist took this further – in a letter to the Australian Prime Minister in March 1947, he asked: 'how may Australia, which entered the war for idealist reasons solely and was not invaded by German troops, now ask for reparations? Therefore, I recommend you, not to take slaves for the progress of Australia, but free men willing to give their best.' Despite these reservations, the Australian recruitment scheme was modestly successful, resulting in the employment of around 150 scientists in total, with 64 of these going to the private sector (including universities), and the rest being hired by public bodies. The professions of the recruited individuals varied widely but included engineers, surveyors, chemists, physicists, metallurgists, food technologists, cartographers, geodetic engineers, geologists, electrical engineers, ceramicists, and aeronautical engineers.[145]

Canada was, on the whole, considerably less successful. They lacked a programme with the same internal cogency and government support as Australia's ESTEA scheme. The Canadian government only decided to follow the British recruitment model in October 1946, and the policy did not come into force until May 1947. Many Canadian investigators were frustrated at this delay, especially if they had encountered desirable German specialists during their time in the field but had been unable to make concrete offers.[146] The implementation of policy did not immediately result in a steady flow and Canadian uptake of German expert personnel was remarkably slow and fitful – much of the impetus actually came from the British who were trying to use Canada as a dumping ground for the German scientists which they did not want, but which they did not want the Soviets to have either. In the end, the Canadians admitted 15 German scientists to Canada by mid-1949, and a further 26 by 1951. Most of these men returned to Germany after the expiration of their contracts, although several did elect to stay and become Canadian citizens.[147] In any case, 41 was a small number and not reflective of Canada's expansive scientific resources or growing status on the world stage – it amounted to less than a third of the Australian recruitment total. One small Canadian success story concerns the Beavers Dental Company at Morrisburg, Ontario – through the Darwin Panel scheme, they were able to hire a number of German scientists and technicians, and within two decades they became one of the largest dental burr manufacturing enterprises in the world; this is especially remarkable as Canada had had no dental burr industry of any consequence prior to 1945.[148] This goes to show that the influx of even a small number of German experts could decisively shape a nation's industrial and technological base within a relatively short period of time.

It was certainly not just the Dominions which sought to profit from this great migration of capable scientists and technicians from Germany at the end of the Second World War. In terms of aeronautics, Michael J. Neufeld, in his assessment of existing scholarship on the phenomenon, has described a 'Nazi aerospace exodus' and highlights how 'cross-border flows of people, information, and technology grew beyond the bounds of the victorious powers' national programs for exploiting German knowledge'.[149] In many cases, this relied on the agency and energy of the German experts themselves – for example, with aircraft design and development prohibited in Germany, many aeronautics specialists proactively sought work overseas.[150] This was particularly true of those individuals who were unwanted by the main occupying powers on professional or political grounds. German rocket technology and expertise also spread throughout the world after 1945 and played a critical role in the post-war proliferation of guided missiles.[151] While it is far beyond the scope of this study to provide a comprehensive account of this far-reaching and worldwide diaspora of German scientists and technicians, there are a handful of interesting case studies which showcase the scope and ramifications of this mass movement.

At the end of the Second World War, Egypt was nominally independent but was still treated in many ways like a British protectorate. It was not, however, considered by Britain to be an ideal place for German scientists to be employed and operated no form of recruitment scheme during the initial occupation period. Following the revolution in 1952, and especially the accession of Gamal Abdul Nasser to the Presidency in 1956, Egypt began looking for ways to expedite technological progress so that it could compete with its rivals. They recruited Willy Messerschmitt from Franco's Spain in 1960 and he took on a supervisory role at the Helwan aircraft factory, where he oversaw the development of the HA-300 light supersonic interceptor – the last aircraft he ever designed.[152] Nasser was also determined to build a domestic guided missile capacity in Egypt and to this end recruited a number of former Peenemünde employees, many of whom had spent the initial post-war period in France, such as Eugen Sänger.[153] In order not to draw the ire of international observers, attempts were made to conceal this Egyptian missile development behind the guise of a space programme. These attempts failed and the revelations about Egypt's rockets caused great consternation and criticism abroad. Nasser's response to this uproar was to disingenuously ask 'if the Russians and Americans could have their German scientists, why couldn't the Egyptians have theirs'?[154]

Argentina was another country which had not been involved in the Second World War (aside from an opportunistic declaration of war on the Axis powers in March 1945) and yet sought to utilise German specialist know-how to develop in the post-war era, under President Juan Perón. One of their first recruits, however, was not German but French – the aviation expert, Émile Dewoitine, who fled France at the end of the war under accusations of collaboration with the Nazis and travelled to Argentina, via Spain, where he designed the *Pulqui* jet fighter.[155] He was followed in 1947 by Kurt Tank and his team, who had been pursued by Britain, the Soviet Union and Sweden but, for various reasons, had not ended up working for any of them. While in Argentina, he oversaw the development of the *Pulqui II* jet fighter, but he lacked the facilities to test it properly and the industrial base for mass production; as a result, only a few were ever made.[156] Tank, and some of his colleagues, moved on to India in the 1950s at the request of President Jawaharlal Nehru and then, in 1967, he joined Willy Messerschmitt in Egypt, where he served as his chief of staff.[157] It was not just aeronautical expertise which Argentina desired and, in 1947, they attempted to recruit the atomic expert Werner Heisenberg who was then residing at Göttingen in the British zone. At the behest of the Americans (who were reluctant to see atomic science taking place in Latin America), the British refused him an exit permit, and nothing more came of this putative move.[158]

As these cases suggest, the movement of German specialists to countries around the globe, especially those in the Third World (often undergoing a nationalist resurgence), was a highly charged and deeply controversial process. The occupying powers, especially Britain and the United States, were often very concerned about this unrestricted movement of German experts to countries with potentially hostile intentions and did what they could to interfere, despite continuing their own recruitment schemes. They knew, however, that if this hypocrisy was revealed, 'their faces would be very red and considerable embarrassment might be caused to them.'[159] In short, the global diaspora described here just goes to show the sheer desirability of German expertise in the aftermath of the Second World War and the lengths to which countries would go to acquire a portion of it for themselves, or to deny a portion of it to their rivals.

To conclude, British exploitation of German science and technology was decisively shaped by the international environment in which it operated. Parallel schemes, especially those run by the United States and France, frequently clashed with that of Britain and the element of competition both hampered British efforts and forced them to be more proactive and creative when seeking solutions. That said, there were also numerous examples of compromise and even co-operation between the national programmes and the collaborative spirit which imbued exploitation during the early, multilateral period of CIOS never disappeared entirely. Meanwhile, the global diaspora of German expertise outlined in the final section of this chapter, though it took place mostly after the initial occupation period, serves as a useful reminder that the processes of personnel exploitation were not neat and discretely contained within the immediate aftermath of the war, or within the control of the four leading Allies. Instead, this was a hugely complex and far-reaching transnational phenomenon that influenced worldwide scientific and technological development, and international relations, long after the formal occupation was concluded in 1949.

One element, however, which remains consistent throughout the schemes and incidents discussed here and which recurs again and again as a motivation or driving force behind numerous policy decisions, is the fear of the Soviet Union. This forced Britain and the United States to collaborate more than they might have done otherwise; it prompted initial suspicion of France, but this gave way to acceptance as another valuable ally in the struggle against possible Soviet aggression; and it made Britain open up exploitation to other countries which it judged to be hostile to the USSR, while also closely monitoring those which it feared might fall under its influence. Understanding the perceptions and impact of the Soviet Union is therefore critical to understanding British exploitation policy, as their actions came to be shaped increasingly by the need to combat the ambitions of this new adversary.

Notes

1 Kori Schake, *Safe Passage: The Transition from British to American Hegemony* (Cambridge, MA: Harvard University Press, 2017).
2 Bower, *Paperclip Conspiracy*, 223; Cairncross, *Price of War*, 16; Hoffmann, 'Germany is No More', 606.
3 Sabine Lee, *Victory in Europe: Britain and Germany since 1945* (London: Pearson, 2001), 7.
4 John Baylis, *Anglo-American Relations since 1939: The Enduring Alliance* (Manchester: Manchester University Press, 1997), 18.
5 Zimmerman, *Top Secret Exchange*, 167, 176.
6 TNA, WO 219/1669, Col. G.G. Vickers to Maj-Gen. K.W.D. Strong, 7 September 1944.
7 TNA, WO 193/432, 'Notes for a Chiefs of Staff Committee Meeting', 13 June 1944.
8 TNA, FO 942/27, A. Balfour to Lt-Col. A. Geddes, April 1944.
9 Hathaway, *Great Britain and the United States*, 13.
10 George L. Bernstein, *The Myth of Decline: The Rise of Britain since 1945* (London: Pimlico, 2004), 84.
11 Zimmerman, *Top Secret Exchange*, 190.
12 John Dumbrell, *A Special Relationship: Anglo-American Relations in the Cold War and After* (Basingstoke: Macmillan, 2001).
13 Roy MacLeod, introduction to Roy MacLeod (ed.), *Science and the Pacific War: Science and Survival in the Pacific, 1939-45* (Dordrecht, Holland: Kluwer, 2000), 5.
14 TNA, CAB 122/343, Chiefs of Staff to Joint Staff Mission, 14 June 1945.
15 TNA, CAB 122/342, 'Minutes of DCOS Meeting', 6 June 1945.
16 TNA, CAB 122/343, Chiefs of Staff to Joint Staff Mission, 4 July 1945.
17 TNA, CAB 122/343, 'Minutes of 6th DCOS Meeting', 5 July 1945.
18 Gimbel, *Science, Technology, and Reparations*, 140, 145.
19 TNA, CAB 122/357, Sir John Magowan to Brig. A.T. Cornwall-Jones, 17 January 1946.
20 TNA, CAB 122/357, Derek Wood to Sir John Magowan, 26 January 1946.
21 TNA, FO 1031/50, 'Minutes of 7th BIOS Meeting', 8 May 1946.
22 TNA, FO 1031/50, 'Minutes of 6th BIOS Meeting', 17 April 1946.
23 TNA, FO 1032/179, 'The Handling of German Documents and Archives', 6 September 1945.
24 Maddrell, 'British-American Scientific Intelligence Collaboration', 81ff.
25 TNA, FO 1031/49, 'Concluding Remarks', 1945.
26 TNA, FO 1032/1459, 'FIAT', 10 December 1946.

27 TNA, FO 1065/12, 'Investigation of T-Force: Report No. 47', 12 December 1946.
28 TNA, FO 1032/177, 'Entrance of Intelligence Investigators into European Theatre, American Zone of Occupation', 23 July 1945.
29 TNA, FO 1031/7, 'Notes for Investigators in US & French Zones', 12 June 1946.
30 Jones, *Most Secret War*, 612.
31 TNA, FO 1032/166, 'Visits by Reparations Assessment Teams to Factories in US Zones', 18 January 1946.
32 Bessel, *Germany 1945*, 200–01.
33 John Krige, *American Hegemony and the Postwar Reconstruction of Science in Europe* (Cambridge, MA: MIT Press, 2008), 52.
34 Bower, *Paperclip Conspiracy*, 158.
35 TNA, FO 1032/302, 'Policy for Families', 2 January 1946.
36 TNA, FO 1031/2, 'Minutes of a JEIA Meeting', 24 February 1948.
37 Tucker, *War of Nerves*, 83.
38 Lasby, *Project Paperclip*, 112.
39 Gallup, *Gallup International Public Opinion Polls*, 179.
40 TNA, CAB 122/363, 'Allocation of Samples of German Material for Intelligence and Research Purposes', 30 June 1945.
41 Bar-Zohar, *Hunt for the German Scientists*, 119–20.
42 O'Reagan, 'Science, Technology, and Know-How', 51.
43 Barbara Marshall, 'German Attitudes to British Military Government, 1945–47', *Journal of Contemporary History*, 15 (1980), 655.
44 Nahum, 'I believe the Americans have not yet taken them all!', 104.
45 TNA, AVIA 54/1294, P.N.N. Synott to G.H. Curtis, 12 June 1946.
46 TNA, FO 1031/9, 'Minutes of 6th BIOS Reception Centre Panel Meeting', 19 June 1947.
47 TNA, FO 1031/69, P.M. Wilson to R.J. Maunsell, 9 November 1946.
48 Lasby, *Project Paperclip*, 112.
49 Nahum, 'I believe the Americans have not yet taken them all!', 109.
50 Reuben Steff, *Strategic Thinking, Deterrence and the US Ballistic Missile Defence Project: From Truman to Obama* (Farnham: Ashgate, 2013), 36–7.
51 TNA, FO 1031/85, USFET to Third and Seventh Armies, 8 August 1945.
52 McGovern, *Crossbow and Overcast*, 200-201.
53 TNA, FO 1031/85, Maj-Gen. A.M. Cameron to HQ, USFET, 13 August 1945.
54 TNA, FO 1031/85, 3 September 1945.
55 TNA, CAB 82/6, 'Deputy Chiefs of Staff Committee', 27 August 1945.
56 TNA, CAB 82/6, 3 September 1945.
57 TNA, FO 1031/85, 'German V-2 Technicians', 2 September 1945.
58 Martin A.L. Longden, 'From "Hot War" to "Cold War": Western Europe in British Grand Strategy, 1945–1948', in Michael F. Hopkins, Michael Kandiah, and Gillian Staerck (eds.), *Cold War Britain, 1945–1964: New Perspectives* (Basingstoke: Palgrave Macmillan, 2003), 124–5.
59 Neufeld, *Rocket and the Reich*, 269.
60 TNA, CAB 122/349, FIAT Main (Br) to Intelligence Division, 4 October 1946.
61 Bower, *Paperclip Conspiracy*, 321. In addition, 25 per cent were in Russia, 8 per cent were in France and the rest had remained in Germany.
62 Ciesla and Trischler, 'Legitimation through Use', 174.
63 Neufeld, 'Nazi Aerospace Exodus', 54.
64 TNA, FO 1032/475, 'The Future of CIOS', 13 April 1945.
65 William I. Hitchcock, *France Restored: Cold War Diplomacy and the Quest for Leadership in Europe, 1944–1954* (Chapel Hill, NC: UNC Press, 1998), 12.
66 Tony Judt, *Postwar: A History of Europe since 1945* (London: Pimlico, 2007), 113.
67 Elisabeth Barker, *Britain in a Divided Europe, 1945–1970* (London: Weidenfeld and Nicolson, 1972), 20.

68 Winston Churchill, *The Second World War*, vol. VI (London: Cassell, 1953), 308–9.
69 G. Maguire, *Anglo-American Policy towards the Free French* (Basingstoke: Macmillan, 1995), 146–47. See also: K.H. Adler 'Selling France to the French: The French Zone of Occupation in Western Germany, 1945–c.1955', *Contemporary European History*, 21 (2012), 575–95.
70 Bronson Long, *No Easy Occupation: French Control of the German Saar, 1944–1957* (Rochester, NY: Camden House, 2015), 20–1.
71 Judt, *Postwar*, 115.
72 Sean Greenwood, 'Coal and the Origins of the Cold War: the British Dilemma over Coal Supplies from the Ruhr', in Hopkins, Kandiah and Staerck, *Cold War Britain*, 144.
73 Michael Creswell and Marc Trachtenberg, 'France and the German Question, 1945–55', *Journal of Cold War Studies*, 5:3 (2003), 9.
74 Aldrich, *Hidden Hand*, 200.
75 Alex May, *Britain and Europe since 1945* (Abingdon: Routledge, 2014), 9.
76 Doris T. Zallen, 'Louis Rapkine and the Restoration of French Science after the Second World War', *French Historical Studies*, 17 (1991), 6.
77 Bessel, *Germany 1945*, 381–2.
78 Hitchcock, *France Restored*, 45.
79 Corine Defrance, 'La mission du CNRS en Allemagne (1945-1950): Entre exploitation et contrôle du potentiel scientifique allemand', *Revue pour l'Histoire du CNRS*, 5 (2001), 54–65.
80 Douglas O'Reagan, 'French Scientific Exploitation and Technology Transfer from Germany in the Diplomatic Context of the Early Cold War', *The International History Review*, 37 (2014), 366–385.
81 Roy F. Willis, *The French in Germany, 1945-1949* (Stanford, CA: Stanford University Press, 1962), 112.
82 Neufeld, 'Nazi Aerospace Exodus', 52, n.10.
83 Aldrich, *Hidden Hand*, 189.
84 TNA, FO 1031/51, 'French Participation in CIOS', 9 April 1945.
85 TNA, FO 1031/50, 'Minutes of 2nd BIOS Meeting (1945)', 29 August 1945.
86 TNA, FO 1031/5, 'British Investigators Visiting the French zone', 31 July 1945.
87 TNA, FO 1065/12, 'Investigation of T-Force: Report No. 47', 12 December 1946.
88 IWM, 99/76/1, Private Papers of Monica Maurice, 18 May 1947.
89 TNA, FO 1031/7, 'Non-arrival of Investigators', 21 June 1946.
90 TNA, FO 1031/6, MG Econ 11 to MG Econ 1, 28 December 1945.
91 O'Reagan, 'French Scientific Exploitation', 367.
92 TNA, FO 1031/5, CCG Intelligence Group to JIC, 9 February 1946.
93 TNA, FO 1031/5, 'French Participation in the Collection of Technical Intelligence in Germany', 15 June 1945.
94 TNA, FO 1031/5, Brig. Pennycook to R.J. Maunsell, 12 August 1945.
95 TNA, FO 1031/7, 'Advance Notes for Investigators', 2 September 1946.
96 TNA, FO 1031/7, 'Minutes of a Panel Meeting', 23 September 1946.
97 O'Reagan, 'French Scientific Exploitation', 368.
98 TNA, FO 1031/5, Brig. R.J. Maunsell to MG Intelligence Division, 25 August 1945.
99 TNA, CAB 82/6, 'General Allocational Policy on Secret Weapons', 1 May 1945.
100 TNA, CAB 126/333, 'German TA (Tube Alloy) Activities', 26 September 1945.
101 Groves, *Now It Can Be Told*, 234.
102 Hitchcock, *France Restored*, 43–5.
103 Krige, *American Hegemony*, 49, n.118.
104 TNA, CAB 81/133, 'JIC(46)51(0)', 24 May 1946.
105 TNA, FO 1031/59, 'Attachment of Personnel of Enemy Personnel Exploitation Section to FIAT Forward (British)', 14 August 1946.
106 TNA, FO 1031/65, Lt.-Col. P.M. Wilson to Brig. R.J. Maunsell, 20 August 1946.

107 TNA, FO 1031/65, Lt.-Col. P.M. Wilson to Brig. R.J. Maunsell, 20 August 1946.
108 TNA, FO 1031/9, 'Minutes of the 6th BIOS Reception Centre Panel Meeting', 19 June 1947.
109 Bar-Zohar, *Hunt for the German Scientists*, 125.
110 Jeffreys, *Hell's Cartel*, 298–9.
111 Bower, *Paperclip Conspiracy*, 268.
112 Neufeld, 'Nazi Aerospace Exodus', 53, 59; Jacques Villain, 'France and the Peenemünde Legacy', in Phillipe Jung (ed.), *History of Rocketry and Astronautics: Proceedings of Twenty-Sixth History Symposium of the International Academy of Astronautics, Washington, D.C., U.S.A., 1992* (San Diego, CA: Univelt, for the American Astronautical Society, 1997), 119–62.
113 TNA, FO 1050/1419, 'Combined Intelligence Priorities Committee – Black List of Targets', 18 August 1944.
114 TNA, CAB 81/24, 'Investigation of German Research and Development', 25 April 1944.
115 Šimůnek, Michal and Miloš Hořejš, 'Exploatace (okupovaného) spojence. Aktivity tzv. Sdruženého podvýboru pro zpravodajské úkoly (C.I.O.S.) a Britského podvýboru pro zpravodajské úkoly (B.I.O.S.) v Československu na příkladě Škodových závodů v Plzni, 1945–47' [Exploiting an (Occupied) Ally: The Activities of CIOS and BIOS in Post-war Czechoslovakia as Exemplified by the Škoda Pilsen Company, 1945–1947], in I. Janovský, J. Kleinová and H. Stříteský (eds), *Věda a technika v Československu v letech 1945–1960* (Prague: NTM, 2010), 385–406.
116 TNA, FO 1031/6, 'Visit of Allied Technical Specialists', 11 July 1946.
117 TNA, FO 1031/5, FIAT Main to FIAT London, 14 August 1946.
118 TNA, FO 1031/6, H.D. Greenwood to HQ, T-Force, 13 August 1946.
119 TNA, FO 1031/50, 'Minutes of 10th BIOS Meeting', 20 December 1945.
120 TNA, FO 1031/50, 'Minutes of 8th BIOS Meeting', 21 November 1945.
121 TNA, FO 1031/5, CCG Intelligence Group to JIC, 9 February 1946.
122 TNA, CAB 131/5, 'Minutes of 13th Defence Committee Meeting', 14 May 1947.
123 TNA, AVIA 54/1403, 'Draft Paper for Ministers on the Emigration of German Scientists and Technicians', 21 July 1947.
124 Bower, *Paperclip Conspiracy*, 171.
125 TNA, AVIA 54/1403, 'Employment of German Scientists in Countries other than the UK, Dominions or USA', 8 May 1947.
126 TNA, FO 1032/164, 'Minutes of 14th Darwin Panel Meeting', 10 January 1947.
127 TNA, AVIA 54/1403, 8 May 1947.
128 Albert Presas i Puig, 'Technological Transfer as a Political Weapon: Technological Relations between Germany and Spain from 1918 to the early 1950s', *Journal of Modern European History*, 6 (2008), 229–30.
129 Neufeld, 'Nazi Aerospace Exodus', 55, 58; Hans J. Ebert, Johann B. Kaiser and Klaus Peters, *Willy Messerschmitt: Pioneer of Aviation Design* (Atglen, PA: Schiffer, 1999).
130 Roy MacLeod, '"All for Each and Each for All": Reflections on Anglo-American and Commonwealth Scientific Cooperation, 1940–1945', *Albion*, 26 (1994), 80.
131 Ibid., 106.
132 TNA, FO 1032/177, 'BIOS: Organisation', 18 July 1945.
133 TNA, FO 1031/50, 'Minutes of 7th BIOS Meeting', 8 May 1946.
134 TNA, BT 211/541, 'Meeting Papers', 6 September 1946.
135 TNA, AVIA 22/940, P.A. Clutterbuck to George W. Turner, 12 October 1945.
136 Steven T. Koerner, 'Technology Transfer from Germany to Canada after 1945: A Study in Failure?', *Comparative Technology Transfer and Society*, 2 (2004), 111.
137 TNA, AVIA 22/940, Frederic Hudd to Viscount Addison, 4 July 1946.
138 Koerner, 'Technology Transfer from Germany to Canada', 105–06.
139 MacLeod, 'All for Each and Each for All', 107.
140 TNA, DEFE 10/66, 'Minutes of 1st IDCGS Meeting', 16 January 1947.

141 TNA, FO 1032/1231A, Lt-Col. W.H.A. Bishop to Maj-Gen. Lethbridge, 16 July 1947.

142 TNA, FO 1032/1231A, Lt-Col. W.H.A. Bishop to COGA, 12 April 1947.

143 TNA, FO 1065/12, D. Wood to Zonal Executive Office, CCG, 28 July 1948.

144 Maddrell, *Spying on Science*, 34.

145 Evan Jones, 'The Employment of German Scientists in Australia after World War II', *Prometheus*, 20 (2002), 305-321; Uta v. Homeyer, 'The Employment of Scientific and Technical Enemy Aliens (ESTEA) Scheme in Australia: A Reparation for World War II?', *Prometheus*, 12 (1994), 77–93.

146 Koerner, 'Technology Transfer from Germany to Canada', 111.

147 Howard Margolian, *Unauthorized Entry: The Truth about Nazi War Criminals in Canada, 1946–1956* (Toronto: University of Toronto Press, 2000), 120.

148 Koerner, 'Technology Transfer from Germany to Canada', 112.

149 Neufeld, 'Nazi Aerospace Exodus', 49.

150 Edgerton, *Shock of the Old*, 124.

151 Neufeld, *Rocket and the Reich*, 271.

152 Ebert, Kaiser and Peters, *Willy Messerschmitt*, 351; Neufeld, 'Nazi Aerospace Exodus', 54.

153 Theo Pirard, 'German Rockets in Africa: The Explosive Heritage of Peenemünde', *Acta Astronautica*, 40 (1997), 885–98.

154 Owen L. Sirrs, *Nasser and the Missile Age in the Middle East* (Abingdon: Routledge, 2006), 55–67.

155 Edgerton, *Shock of the Old*, 124–5.

156 Neufeld, 'Nazi Aerospace Exodus', 54.

157 Ebert, Kaiser and Peters, *Willy Messerschmitt*, 351.

158 Ruth Stanley, 'German-speaking Armaments Engineers in Argentina and Brazil, 1947–1963', in Oliver Rathkolb (ed.), *Revisiting the National Socialist Legacy: Coming to Terms with Forced Labor, Expropriation, Compensation, and Restitution* (New Brunswick, NJ: Transaction, 2004), 213.

159 TNA, CAB 122/357, Sir John Magowan to Brig. A.T. Cornwall-Jones, 17 January 1946.

8   A new adversary

In the improbable wartime alliance, the Soviet Union was certainly the most anomalous element. Since the Russian Revolution of 1917 and the subsequent civil war, relations between the Soviet Union and the West had been largely unfriendly and characterised by suspicion and rivalry. They were only forced to collaborate in the Second World War by Nazi Germany's surprise invasion of the Soviet Union in June 1941. Under these inauspicious circumstances, a marriage of convenience was reached, where parties set aside, or at least temporarily veiled, ideological differences, racial stereotypes and varying strategic aims in order to unite for a common cause – the defeat of Nazi Germany.[1] Once that aim had been achieved, or even before, when it looked increasingly likely, cracks appeared in the unsteady foundations of the alliance and gradually a gulf reopened between East and West.[2] In September 1946, a Gallup poll revealed that 61 per cent of British respondents felt that the alliance between the US, Britain and the Soviet Union had disappeared, while less than a quarter thought it was still intact.[3]

The British, for their part, reacted to the breakdown in Anglo-Soviet relations swiftly and prudently.[4] This was partly because some British intelligence and military experts had never stopped thinking of the Soviet Union as the real enemy, even while the war was still being fought. On 27 July 1944, Chief of the Imperial General Staff, Sir Alan Brooke, wrote in his diary, 'Germany is no longer the dominating power of Europe, Russia is. … She has vast resources and cannot fail to become the main threat in 15 years from now.'[5] While his timescale may seem a little naïve in hindsight, Brooke was generally prescient about the threat posed by the Soviets. Others were slower to adjust to these new geopolitical circumstances, or at least slower to outwardly reveal their adjustment. Britain's new post-war Labour leadership publicly displayed a willingness to continue working with the Soviets, even if, in private, they began to seriously doubt whether such collaboration was possible.[6] In this way, senior British policymakers were ahead of the curve when compared to their American counterparts. The idea that the wartime alliance could persist after 1945 and be the core of a lasting global peace died hard in Washington and there were members of Harry Truman's administration who felt that the USA could straddle Soviet and British imperialism and mediate between them.[7] For the British, the experience of the occupation of Germany simply deepened existing mistrust and suspicion of Russia, which dated back to the nineteenth century and to the Bolshevik seizure of power.[8] This was gradually reflected in the attitudes of the British public too – by July 1948, a remarkable 70 per cent of British respondents felt that the goal of the Soviet Union was world domination.[9]

Away from the public domain, where geopolitical realignment happened more gradually, British military planners rapidly accepted the new paradigm and began acting on the basis of it. Very soon after the war, the Joint Intelligence Sub-Committee (JIC) predicted that, before long, 'Russia will contain within her own frontiers such military and economic resources as would enable her to face without serious defeat even a combination of the major European powers'.[10] As such, these planners even began conducting assessments of what military strategy would be most appropriate for defeating the Soviets should it come to war between East and West, concluding that the use of weapons of mass destruction, especially atomic bombs, would be necessary to counteract the Soviet Union's overwhelming superiority in conventional land forces.[11] This built upon an earlier staff study, conducted at Winston Churchill's request mere days after the war's end, on how a campaign could be waged against the Soviets. Tellingly codenamed 'Operation Unthinkable', it envisaged hundreds of thousands of British and American troops, augmented by 100,000 rearmed German soldiers and provided with air support by the RAF, launching a pre-emptive strike on the war-weary Soviet Union.[12] While such a recommendation is horrifying and fascinating in equal measure, one particular point of interest is the proposed involvement of German forces. This reflected the fact that, just as the Soviet Union was being recast as a foe, Germany was starting to be seen as, if not quite an ally, then at least as a tool to help secure British interests in Europe.[13] In fact, in an interesting parallel to scientific and technical exploitation, the US military asked General Alfred Jodl – Chief of the Operations Staff of the Wehrmacht High Command during the war – to set out his plans for an effective Western strategy to attack the Soviet Union in 1946.[14] A few months later, he was executed for war crimes and crimes against humanity.

If British and American utilisation of German resources against the Soviets was one side of the coin, the other was widespread fears that Germany would end up firmly in the Soviet orbit. Even as early as summer 1944, the Foreign Office worried about a revived and remilitarised Germany entering into alliance with the Soviet Union, something which it described as a 'most formidable combination'.[15] In this, British policymakers were haunted by the spectre of the Treaty of Rapallo, which had been signed between Germany and Russia in 1922 and saw them renounce all claims against one another following the First World War and build normalised, and even friendly, diplomatic relations.[16] This had, in turn, led to secret military co-operation, allowing Weimar Germany to circumvent the restrictions imposed by the Treaty of Versailles and begin rearming with Soviet assistance.[17] In this sense, suspicions of a resurgent Germany or an openly hostile Soviet Union ceased to be mutually exclusive and it was instead considered possible that the two might occur in tandem. These anxieties crystallised in the immediate post-war period and began to inform policy; by 1946, it has been argued that

British strategy had evolved from containing Germany to containing the Soviet Union in Germany.[18]

These concerns extended to the exploitation of science and technology, too. As has been noted, the British and Americans believed that their preponderance of advanced weaponry could counter the Soviet Union's larger conventional forces, so it became a major priority to prevent the Soviets achieving parity in so-called 'scientific strength', especially if they tried to expedite this by utilising German expertise.[19] Indeed, in May 1946, the Joint Intelligence Sub-Committee stated its anxious belief that 'the alliance of German brainpower and Russian resources may well prove to be the most important outcome of the occupation of Germany'.[20] In his work on the Russian occupation of Germany, Norman Naimark asserts that British and American exploitation can *only* be understood in terms of the relationship with the Soviet Union. Certainly, one of the key aspects of the Cold War was the arms race between East and West, as both sides quickly established that an advantage in the science and technology of warfare might give them a crucial edge at the negotiating table as much as on the battlefield. This arms race was christened in the struggle for the spoils of Germany, with all participants realising that a shortcut to technological superiority might be found among the ruins of the Third Reich. This lends considerable credence to Naimark's point, in that exploitation would never have ranged so widely nor lasted so long had the prospect of another global conflict not been looming on the horizon.[21]

Working with the Soviets

To fully understand Anglo-Soviet relations regarding exploitation, it is necessary to establish what the broader aims of the Soviets were in Europe during this period. Ultimately, following two brutal German incursions onto Russian soil in the past 30 years, they sought to do whatever was necessary to prevent these sufferings from being repeated.[22] This guided much of their policy in Eastern Europe where the creation of loyal 'buffer states', for example in Poland, was considered a high priority.[23] With regard to Germany itself, they aimed to ensure that it would never again be in a position to launch a war of aggression or threaten their territory. The Western Allies shared this objective, but the two sides differed on how to achieve it – a disparity which meant that, though none of the Allies actually wanted a divided Germany, the split between East and West gradually became inevitable. Britain (along with the United States) sought to rebuild an independent, self-sufficient and democratic West Germany which could contribute to a general Western alliance and act as a bulwark against threats from further east. This explains British support for West German rearmament and admission to NATO in the early 1950s.[24] The Soviet Union, meanwhile, wished East Germany to be a subservient socialist client-state which would grant them a foothold in central Europe and help serve their broader diplomatic purposes.[25]

In keeping with this conception, and because they had invested so much, not just in terms of human lives but also economically, in the war against the Nazis, the Soviets felt they had a right to strip Germany of all it had to offer.[26]

This belief underpinned Soviet exploitation of German science and technology, as part of their extensive and punitive dismantling strategy – for example, one-third of all railway track in the Soviet zone was simply torn up and transported east, along with large quantities of agricultural, industrial and commercial stock. But the Soviet gain was much less than what East Germany lost.[27] This was because dismantling and exploitation were conducted in such a chaotic and careless fashion that much which was of value was damaged or destroyed, especially in terms of delicate scientific equipment.[28] In addition, the transport available to move the materials back to the Soviet Union was insufficient and large quantities were simply abandoned at railway sidings, to the double fury of the wastefully deprived German population.[29] In many ways, early Soviet exploitation was little more than an extension of the widespread looting conducted by Red Army soldiers, who were the worst perpetrators of crimes against the German population, with incidents of rape, pillage and murder horrifyingly commonplace.[30] George F. Kennan, the senior American diplomat, later wrote in his memoirs that the Russians 'swept the native population clean in a manner that had no parallel since the days of the Asiatic hordes'.[31] British exploitation teams even heard first-hand stories of the horrors of Soviet occupation – the works manager at the *Dominitwerke* in Brilon, Westphalia, described the effect of the presence of 1,500 Russian soldiers for six weeks as being 'as good as an air raid except that the ceilings remained intact'.[32]

However, the Soviets soon realised the shortcomings of this haphazard and wasteful approach to exploitation and resolved to develop a more effective and better-co-ordinated policy in its place. To this end, in August 1945, the Soviet exploitation programme was placed under the jurisdiction of the People's Commissariat for Internal Affairs (NKVD), directed by Stalin's powerful secret police chief, Lavrenti Beria. This afforded it impressive latitude and access to resources in pursuit of the scientific and technological spoils of war available in occupied Germany.[33] Furthermore, Soviet exploitation efforts were never officially conducted on an inter-Allied basis (there was, for instance, no Soviet membership of CIOS, nor was it ever seriously contemplated), which granted a certain level of freedom to their endeavours. That said, and despite the wider breakdown in relations between the wartime Allies which was occurring in this period, the early stages of exploitation saw a general spirit of civility and even co-operation persist between agencies and operatives from East and West.

In Britain, some advocated a strongly collaborative approach, such as the Field Information Agency, Technical (FIAT) chief, R.J. Maunsell, who called for 'basically, full co-operation',[34] albeit with some particularly sensitive topics, such as chemical warfare, excluded. However, the prevailing view in Whitehall in the months following the end of the war was that some form of reserved, partial co-operation was the best option, including an insistence on reciprocity for any exchanges. For example, Admiral Sir Harold Burrough, the British Naval Commander-in-Chief in Germany, replied to Maunsell's suggestions by saying that his 'past experience has shown that Russians are prepared to take everything and give nothing' and suggesting a firm reciprocal basis, with all 'requests and proposals to be initiated by Russians'.[35] Such sentiments were shared by the

Foreign Secretary, Ernest Bevin, who, following a meeting of the Council of Foreign Ministers in Moscow in 1947, wrote to Prime Minister Clement Attlee, that the Soviet aim was to 'rehabilitate their own zone at our expense and then on top of that get reparations from current production. ... The result is that it is impossible to reconcile [British objectives] with the desires and determination of Russia to loot Germany at our expense.'[36]

The official policy which was handed down by the British Chiefs of Staff in September 1945 predated Bevin's letter but reflected many of his sentiments. Building on Admiral Burrough's appraisal, it stated that Britain 'should have reasonable latitude in permitting conducted Russian visits to German intelligence targets within our zone for strictly limited periods, subject to the exclusion of certain specific targets'.[37] Areas which were not open for exchange with the Soviets included chemical and bacteriological warfare, applied nuclear physics, supersonic aerodynamics, control of guided missiles, the work of IG Farben and all diplomatic and political documents. Problems were found with this scheme almost immediately. The economic intelligence division of the Military Government wrote to the JIC, describing the exclusions as 'somewhat unrealistic' and listing several particularly telling examples:

> In particular, the exclusion of Russians from all IG Farben plants and the ban on any reference to this cannot be effected while there is a quadripartite enquiry into the ramifications of the IG Farben. It is also going to be very difficult to avoid discussion about the control of guided missiles when the Russians have been asked to attend the Backfire demonstrations. Again, there seems little point in restricting information on the new poison gases Sarin and Tabun when the plant for their manufacture is in the Russian zone. Other examples could be mentioned, but these are typical.

It was also noted that the list was so comprehensive that if it was adhered to literally, 'we are bound to create a feeling of suspicion in the minds of the Russians and all to very little purpose'.[38]

Furthermore, on the British side, there was also a feeling that any push for reciprocity was pointless as 'it is likely that worthwhile intelligence targets within the Russian zone are few' though it was still considered 'desirable to learn what does in fact remain in the Russian zone'.[39] This reflected a similar point which had been raised by the British at the Potsdam Conference, that 'even if a more general undertaking for reciprocal exchange of all information were made and loyally observed by the Russians, we should not expect to obtain much valuable information from them'.[40] Moreover, early experience showed that Admiral Burrough's predictions had been correct and that the Soviets were not particularly willing to loyally observe the terms of reciprocity. For example, in February 1946, representatives of the Chemical Industries Branch of the Control Commission for Germany (British Element) (CCG(BE)) complained that 'the position with regard to Russian visits to our zone is extremely unsatisfactory', largely because the Soviets kept sending new teams without allowing for British return visits, thus removing any chance that 'a proper give and take basis would be established'.[41]

Sometimes obstructions were not created by the Soviets, but rather by the security-conscious British authorities – Monica Maurice, who led a BIOS (British Intelligence Objectives Sub-Committee) trip to Germany in 1947, recorded ruefully that one member of her team was offered a 'heaven sent opportunity' to visit the Soviet zone 'but we are not allowed to do that'.[42] However, when her team travelled through the Soviet zone to get to Berlin, they found they were not permitted to stop anywhere *en route* – as a result, they covered 100 miles in just under two hours, despite the poor condition of the roads.[43] They were also not allowed to travel through the Soviet zone after dark and there were Red Army sentries posted every two or three miles along the roads to make sure these rules were acknowledged.[44] Nevertheless, British investigators did have some positive experiences – in April 1946, although 'some difficulty had been experienced in making the initial arrangements', the leader of a BIOS team which had travelled to the Soviet zone emphasised the 'courtesy and good treatment' he and his men had received, and even suggested that a letter of thanks be sent to the Russian Liaison Officer in Berlin.[45]

Investigations into IG Farben offered another occasion for greater co-operation, or at least the outward experience of it, and, in January 1946, British officials chose to visit three IG facilities under Soviet control (at Auschwitz, Staßfurt and Bitterfeld) in return for Soviet visits to three facilities in the British zone (at Düsseldorf, Uerdingen and Leverkusen).[46] This was, in part, the legacy of a proposed Quadripartite IG Investigation Working Committee which had been plagued by a number of issues, including the lack of a dedicated detention centre and conflict with the Nuremberg Trials and, in December 1945, was dissolved as it 'has not now and never has had official sanction, and in fact has never acted on a quadripartite basis'.[47] Furthermore, by 1947, BIOS had lodged 70,000 patents for German technology at the British Patents Office, four-fifths of which carried no security rating and were therefore available for Soviet officials to consult; there was, however, no equivalent arrangement for British officials to access Soviet reports.[48] As these cases suggest, the spirit of co-operation between East and West was often little more than an optimistic ideal, and examples of genuine collaboration were the exception rather than the rule.

As with all international interactions on exploitation, the biggest sources of contention were linked to the most significant technological advances achieved by German researchers. The Soviets were not allowed access to any 'Top Secret, Secret or Confidential' British documents relating to the V-weapons or rocketry in general,[49] and in terms of the allotment of specimens of these secret weapons to the Soviets, it was decreed that they should not be handed over until specifically asked for and even then 'should always be subject to reciprocal action, not necessarily in kind but in equitable exchange of information, materiel or visits'.[50] The Americans even considered destroying the Mittelwerk underground V-2 factory near Nordhausen before the area was handed over to the Soviets, in order to 'preclude resumption of production within a comparatively short time'.[51] However, it was deemed that such action could have 'unfortunate repercussions' so it was called off, though as much specialist equipment as possible was removed before the handover and relevant German experts were relocated to Cuxhaven, deep within the British zone.[52] In spite of this, the British did invite three Soviet observers to attend the third test-firing of a V-2 conducted as part of Operation Backfire – the Soviets brazenly sent six men (including Sergei Korolev, future Chief Designer of the Soviet space programme), three of whom were refused entry, meaning they had to try and catch a glimpse of the launch from outside the facility.[53]

Unsurprisingly, another area which gave rise to considerable conflict and competition was the German atomic research programme. Britain and the US were obviously very concerned that the West should maintain a monopoly on atomic weapons in order to give them a greater edge at the negotiating table with the USSR, but intelligence on the Soviet atomic bomb project was notoriously difficult to gather.[54] As they had little faith in Soviet science to develop an atomic bomb of their own accord, the withholding of any German material, equipment or personnel with connections to atomic research became of paramount importance.[55] Brigadier Leslie Groves, head of the Manhattan Project, described prominent German atomic physicist Werner Heisenberg as of greater worth than ten divisions of German soldiers and predicted that if he fell into Soviet hands, he would prove 'invaluable' to them.[56]

Alsos was at the forefront of the efforts to prevent the Soviets deriving any benefit from German atomic science, an endeavour which included denying them access to any relevant substances as well as personnel. In March 1945, following intelligence gathered by Alsos operatives, the US Air Force bombed a thorium and uranium processing plant at Oranienburg, while in April, an Anglo-American team removed some 1,200 tons of uranium ore from a salt mine near Staßfurt and shipped it back to Britain – both locations were due to fall within the Soviet zone.[57] Similarly, in September 1946, a T-Force unit was sent to the Krupp Works at Essen to discreetly remove some 6,000 kilograms of highly refined uranium ore before a quadripartite team, including four Russians, came to inspect the facility two days later. Its removal prevented the Soviet authorities from making a bid for this material as part of their reparations demands.[58] Tactics such as these marked the first incidence of the denial of fissile materials being employed as an atomic non-proliferation measure – a tactic which has persisted, but not proved widely successful, throughout the Cold War and up to the present day.[59]

It is interesting to note that, while the British government was working hard to prevent German technology from falling into Soviet hands, they were quite content to hand over British technology. This is seen in the government-sanctioned sales of Rolls-Royce Nene and Derwent jet engines to the Soviet Union (these engines powered the MiG-15 aircraft which later outperformed British-made aircraft in the Korean War), despite strong opposition from President Truman.[60] However, as the case studies of atomic physics and rocketry reveal, general British policy was to continually seek ways to thwart the Soviet exploitation of German science and technology. This did not just extend to the material spoils of war; instead, just as in their relations with the Americans and French, the area which drew the British authorities' particular attention was the Soviet utilisation of German expert personnel.

'A completely open race'

In the conclusion to a FIAT intelligence report from August 1946, the unnamed author expressed his feeling that 'we may just as well acknowledge the situation for what it is between Russia and the Western powers: a completely open race for the best talent and skill Germany has to offer.'[61] The British and American side of this race was spurred on by an overwhelming concern about what could happen if the Soviet Union was able to maximise the benefits of personnel exploitation. In reality, the extent to which German experts contributed in a significant way to Soviet technological development after the war is unclear, muddied as much by contemporary secrecy and national pride as by subsequent historiography, much of which has been based on only limited access to the pertinent files.[62] However, in this case the reality was less important than the imagery conjured up by the fertile imaginations of Western military planners – their visions of a Soviet war machine which combined the expansive resources and manpower of the USSR with the scientific and technological prowess of the Third Reich, inspired fear and drove Britain and the USA into fierce competition with their wartime ally.

This process began in the immediate aftermath of the war. When Germany surrendered in May 1945, British and American troops had captured parts of Saxony and Thuringia which were due to fall under Soviet occupation. In the weeks before these areas were handed over, much of scientific and technical value, including expert personnel, was evacuated from them back to Britain and the USA. At the Potsdam Conference, Stalin directly challenged President Truman on these illicit removals.[63] Truman had come prepared and responded that the removals 'were not made under instructions of the American government and that they would be accounted for... He added that no people had been removed by the American Army.'[64] He also promised to have a full investigation into these removals conducted by the US Military Governor, Dwight D. Eisenhower, which duly took place and concluded that, 'with certain exceptions, we did evacuate equipment and personnel from the Russian zone as claimed.'[65] Truman unsurprisingly felt no need to share these findings with Stalin or the Soviets. An EPES (Enemy Personnel Exploitation Section) report filed a year later, in August 1946, went into greater detail and noted that the British evacuated 250 German experts and their families from the future Soviet zone, while the number handled by the Americans was closer to 2,000, surmising that 'the whole operation is now regarded with favour by British and American authorities, especially in view of the valuable results which have been obtained in the exploitation of German evacuated scientific and technical personnel.'[66]

A note appended to this report by R.J. Maunsell of FIAT warned of the 'strenuous efforts' which the Soviets were making to induce the evacuated specialists to return to their zone. It detailed that:

> Every effort is now being made by the Russians to persuade the evacuees to go back. Russian methods of persuasion include the offer of lucrative terms of employment and if these are not accepted the victimisation of the families of evacuees still in the Russian zone and the confiscation of their property.[67]

These carrot-and-stick tactics would later become characteristic of the whole Soviet recruitment effort, but their strong protests about these British and American removals concealed, perhaps deliberately, the fact that they had done something very similar in parts of Berlin before they were handed over to Western occupying forces. In the district of Dahlem, for instance, the Soviets lured away the bulk of the scientific workforce at the Kaiser Wilhelm Institutes for Biology, Biochemistry, Chemistry and Anthropology, using offers of lard to prove that they were serious about looking after these men and their families.[68]

These initial mutual poaching attempts were also replicated in some of the high-priority areas of German science and technology. For example, in his memoirs, Albert Speer recounted a rumour that the Soviets had contrived to use the kitchen staff at the US Army's camp at Garmisch-Partenkirchen in southern Bavaria to pass a secret offer of employment to Wernher von Braun, while he was briefly detained there after the war.[69] Unable to secure von Braun, the Soviets had to settle for some of his Peenemünde subordinates, the most senior of whom was Helmut Gröttrup, whose left-leaning politics made him more amenable to Soviet offers and who also bore considerable personal resentment towards von Braun.[70] Under the auspices of the so-called Institute Rabe, Gröttrup spent a year at Bleicherode, not far from the Mittelwerk, overseeing the construction of more V-2 rockets for Soviet use.[71] During this time, the British attempted to lure him to work in Britain by way of a top-secret letter – Gröttrup remained loyal to his Soviet masters though and handed the letter over to them.[72] Even German rocket scientists who had taken up employment in other Allied nations after the war were not safe from Soviet intrigues – in 1947, Stalin dispatched his son, Vasily, along with two NKVD operatives – Grigory Tokaev and Ivan Serov – to France to bring Eugen Sänger and his engineer wife, Irene Bredt, back to the Soviet Union in a 'voluntary-compulsory manner', so they could help develop a rocket-powered bomber capable of reaching the United States. Unfortunately for Stalin, the mission failed.[73]

In addition, the field of atomic physics was also subject to fervent activity with regard to personnel exploitation. The Soviet atomic investigative organisation in Germany – which has been dubbed 'Russian Alsos' – was led by an NKVD lieutenant-colonel, Avraam Zaveniagin, who brought an approach more driven by intelligence-gathering than by scientific curiosity. In fact, many Soviet scientists were reluctant to participate, fearing that they would be replaced by the very German experts whom they helped to recruit. Although the Soviets soon discovered that they had missed out on the most talented of the German physicists (who by this stage were interned at Farm Hall, near Cambridge), they were able to benefit from the fact that many others did not wish to go to the USA as they felt that they had nothing to offer to the vastly advanced American project and did not want to rely on charity. The Soviet bomb project, meanwhile, was only slightly ahead of German research and therefore the German experts felt they could contribute more to it.[74] Certainly, this competition for German expertise was the first phase in the close relationship between espionage and atomic physics which existed throughout the Cold War, and gave both sides ample experience.[75]

Indeed, concerns about atomic secrecy did not disappear after the initial mad rush for spoils which took place in 1945. For instance, when the Farm Hall scientists returned to Germany after their period of internment in Britain had ended, they were placed under 'special surveillance', codenamed Operation Scrum Half. This continued and expanded in 1947 and 1948, as fears of these scientists being kidnapped, murdered or swept up by the Soviets in the instance of a land invasion of the western zones of Germany grew. However, it also came in for considerable criticism, including by the US Military Governor Lucius Clay, who felt it was foolish to have these vital individuals living freely in Germany but under almost prohibitively expensive surveillance. It was suggested that it would be better to either incarcerate them or move them permanently to Britain or the US.[76]

During the initial post-war period, the British and Americans began to take a greater interest in exactly how the Soviets were conducting their own personnel exploitation programme. As early as July 1945, the Deputy Chiefs of Staff were informed that 'public offers of employment of German scientists have already been made over the Russian-controlled radio',[77] while in March 1946, the US branch of FIAT produced a report entitled 'Soviet Sponsored Research Organisations Currently Active in Berlin'. This report comprehensively detailed the way in which these organisations contributed towards Soviet exploitation and commented that 'interested Russian agencies largely dominate scientific and technological life in Berlin. The three Western powers, for their own part, are apparently unaware of the nature and extent of this domination.'[78]

Despite this, British and American exploitation agencies took solace in their belief that the German people had a fundamental dislike of the Soviets and were therefore generally reluctant to work for them. In January 1946, Brigadier C.F.C. Spedding of the Research Branch of CCG(BE) dismissed claims that it was risky to let desirable German specialists live too close to the border of the Soviet zone, 'since popularity of [the] Russian zone is inversely proportionate to its proximity'.[79] A Civil Censorship intercept from August 1946 revealed that, upon receiving an offer to go and work for the Soviets, German rocket scientist Helmut Reichstein felt that although he 'would have immediately acquiesced for the Americans, the matter requires some real deliberation when it concerns the Russians'.[80] Reports such as these gave the Western powers an inflated sense of confidence, leading one FIAT intelligence assessment to conclude that, in terms of the majority of German scientists and technicians, 'most of them are ours for the asking – if we ask'.[81]

However, the Soviets actually had many ways to make their offers attractive to a wide range of German experts. By the end of 1946, the British began to recognise that their own commitment to fairly thorough programmes of denazification and disarmament was 'having the effect of driving German scientists and technicians to work for the Russians ... who have no such scruples'.[82] The extent of Soviet commitment to denazification has been disputed,[83] but there are clear examples of their active recruitment of fairly obvious and committed Nazis, such as physical chemist Peter Adolf Thiessen, who had been a senior figure in the Nazi scientific hierarchy, the holder of several Nazi Party awards and had been a member of the Party since 1933 – in short, he was no mere 'fellow traveller' nor one of the so-called 'apolitical' scientists which the British and Americans claimed to exclusively recruit.[84] When the British tried to understand how such a man as Thiessen could be happy to go and work for the Soviets, the conclusion they drew was that he had done so 'probably to contribute to Germany's renewed strength and greatness with the help of the country which made a pact with Hitler against Britain in August 1939'.[85] Such a move certainly did Thiessen's career no harm – while working in the Soviet Union, he was even able to add the Stalin Prize to his collection of Nazi accolades.[86]

For other experts, especially those with a less tainted political background, there were additional elements of the Soviet offers which they found appealing. The Soviets offered salaries ranging from RM 800 to 8,000 a month, which completely dwarfed the average British offers of RM 400, and they augmented this with generous double ration packages.[87] The Soviets also used a system of *payoks* to entice their targeted specialists – these were variously sized parcels of much sought-after items used to sweeten the deal, ranging from five cigarettes at one end of the scale to two cases of foodstuffs at the other.[88] As the food shortages in the British zone worsened in 1946, more and more German experts began looking eastwards for a more secure future, though the British wondered 'whether the prospects of physical starvation weigh as heavily with these men as the virtual certainty of mental starvation if they remain in western Germany'.[89] This was in reference not only to the Anglo-American policy of picking a German's brain, leaving him in uncertainty and often not offering him any financial recompense or job prospects in return, but also in reference to the ban on any warlike industries in the western zones, which included fields such as aeronautics and rocketry, in which many of the relevant experts specialised.[90] This discrepancy was not lost on the German specialists – in March 1946, Heinrich Waas, a German naval technician, compiled a report for the JIC, in which he sardonically reflected German views on Allied recruitment policies: 'one can often think that an agreement exists between the British and Americans on the one hand, the Russians on the other, to drive all valuable technicians out of the western zones into the Russian.'[91]

The British and Americans consoled themselves by suggesting that the German scientists which the Soviets were able to recruit were fairly insignificant individuals. Some government officials felt that Britain had already secured some of the truly outstanding German researchers, too much emphasis was being placed on the 'aiders and abettors', and that Britain should not 'really mind if these lesser lights do go to the Russians'.[92] EPES, meanwhile, noted that the Soviets had taken 'chiefly technicians and engineers … and left behind many of those who specialised in construction and planning', which it was hoped would limit the amount of long-term benefit the Soviet Union could derive from exploitation.[93] It was also recognised that, being mostly younger men, 'assistants are usually more willing to take the risk with the Russians', but sufficient foresight was shown to acknowledge that as 'the assistants will normally be the professors in about ten years, it is considered just as important to keep them in work and in good will towards us'.[94]

The Soviets, however, were not content with simply siphoning off a sizeable number of able German scientists and technicians of all levels using attractive offers of continued work in their particular field, good pay and rations allowances, real opportunities for professional development and a working environment characterised by respect and good relations with their supervisors. Under the Cold War conditions of heightened paranoia, they began contemplating a more drastic way to secure a large number of Germany's best and brightest for themselves. In the autumn of 1945, the German press in the British and American zones ran numerous sensationalist stories about the Soviet kidnappings of countless German specialists; much of which was little more than a thinly veiled propaganda attempt to counter the many positives of Soviet recruitment.[95] Nonetheless, these fears were felt very acutely by British exploitation officials, as shown by the continued monitoring of the Farm Hall scientists upon their return to Germany, mentioned above.

The fears were also not totally groundless. In the autumn of 1946, the Soviets began moving small groups of German specialists forcibly from the eastern zone to the Soviet Union proper. In many cases, they did so covertly, in order not to incite their targeted men to flee or to arouse too much suspicion in the West. In one example, the British Scientific and Technical Intelligence Branch (STIB) recorded that during deportations from the Junkers works around Magdeburg and Dessau, the presence of German police and Red Army soldiers on the streets was explained away as 'a drive … being made against Black Market racketeers'.[96] However, it was at 4 a.m. on Tuesday 22 October 1946 that the real extent of the Soviet deportation plan came to light. This was 'zero hour' for Operation Osoaviakhim, a well-planned and neatly executed mass forced evacuation scheme. It was co-ordinated and led by General Ivan Serov, Deputy Commissar of the NKVD under Beria and future Chairman of the KGB, once again reinforcing the image that Soviet exploitation fell very much in the domain of intelligence and espionage as opposed to the civil service, as was the case in Britain.[97]

The immediate goal of Osoaviakhim was to move huge aviation, rocketry, and other weapons research and production facilities from Saxony and Thuringia to the Soviet Union. These Nazi-era facilities had been rebuilt and the staff was primarily German, under the supervision of the Soviets, who were well aware of the perils of conducting military research in Germany, given the relatively porous frontiers between the various zones of occupation and the supposed four-power prohibition of such research.[98] The aim, therefore, was to relocate these men from Germany, where, despite already being in Soviet employ, they were at risk of poaching by another occupying nation, to the USSR where they were almost completely safe. Among those forcibly relocated were Helmut Gröttrup and his fellow rocketeers who had been working for the Soviets at Bleicherode – on 22 October, they were entertained at a banquet by the Soviet officer in charge, General Lev Gaidukov, while their families and personal possessions were gathered up and put on trains by Red Army soldiers. By the time they were informed that they were being taken to the Soviet Union, it was too late, and they were too inebriated, to effectively protest.[99] Others who were taken were experts in aviation, nuclear physics, electronics, optical science, radio, and chemical engineering, and included former employees of BMW, AEG and Junkers, among many more.[100]

The process for each individual who was included in Osoaviakhim was much the same across the board:

> The man concerned was awakened by Russian soldiers in the early hours of the morning and informed that he would be leaving for Russia immediately. In many cases the man was permitted to take with him his family and as much of his furniture as could be loaded into one third of a railway freight wagon.[101]

The men were told that they would receive a contract for five years employment on arrival in the Soviet Union and that their salary would match that of equivalent Soviet experts.[102] These men and their families were then moved by lorry and private car to the eastern outskirts of Berlin, where they were loaded onto 92 trains, totalling some 700 coaches, at the stations of Friedrichshagen and Köpenick. The destinations of these trains were major cities and industrial centres in the USSR, including Moscow and Odessa, and the Germans aboard were vaguely told that their journeys would last from three to seven days.[103] The deportations continued throughout 22 October and were still ongoing at 5 p.m. that evening, with trucks loaded with scientists, their families and their household possessions arriving at the railway stations every three to four minutes. The scale of the operation was unprecedented, involving 2,552 German specialists – a number which rises to 6,560 once family members are factored in. Within two weeks, these German deportees were spread among 31 different defence and industrial institutions across the Soviet Union.[104]

There is no doubt that Osoaviakhim was meticulously well planned and prepared for. In the months leading up to it, the Soviets had lured many German experts who worked in their zone or sector of Berlin, but lived elsewhere, to relocate closer to their workplace, by offering much larger and more comfortable accommodation at a fraction of the cost. With the dire housing shortage in Germany at this time, especially in Berlin, it was considered 'small wonder' that few could bring themselves to refuse such a tempting offer.[105] Nonetheless, certain 'rugged individualists' still 'refused to budge, and stuck in their homes in the British or US Sectors', thus forcing the Soviets to make some very risky, and largely fruitless, raids outside of their own zone, sometimes using German police officers to bring the specialists in question from their homes to their workplaces where they were promptly handed over to the Soviet commissars.[106] In addition, Red Army commando units and troops with trucks were posted to street corners and important locations during the night, in order to pre-emptively deter any resistance which might be provoked.[107] Irrespective of these tactics, 90 per cent of specialists resisted their deportation or pleaded for their families to be left behind but, in most cases, these appeals fell on deaf ears.[108]

Some individuals still remained undaunted, such as Dipl.-Ing. Zumpe, chief of the flak rocket department at GEMA in Berlin, who turned up to work as usual on the morning of 22 October, where the Russian director informed him that he was to go to the USSR for work. Zumpe immediately acquiesced, knowing that to refuse could prove fatal, and arranged for Soviet transport to fetch him, his wife and their possessions from their home in the British Sector the following morning. Before they could do so, the Zumpes concealed themselves at a friend's apartment nearby and watched as a succession of different Soviet officials and soldiers attempted to locate them. In the meantime, he managed to contact the British element of EPES, who arranged to evacuate them, by air, to Frankfurt, just over a week later. EPES also recorded the story of Dr Ulrich Capeller, a physicist from Jena in Thuringia, who was loaded onto a train by the Soviets but managed to jump off while it was moving during the night and make his way back to Berlin, where he immediately made himself known to the British authorities.[109]

Unsurprisingly, despite its secretive origins, the full extent of Osoaviakhim soon came to the attention of the shocked British and American exploitation agencies and to the wider public, too. Horror stories appeared in the Western press, which the Soviets dismissed as 'calumnious attacks', and they attempted to mitigate the damage which the deportations threatened to wreak to their public image in Germany by arguing that their operation was no worse than the removals made by the Americans and British from areas due to be handed over to the Soviets in the summer of 1945. One story from the time runs that Marshal Vasily Sokolovsky, the head of the Soviet Military Administration in Germany, snidely told Colonel Frank Howley, the American commandant of Berlin, 'I am not asking the Americans and British at what hour of the day or night they took their technicians – why are you so concerned about the hour at which I took mine?'[110]

Despite the uproar from the Western powers, Osoaviakhim was only really concerned with the evacuation of German specialists already in Soviet hands and its repercussions for Britain and the US were, in reality, predominantly positive. The majority of German scientific and technical experts were so shocked by Osoaviakhim that EPES was almost immediately swamped by a great number of 'callers, correspondents and other enquirers, all with the same aim in view' – to escape the possibility of deportation and remove themselves as quickly as possible to the western zones, the United Kingdom or the USA. 'One man went so far as to ask to be arrested for his own safety.'[111] The FIAT Forward office predicted that if the threat of deportation continued, this stream of applicants was 'liable to become a flood'.[112] Desperation was truly commonplace – at a Zeiss plant in Jena, the removal of so many personnel, as well as nine-tenths of the equipment, led to a spate of suicides.[113] More generally though, this was a golden opportunity for the Western powers, especially Britain, to reverse the flow of German specialists heading eastward on account of generous Soviet inducements and maximise their own exploitation potential.[114]

Denial policy

What Osoaviakhim allowed Britain and the US to finally realise was a policy of denial, which had been in the pipeline, albeit putatively, since the end of the previous year. In December 1945, the Joint Intelligence Committee of the Control

Council for Germany (JIC-CCG) had received reports from the Naval Intelligence Division, which cast a 'somewhat sinister light on Russian activities *vis-à-vis* German scientists' and prompted them to call for 'policy guidance at a high level ... as to whether HMG [His Majesty's Government] would wish strenuous efforts to be made to deny scientists and technicians to the Russians'.[115] It was May 1946 before the main JIC considered these reports in full and they concluded that, as a result of the efficiency of Soviet recruitment and the lack of British counter-measures, 'by the end of 1946 a large proportion of German brainpower will have gone to the Russians and there will be no looking back'.[116] In August, they added that 'this movement, if unchecked, will increase significantly Russian war poten-tial.'[117] Osoaviakhim lent considerable credence to these fears and in December, almost exactly one year after the JIC-CCG had submitted its initial report, the Defence Committee of the Cabinet, chaired by Attlee, 'agreed in principle that it was necessary to deny to the Russians those German scientists and technicians, within our influence, who could contribute substantially to the building up of Russian war potential'.[118] Thus began the 'scientific containment' of the USSR, bringing exploitation into line with wider diplomatic thinking on the West's strat-egy towards the Soviet Union, as espoused in George Kennan's 'Long Telegram' and the Truman Doctrine – that is, to prevent Soviet expansion and advancement at every opportunity.[119]

The practical implications of this new development were almost immediately evident. While previously the only criterion for securing a German scientist or technician had been whether he had some contribution to make to British sci-ence, this was now expanded to include any expert who could offer something of value to the Soviet Union. In April 1947, the newly formed Ministry of Defence estimated that there were approximately 290 such scientists within the British zone, but this figure got considerably larger as time wore on and the definitions broadened.[120] The most obvious manifestation of this new policy was the creation of the transit hotels and holding centres, opened under the auspices of Operation Matchbox.[121] The first of these, at the spa resort of Bad Hermannsborn, opened on 16 January 1947 and two days later it already had 40 Germans in residence; by April, that number was 191 and, by August, it was 280 (of which 119 were scientists or technicians and the remainder were family members).[122] Very often, the demand for Matchbox exceeded its capacity, which EPES considered deeply regrettable 'since it means that many useful subjects will be irretrievably lost to us, and when the news of their fate goes round on the grapevine it will be yet another blow to British prestige'.[123] As a result, another hotel was opened, in nearby Bad Driburg, in October 1947 and other smaller facilities followed thereafter. All those who stayed at the hotels were afforded considerable luxuries, including a heavy worker's ration allowance, additional fuel and 'all amenities normally enjoyed by families of scientists and technicians'.[124] In addition, the scientists received a salary of RM 200, though this compared very unfavourably with the amounts offered by the Soviets.[125]

This last point is surprising as the ultimate objective of Matchbox was 'to remove from Russian influence and control, scientists and technicians eminent in certain warlike subjects who were materially contributing, or could materially contribute, to Russian war potential'.[126] This goal was reflected in the criteria which German experts had to meet before they were considered for inclusion in Matchbox. The three categories fit for inclusion were:

a Scientists and Technicians whom it is desired to deny to the Russians on account of their scientific or technical eminence in certain warlike subjects.
b Scientists and Technicians who, while not to be classed in Category (a), would nevertheless have a serious effect on Russian sponsored development and research should they be removed from, or denied to, the Russians.
c Scientists and Technicians who are valuable, not for their professional competence, but because they can give intelligence of value to us about Russian sponsored research and development.[127]

Unfortunately, the breadth of these guidelines meant that they were very open to manipulation by ambitious and enterprising German individuals of dubious exploitation value.

For instance, EPES were bombarded by appeals from characters such as Ernst Schnubel, who claimed he had invented a 'Death Ray Transmission Apparatus', which could be used as a battlefield weapon, a defence against bombs (including atom bombs), and in peacetime against garden pests, vermin, lice and, in the inventor's own words, 'gangsters, terrorists, demonstrationists, rebels, etc.'![128] Schnubel was obviously too eccentric to ever be taken seriously, but several others did slip through the net and were able to take advantage of the amenities at Matchbox which were often so hard to come by elsewhere in post-war Germany. This formed the basis of much of the criticism directed at Matchbox, such as that of Dr Bertie Blount, the Director of Research Branch and one of its harshest critics, who considered that it had become 'a place of permanent residence' rather than 'a place of transit'. This was especially true for individuals such as Gotthilft von Studnitz 'who pretends to be a physiologist but is universally regarded as a quack both by physiologists in other countries and by German scientists'.[129]

Certainly, denial policy gradually became something of an obsession for the British exploitation agencies and it became unthinkable to let any German scientist of any calibre slip through the net lest they turn towards the East. This, in turn, opened the whole policy up to even greater criticism, including another polemic from Blount. In March 1948, he expressed his feeling that Matchbox was 'one of those unfortunate projects which are thoughtlessly entered into and leave a trail of difficulties behind them'. He wondered 'whether the financial results of setting up Matchbox were ever envisaged' and stated his belief that it was quickly becoming 'quite a big and expensive show which directly or indirectly must fall on the British taxpayer', foreseeing that 'sooner or later the cost ... will be queried and there may be a gigantic row'. On the other hand, he worried that if they tried to save money by being more restrictive on who was admitted to Matchbox, they would be blamed 'for almost every scientist and technician who crosses over to the Russians'.[130]

Furthermore, accommodation at Matchbox alone was not always enough. It only constituted a short-term delaying strategy and STIB worried that offers of full employment were not always suitably forthcoming from potential British employers for the German experts being held there. While STIB acknowledged that before employment could be offered, the relevant departments had to be satisfied that no British individual could fill the position adequately, they countered that 'it should be remembered that unless Matchbox scientists and technicians can be suitably employed ... there is every possibility that they will turn to their late Eastern masters'.[131] Naturally, the situation which emerged from these circumstances was that numerous German scientists were held under Matchbox auspices lest they be seized by the Soviets, but were offered no employment and so remained in an unenviable state of limbo. This was the case of Heinz Peukert, an aeronautical scientist who had participated in Operation Surgeon at Völkenrode. When that commitment ended in February 1947, he had been instructed by the Ministry of Supply that he could undertake no further work without their permission and had returned to his home in the French zone of Germany. Seven months later, he wrote to the British authorities, restating his willingness to work in Britain, South Africa or Canada, and asking for a speedy resolution to his predicament, as he had no income and was encountering difficulty in obtaining a ration card. His case remained unresolved as late as December.[132]

As such, the British realised that Matchbox could not function as the only element of denial policy. Instead, a much wider initiative was established, incorporating efforts back in Britain (as well as in the Dominions and other 'friendly' nations) to create employment for German scientists and technicians, and to thus deter them from seeking work with the Soviets. This approach had the potential to be highly effective, as shown in the case of the Linke team. This was a group of six guided missile specialists evacuated from Berlin, brought through Matchbox, interrogated, and then all offered permanent employment at the Royal Aircraft Establishment (RAE), Farnborough.[133] Intelligence Division believed that their removal from the eastern zone 'seriously ... affected Russian exploitation of German guided missile research'.[134] Elsewhere, the Darwin Panel, which was only concerned with recruitment for civil-industrial work, looked to bring over 'German technicians and scientists possessing knowledge of secret processes ... in order to deny their services to the Russians'.[135] The needs of denial were also a strong positive argument when the Inter-Departmental Committee on German Scientists was pushing to allow private firms to employ German specialists exclusively.[136] Even in Operation Bottleneck – a scheme to outsource some of the work of British firms to the surplus labour force in Germany – it was hoped that 'by providing employment for Germans, [it would] help to arrest their drift to employment with the Russians'.[137]

Unsurprisingly, this ongoing wrangling over the fates of innumerable German scientists and technicians, which became a major post-war preoccupation for British exploitation officials both in Britain and in Germany, gave rise to a considerable increase in the use of espionage and subterfuge – a development which would come to characterise scientific intelligence throughout the secrecy-heavy years of the Cold War.[138] One suggested tactic for British agencies wanting to contact German experts living in the eastern zone was to write a letter on German stationery 'under a false German name such as Muller' (or Schmidt or Wolff) and send it to a private address in the British or French sector of Berlin – 'the owners of such houses should be selected for trustworthiness and should be offered cigarettes etc. as an inducement to co-operate.' The letter would state that 'Muller' had been offered work in an Allied country and that they were looking for other men for this same work, with emphasis to be laid on the 'excellent conditions and good and fair treatment offered, and the fact that there will be an opportunity to work outside Germany'. This letter would then be forwarded on to the desired German expert to await his response.[139]

Another option was to send a loyal German to enquire directly with the targeted specialist but as a travel permit, including stated purpose of journey, was needed for a German civilian to enter the Soviet zone, this was not always practicable.[140] Moreover, it could be quite hazardous for ordinary German citizens to aid the British exploitation efforts. Henry Mecklenburg, who ran a Matchbox transit hotel in the British Sector of Berlin, had several close encounters with the Soviet security services. His night-porter was detained by the police, questioned by a Soviet agent and told to report back on the British officers who visited the hotel, with the threat of 'Red Army disciplinary action' if he did not comply. Mecklenburg himself felt he was about to be attacked by two uniformed Russian men on one occasion when walking home late at night with his wife, but the timely arrival of a British military Volkswagen scared them off.[141] The Soviets often acted with remarkable impunity in their attempts to interfere with British exploitation. On the night of 18 October 1946, the British Military Train from Berlin to Hannover was halted while passing through the Soviet zone and, despite the armed guard, a number of German passengers were removed from a sealed coach, often used to transport scientists recruited by FIAT and EPES. On this occasion, there were no such scientists aboard, but the Soviets had obviously hoped there would be, as they had turned up with enough men to leave the train guards 'heavily outnumbered'. Only the guard commander's 'anxiety to avoid an international incident coupled with his uncertainty as to how to act in these extraordinary circumstances' maintained calm and the train's security cohort was strengthened thereafter.[142]

Altogether, the impact of denial policy is hard to judge. In terms of scale, by April 1948, 321 German specialists had passed through the Matchbox machinery, of whom 286 had been taken on as consultants, paid by the British government. Within this number were numerous guided missile experts, colour film specialists from the Agfa corporation and the entire 15-member Technical Directorate of the Brückner-Kanis company, which developed high-speed underwater-propulsion turbines.[143] Also in April, the CCG(BE) Intelligence Division produced a report which suggested that the work of the design and development departments of a number of important aircraft firms, including Junkers, Heinkel and BMW, reconstituted under Soviet administration, had been 'retarded by the evacuation of some good specialists from each Establishment'.[144] FIAT, meanwhile, felt about the best that could be said of denial policy was that, while it 'may have delayed Russian developments, it has hardly prevented them' and that 'the main value of [securing] a first rate man, at the moment anyhow, consists in saving time.'[145] As Paul Maddrell has argued, the enormous scientific resources of the Soviet Union doomed the Western Allies' non-proliferation measures to failure and the best they could hope for was to slow the pace of certain development projects.[146] Nonetheless, Matchbox lived on. In fact, in March 1950 it acted to prevent Paul Schröder, who was described by British scientists as 'the greatest mathematical authority on rockets alive' but who had fallen on hard times after an initial period of employment with the British, from drifting into Soviet employment by offering him a two-year contract as a Matchbox 'consultant'.[147] As this suggests, the fear of effective Soviet utilisation of German expertise continued to motivate British policy in Germany even after their formal military occupation had ended.

In conclusion, the British exploitation programme can only be fully understood through the lens of Anglo-Soviet relations, specifically the suspicion and hostility which characterised the nascent Cold War. As soon as the war with Germany ended, and in fact even while it was still being fought, British intelligence operatives became very aware that the new enemy was likely to be the Soviet Union and this was reflected in a whole raft of policies designed to contain and neutralise the Soviet threat. The new ideological divide and the absorption of both Britain and western Germany into a US-dominated Western alliance simply brought the conflict into sharper contrast. Denial policy and Soviet actions like Operation Osoaviakhim are both examples of the two camps focusing all their attention on the next war, and not the last. Shaped by the significant role which new weapons and forms of warfare had played during the Second World War, it was evident that any future conflict would be decisively affected by the technological fruits of scientific labour. Therefore, exploitation of German equipment and expertise can be seen as the first phase of the Cold War arms race.

As has been indicated, all the efforts expended by Britain to deny this vital material and know-how to the Soviet Union largely turned out to be futile. The Soviet recruitment programme certainly far outstripped its British or even American equivalents – rough estimates suggest that around 3,000 German specialists went to the Soviet Union after the war, though many of these were only low-skilled technicians or were taken as POWs.[148] Furthermore, the Soviet Union very quickly combined its superiority in conventional land forces with an extensive and advanced technological arsenal, as demonstrated by their detonation of an atomic bomb in 1949 and a thermonuclear bomb in 1953, the first test-firing of an intercontinental ballistic missile in August 1957, and the launch of the first satellite, Sputnik I, into space just over six weeks later.[149] German involvement can be detected in all of these projects, though the exact extent of their contribution is more difficult to establish. Certainly, the Western Allies considered it to have been pivotal – in a press conference less than a week after Sputnik's launch, President Eisenhower attributed the Soviet success to their use of German scientists.[150] It should be remembered, however, that British exploitation was not solely concerned with preventing the Soviets from gaining access to the scientific spoils of war, it was also about obtaining a genuine benefit for Britain too, so that it would be better placed to fight a future war against any opponent, including the Soviet Union. Either way, it was this spectre of another global conflict on the horizon which made the exploitation of science and technology such a critical and central part of British occupation policy in Germany, at the dawn of the Cold War.

Notes

1   Geoffrey Roberts, *Stalin's Wars: From World War to Cold War, 1939–53* (New Haven, CT: Yale University Press, 2006), 296.
2   Richard Overy, *Russia's War* (London: Penguin, 1998), 282–3.
3   Gallup, *Public Opinion Polls*, 137.
4   Lewis, *Changing Direction*, xcvii.
5   Alex Danchev and Daniel Todman (eds.), *Field Marshal Lord Alanbrooke: War Diaries 1939–1945* (London: Weidenfeld & Nicolson, 2001), 575.
6   Anne Orde, *The Eclipse of Great Britain: The United States and British Imperial Decline, 1895–1956* (Basingstoke: Macmillan, 1996), 160.
7   Bernstein, *Myth of Decline*, 78; Hathaway, *Great Britain and the United States*, 12.
8   Knowles, *Winning the Peace*, 184.
9   Gallup, *Public Opinion Polls*, 179.
10   D. Cameron Watt, 'British Military Perceptions of the Soviet Union as a Strategic Threat, 1945–50', in Josef Becker and Franz Knipping (eds.), *Power in Europe* (Berlin: de Gruyter, 1986), 330.
11   Aldrich, 'British intelligence and the Anglo-American 'Special Relationship' during the Cold War', 332–33; Benbow, 'Royal Navy', 379.
12   Aldrich, *Hidden Hand*, 58.
13   Deighton, *Impossible Peace*, 5.
14   Aldrich, *Hidden Hand*, 211.
15   Watt, 'British Military Perceptions', 329.
16   Spencer Mawby, 'Revisiting Rapallo: Britain, Germany and the Cold War, 1945–1955', in Hopkins, Kandiah and Staerck, *Cold War Britain*, 81–94.
17   Gordon H. Mueller, 'Rapallo Reexamined: A New Look at Germany's Secret Military Collaboration with Russia in 1922', *Military Affairs*, 40 (1976), 109–17.
18   Lee, *Victory in Europe*, 24.
19   Jon Agar and Brian Balmer, 'British Scientists and the Cold War: The Defence Research Policy Committee and Information Networks, 1947–1963', *Historical Studies in the Physical and Biological Sciences*, 28 (1998), 209; Audra Wolfe, *Competing with the Soviets: Science, Technology, and the State in Cold War America* (Baltimore, MD: Johns Hopkins University Press, 2013).
20   TNA, CAB 81/133, 'JIC(46)51(0)', 24 May 1946.
21   Norman Naimark, *The Russians in Germany: A History of the Soviet Zone of Occupation, 1945–1949* (Cambridge, MA: Harvard University Press, 1995), 206.
22   Roberts, *Stalin's Wars*, 288.
23   Judt, *Postwar*, 118–19.
24   Saki Dockrill, 'Britain's Strategy for Europe: must West Germany be Rearmed? 1949–51', in Richard Aldrich (ed.), *British Intelligence, Strategy and the Cold War, 1945–51* (Abingdon: Routledge, 1992), 193.
25   Naimark, *Russians in Germany*, 10; Judt, *Postwar*, 118–22.
26   Graham-Dixon, *Allied Occupation of Germany*, 100.
27   Liberman, *Does Conquest Pay?*, 126–7.

28 Filip Slaveski, *The Soviet Occupation of Germany: Hunger, Mass Violence and the Struggle for Peace* (Cambridge: Cambridge University Press, 2013), 28.
29 J.P. Nettl, *The Eastern Zone and Soviet Policy in Germany, 1945–50* (New York: Octagon, 1977), 200–1.
30 Judt, *Postwar*, 19–21.
31 George F. Kennan, *Memoirs 1925–1950* (London: Hutchinson, 1968), 265.
32 IWM, 99/76/1, Private Papers of Monica Maurice, 21 May 1947.
33 Naimark, *Russians in Germany*, 206–7.
34 TNA, FO 1031/5, Brig. R.J. Maunsell to Mil. Gov., 9 August 1945.
35 TNA, FO 1031/5, Admiral H. Burrough to Brig. R.J. Maunsell, 11 August 1945.
36 Ernest Bevin, quoted in Barker, *Britain in a Divided Europe*, 57–8.
37 TNA, FO 1031/5, Lt-Gen. B. Robertson to Lt-Gen. L. Clay, 17 September 1945.
38 TNA, FO 1031/5, Mil. Gov. Econ. 3 to JIC, 15 October 1945.
39 TNA, FO 1031/5, 'Russian Access to Targets in the British Zone of Germany', 20 August 1945.
40 TNA, CAB 122/343, 'Policy for the Exploitation of German Science and Technology', 1 August 1945.
41 TNA, FO 1031/6, M. Zvegintzov to C.G. Wickham, 23 February 1946.
42 IWM, 99/76/1, Private Papers of Monica Maurice, 26 April 1947.
43 IWM, 99/76/1, Private Papers of Monica Maurice, 22 May 1947.
44 IWM, 09/21/1, Private Papers of Gilbert A. Hunter, January 1946.
45 TNA, FO 1031/50, 'Minutes of 6th BIOS Meeting (1946)', 17 April 1946.
46 TNA, FO 1031/6, 'Mutual Visits to IG Farben Plants', 8 January 1946.
47 TNA, FO 1031/53, 'Quadripartite policy', 11 December 1945. For more on the fate of IG Farben in occupied Germany, see Raymond Stokes, *Divide and Prosper: The Heirs of I.G. Farben under Allied Authority, 1945–1951* (Berkeley, CA: University of California Press, 1988), esp. chapters 2–4.
48 Farquharson, 'Governed or Exploited?', 34–5.
49 TNA, CAB 82/6, 'DCOS Papers', 22 June 1945.
50 TNA, CAB 122/363, JIC London to JIC Washington, 10 May 1945.
51 TNA, WO 219/2165, 'Operation Backfire', 19 June 1945.
52 TNA, CAB 122/363, 'Allocation Policy on Samples of Secret Weapons', 26 May 1945.
53 Simons, *Operation Lusty*, 133.
54 Catherine Haddon, 'Union Jacks and Red Stars on Them: UK Intelligence, the Soviet Nuclear Threat and British Nuclear Weapons Policy, 1945–1970', PhD dissertation, QMUL (2008), 66.
55 William H. McNeill, *The Pursuit of Power: Technology, Armed Force and Society since AD1000* (Oxford: Blackwell, 1983), 367.
56 Groves, *Now It Can Be Told*, 245.
57 David Holloway, *Stalin and the Bomb: The Soviet Union and Atomic Energy, 1939–1956* (New Haven, CT: Yale University Press, 1994), 111.
58 Howard, *Otherwise Occupied*, 156–7.
59 Harold A. Feiveson et al., *Unmaking the Bomb: A Fissile Material Approach to Nuclear Disarmament and Non-Proliferation* (Cambridge, MA: MIT Press, 2014), 174.
60 Engel, 'We are not concerned who the buyer is', 48–9.
61 TNA, FO 1031/59, 'Periodic Intelligence Report No. 2', 6 August 1946.
62 Asif Siddiqi, 'Germans in Russia: Cold War, Technology Transfer, and National Identity', *Osiris*, 24 (2009), 122–3.
63 Gimbel, 'US Policy and German Scientists', 433–4.
64 FRUS, 'Minutes of 11th Plenary Meeting', 31 July 1945, 516–7. [Accessed online 27 May 2015, http://digital.library.wisc.edu/1711.dl/FRUS].
65 'Eisenhower to Truman', 24 September 1945, in *The Papers of Dwight David Eisenhower*, vol. VI (Baltimore: Johns Hopkins University Press, 1978), 367–8.

66 TNA, FO 1031/67, 'Evacuation of German Scientists and Technicians from Russian zone', 14 August 1946.
67 TNA, FO 1031/67, R.J. Maunsell to EPES HQ, 14 August 1945.
68 Naimark, *Russians in Germany*, 209.
69 Speer, *Inside the Third Reich*, 674.
70 Neufeld, *Rocket and the Reich*, 268–9.
71 McGovern, *Crossbow and Overcast*, 198–99.
72 TNA, FO 1031/59, Col. P.M. Wilson to Maj. E.C. Malet-Warden, 31 May 1946.
73 James Duffy, *Target America: Hitler's Plan to Attack the United States* (Westport, CT: Praeger, 2004), 124.
74 Holloway, *Stalin and the Bomb*, 109–10.
75 Richelson, *Spying on the Bomb*, 62.
76 Aldrich, *Hidden Hand*, 222–23.
77 TNA, CAB 122/343, 'British Requirements from Germany', 16 July 1945.
78 TNA, FO 1031/65, 'Soviet Sponsored Research Organisations Currently Active in Berlin', 1 March 1946.
79 TNA, PREM 8/373, Brig. C.F.C. Spedding to M.W. Perrin, 17 January 1946.
80 TNA, FO 1031/65, 'Civil Censorship Submission: H. Reichstein to K. Hoertnagel', 31 August 1946.
81 TNA, FO 1031/59, 'Periodic Intelligence Report No. 2', 6 August 1946.
82 TNA, FO 1031/68, 'Russian Activities regarding German Scientists and Technicians', 21 November 1946.
83 Timothy Vogt, *Denazification in Soviet-Occupied Germany: Brandenburg, 1945–48* (Cambridge, MA: Harvard University Press, 2000); Mary Fulbrook, *Dissonant Lives: Generations and Violence through the German Dictatorships* (Oxford: Oxford University Press, 2011), 286–8.
84 Naimark, *Russians in Germany*, 209.
85 TNA, FO 1031/138, Maj. E. Tilley to Lt-Col. P.M. Wilson, 21 May 1946.
86 Pavel V. Oleynikov, 'German Scientists in the Soviet Atomic Project', *The Non-Proliferation Review*, 7:2 (2000), 20.
87 TNA, FO 1032/1231B, 'Matchbox: General Report', 15 April 1948.
88 Naimark, *Russians in Germany*, 218–9.
89 TNA, CAB 81/133, 'Employment of German Scientists by Russians', 7 April 1946.
90 Krige, *American Hegemony*, 46–47.
91 TNA, CAB 81/133, 'Report by Heinrich Waas', 26 March 1946.
92 TNA, AVIA 54/1403, 'PDSR(D)', 20 September 1946.
93 TNA, FO 1031/25, 'Special Intelligence Report No. 3', 31 January 1947
94 TNA, FO 1031/75, 'EPES: policy', 25 April 1946.
95 Naimark, *Russians in Germany*, 219.
96 TNA, FO 1039/672, 'Reports from STIB', 25 November 1946.
97 Naimark, *Russians in Germany*, 220.
98 Dolores L. Augustine, *Red Prometheus: Engineering and Dictatorship in East Germany, 1945–1990* (Cambridge, MA: MIT Press, 2007), 8.
99 McGovern, *Crossbow and Overcast*, 216ff.
100 Siddiqi, 'Germans in Russia', 127–8.
101 TNA, FO 1031/59, 'Special Intelligence Report No. 2', 6 November 1946.
102 Siddiqi, 'Germans in Russia', 127.
103 TNA, FO 1031/59, 'Special Intelligence Report No. 2', 6 November 1946.
104 Siddiqi, 'Germans in Russia', 127–8.
105 Maddrell, *Spying on Science*, 30.
106 TNA, FO 1031/59, 6 November 1946.
107 Naimark, *Russians in Germany*, 220–1.
108 Siddiqi, 'Germans in Russia', 127.
109 TNA, FO 1031/59, 6 November 1946.

110 Naimark, *Russians in Germany*, 226.
111 TNA, FO 1031/59, 6 November 1946.
112 TNA, FO 1031/68, 'German Scientists and Technicians Escaping from Russian Zone', 9 January 1947.
113 Naimark, *Russians in Germany*, 229.
114 Maddrell, *Spying on Science*, 31–2.
115 TNA, CAB 176/8, 'JIC/1907/45', 15 December 1945.
116 TNA, CAB 81/133, 'JIC(46)51(0)', 24 May 1946.
117 TNA, CAB 81/134, 'JIC(46)79(0)', 21 August 1946.
118 TNA, CAB 131/1, 'Minutes of 24th Meeting of Defence Committee', 11 December 1946.
119 Paul Maddrell, 'Operation Matchbox and the Scientific Containment of the USSR', in Peter Jackson and Jennifer Siegel (eds.), *Intelligence and Statecraft: The Use and Limits of Intelligence in International Society* (Westport: Greenwood, 2005), 174.
120 TNA, AVIA 54/1403, 'Russian Enticement of German Scientists', 18 April 1947.
121 Maddrell, 'Operation Matchbox', 189.
122 TNA, FO 1032/1231A, Lt-Col. W.H.A. Bishop to Control Office, 12 April 1947.
123 TNA, FO 1031/25, 'EPES Special Intelligence Report No. 4', 28 February 1947.
124 TNA, FO 1013/373, 'Operation Matchbox', 1947.
125 TNA, FO 1032/1231B, 'Intelligence Operation: Matchbox', 15 April 1948.
126 TNA, FO 1032/1231B, 'Intelligence Operation: Matchbox', 15 April 1948.
127 TNA, FO 1032/1231A, Lt-Col. W.H.A. Bishop to Control Office, 12 April 1947.
128 TNA, FO 1031/25, 'EPES Special Intelligence Report No. 5', 31 March 1947.
129 TNA, AVIA 54/1403, 'German Section, FO, to Berlin', 2 December 1948.
130 TNA, FO 371/71038, B.K. Blount to I. Worsfold, 5 March 1948.
131 TNA, AVIA 54/1403, J.J.K. Graham to F.H. Hollingdale, 22 January 1948.
132 TNA, AVIA 54/1403, Heinz Peukert to British Liaison Mission, French Army of Occupation, 10 September 1947.
133 TNA, AVIA 54/1403, E.V. Marchant to Ivor Worsfold, 27 July 1948.
134 TNA, FO 1032/1231B, 'Intelligence Operation: Matchbox', 15 April 1948.
135 TNA, LAB 8/1198, 'Darwin Panel Scheme', November 1945.
136 TNA, AVIA 54/1403, 'Employment of Germans in Industry in a Private Capacity', 24 January 1947.
137 TNA, BT 211/62, 'Operation Bottleneck: Policy and Arrangements', November 1947.
138 Dylan, *Defence Intelligence and the Cold War*, 157.
139 TNA, FO 1031/59, Col. P.M. Wilson to Maj. E.C. Malet-Warden, 31 May 1946.
140 TNA, FO 1031/59, Malet-Warden to Wilson, 24 May 1946.
141 TNA, FO 1031/25, 'Special Intelligence Report No. 7', 31 May 1947.
142 TNA, FO 1031/59, 'Incidents on the British Military Train', 6 November 1946.
143 Maddrell, 'Operation Matchbox', 191–3.
144 TNA, FO 1032/1231B, 'Matchbox: general report', 15 April 1948.
145 TNA, FO 1031/59, 'Periodic Intelligence Report No. 2', 6 August 1946.
146 Maddrell, 'Operation Matchbox', 174.
147 Ibid., 191.
148 Neufeld, 'Nazi Aerospace Exodus', 53.
149 Maddrell, 'Operation Matchbox', 203–04.
150 'Transcript of the President's News Conference on Foreign and Domestic Matters', *New York Times*, 10 October 1957, 14.

9   Exploitation and the occupation

The British occupation of Germany was a costly and complex undertaking, which forced the occupying authorities to balance and prioritise a number of different demands, and ultimately reconcile the moral imperatives incumbent on a victor over its vanquished foe with the practical necessities of governing a large area of foreign, war-ravaged and potentially hostile territory. Further complicating matters was the fact that, as we have seen, Germany quickly became the frontline of the Cold War, which added shifting geopolitical considerations to the list of concerns, while Britain's parlous economic situation in 1945 presented them with constraints which mattered far less to the two larger occupiers, the United States and the Soviet Union. As such, the relevant officials in Whitehall and on the ground in Germany were forced to develop a 'British way' in occupation, which was distinct from the approaches of the other Allies and which was explicitly shaped by Britain's unique position at the end of the war – as a former Great Power, now in decline, but still convinced that it had a substantial role to play on the world stage. As Adam Tooze has noted, western Germany was where the European dilemma of coming to terms with the past, encouraging economic growth and satisfying the urgent demands of the Cold War was felt most acutely.[1]

It was within this framework that the exploitation programme had to operate. To some extent, it benefitted from the broader occupation strategy, especially when the focus was on extracting reparations from Germany and building defences against potential Soviet aggression. In other instances, it fell afoul of bigger initiatives, particularly once the reconstruction of a stable and prosperous Germany became the overriding objective of the occupation. On a smaller scale, exploitation also coexisted, and sometimes clashed, with other specific policies enacted by the occupation authorities and, therefore, both shaped and was shaped by the wider discourse. Three of these other policies and their relationships with the exploitation scheme will be discussed here in greater depth – control of science, the process of denazification, and efforts to demilitarise, disarm and partially deindustrialise the British zone of occupation – all of which were, in one way or another, geared towards preventing Germany from ever again posing a threat to world peace. What this ultimately resulted in was a delicate balancing act in which Britain tried to learn as much as possible about Germany's technique from the last war while preventing them from waging the next. Overall, this chapter will show how exploitation fit into British occupation policy as a whole and how it was directly affected by the frequently changing goals which the British pursued in their zone of Germany.

The British zone

The division of Germany into four zones of occupation had been decided in principle at the Yalta Conference in February 1945 and was formally ratified at the Potsdam Conference in July and August of that year. The British zone was in the north-west of the country and comprised three states, or *Länder*, demarcated by Britain after the war – Schleswig-Holstein, Lower Saxony and North Rhine-Westphalia. Crucially, it included the major city of Hamburg (while Bremen and Bremerhaven became US-occupied exclaves, to grant American forces access to the North Sea) and the industrial heartland of the Ruhr. The British had pushed to have control of this region, standing firm against American demands, but it soon turned out to be an economic drain rather than an asset. While it was highly industrialised, it had also been badly bombed, with an already high population swollen by an influx of refugees (over a third of the population of Germany were in the British zone in October 1946), and no natural food supply (the bulk of German agricultural land lay in the Soviet zone).[2] Unsurprisingly, it was said at the time that in the allocation of German territory between the Western powers at the end of the war, the Americans got the scenery, the French got the wine and the British got the ruins.[3]

These factors meant that many of the issues faced by all four occupying powers in Germany were magnified substantially in the British zone. Short-term considerations such as feeding the population and carrying out disarmament were made far more difficult and the British authorities struggled to find an efficient mode of administration.[4] Many officials, including the first Military Governor of the British zone, Field Marshal Bernard Montgomery, thus found themselves falling back on colonial government techniques, such as indirect rule. Indeed, such strategies had been adopted in an earlier British military occupation – Lord Rennell, Chief Civil Affairs Officer of the Allied Military Government of Occupied Territories, was a keen advocate of indirect rule and applied it actively in Italy in 1943.[5] In Germany, many former colonial officials were brought in to aid in the administration of the British zone and they utilised several of their well-worn tactics, including collaboration with local elites, the appointment of district commissioners and the appointment of a British Military Governor, followed by the formation of representative local assemblies. Ultimately, the defining characteristic of British policy, in both occupied Germany and their overseas colonies, was an overriding concern to pursue their interests at the lowest possible cost and a willingness to work with local groups that benefitted from British rule.[6]

The colonial approach raised many issues, not least that it came in for serious criticism from the German citizens who found themselves under British occupation. For example, the leader of the German Social Democrats, in a statement welcoming India's independence in 1947, was recorded as saying that he hoped that yet more ex-colonial administrators would not come to Germany. Special clubs for

British officers (such as the tactlessly named Victory Club in Hamburg) provided five-course meals in luxurious surroundings, while ordinary Germans came close to starving. Concerts and cinema showings were initially segregated so that occupiers and occupied did not mix during leisure time. In the words of John Ramsden, 'here again was the hauteur of the Raj, deployed to equally self-defeating effect'.[7] For the British officials with colonial backgrounds, they found the notion of having to 'sell' their policies to the native population completely alien.[8] As this suggests, the colonial approach to British occupation was not particularly tenable in the long-term. Instead, it was steadily replaced with an enormous civilian bureaucracy, comprising some 26,000 officials, five times as many as in the American zone.[9] This was characterised by a desire to control every element of life in British-occupied Germany, a trait which also drew the ire of ordinary Germans, who complained of the 'meddlesome inefficiency' of the British.[10]

In any case, the British zone was not only shaped by the way in which it was administered, but also by the overall goals which drove British policy. At the very beginning of the occupation period, Montgomery declared that 'we are in Germany today to prevent, if we can, a state of affairs arising in Europe which might produce another war'.[11] While this was the overriding aim, there was considerable disagreement over the best way to achieve it. For example, the 1944 Morgenthau Plan, named for its originator, US Secretary of the Treasury, Henry Morgenthau, Jr., proposed the imposition of severe deindustrialisation and pastoralisation measures on Germany. The British were quick to discard this approach once the war ended – not only did they see it as leading to vastly inflated occupation costs, they also felt that it made the re-emergence of a German military threat more, not less, likely.[12] The rhetoric of 'unconditional surrender' followed by a strict and punitive occupation which had dominated Allied post-war planning during the war years, was quickly set aside in favour of milder policies in 1945, a decision driven in part by economic necessity.[13] Certainly, the expense of sustaining and governing the British zone was considerable; the estimated cost of the British occupation for 1946 alone was £80 million (some £2.8 billion in today's money), no small sum as Britain teetered on the brink of economic insolvency in the immediate post-war years.[14]

Therefore, one of the primary aims of British occupation policy was to restore German self-sufficiency while simultaneously ensuring that Germany remained peaceful and amenable.[15] However, this latter priority faded as the target of British enmity shifted from Germany to the Soviet Union and the idea of building Germany up as a bulwark against Communist expansion gained traction.[16] As such, it can be said that the rehabilitation of Germany was directly accelerated by the deepening East-West tensions and very swiftly replaced the harsher approach which had initially been formulated. In seeking to achieve this reintegration of Germany into the West, the British pursued a carrot-and-stick approach but recognised that the stick could not be employed too readily and the carrots had to be succulent enough to fight off the temptations from the East.[17] In spite of all this, policymakers (and large swathes of the British public more generally) still remained fairly mistrustful of Germany and many believed the population there had a natural predisposition to prefer autocracy and extreme nationalism.[18] A British exhibition about the occupation of Germany declared that many German political figures 'think and act with the inherent autocratic mind of the German' and that 'it will be many years before they can learn fully many of the essentials of democracy.'[19]

Certainly, the British were keen to foster a democratic spirit at the grassroots in Germany, including through positive reconstruction, which it was felt would make the German people more receptive to democratic ideas and less likely to be attracted to alternative political systems, such as National Socialism or communism. In terms of the latter, Britain also acknowledged that a strong and profitable Germany was more resistant to communism, which, in their analysis, thrived on hunger, chaos and poverty.[20] In reality, the British authorities need not have worried – while some socialist ideas did gain traction in post-war West Germany, their association with the Soviets meant they never developed a wide foundation of support. In fact, the proximity and threat of the Russians often actually acted as a stimulus for co-operation between occupiers and occupied in the British zone.[21] For some German citizens, the only virtue seen in the British occupation was that it kept the Russians out.[22] Overall, Cold War divisions played a central role in dictating both British policy in their zone of occupation and the nature of relations between British officials and German citizens, in the immediate post-war period. As has been explored in the previous chapter, it also influenced the British exploitation scheme, especially with respect to the recruitment of expert personnel.

Other elements of British occupation policy also had a bearing on the way in which exploitation was conducted. For example, the remarkably draconian early non-fraternisation measures – in March 1945, Montgomery instructed all British troops to 'keep clear of Germans, man, woman and child' and forbade them from shaking hands with German citizens, playing games with them or attending social events with them.[23] These were later eased, not least because they were degrading and savoured of colonial-style repression, but, in the early period, the restrictions made it very difficult for exploitation agents to conduct sufficiently open and frank exchanges with German scientists and technicians. Also, the food and housing shortages in the British zone severely limited the desirability of British offers of employment made to German specialists, especially in contrast to American or Soviet offers. In fact, in 1946, the British zone lurched dangerously close to a starvation crisis – the average official ration for ordinary German citizens during this period was 1,630 calories, two-thirds of what it had been in 1939 and 1940.[24] This even forced British authorities to introduce bread rationing in Britain, a privation which had been avoided during the war but was now deemed necessary so that grain could be diverted to Germany.[25] These larger issues aside, the British occupation policies which most directly and consistently affected the conduct of exploitation were denazification, demilitarisation and control of science.

Control of science

The pursuit of a large-scale post-war exploitation programme was primarily fuelled by the belief that German science was far superior to that of the Allied countries and that its potential military applications were substantial. However, for these very same reasons, the Allies also deemed it necessary to directly control German science, in the interests of preserving peace in Europe.[26] Once again the spectre of Rapallo, which had allowed German science to remilitarise, with Russian aid, after the First World War, reared its ugly head. British policymakers were determined not to repeat the mistakes of the past. In the House of Lords on 29 May 1945, Baron Robert Vansittart, the renowned Germanophobe, bemoaned how inadequate British responses to German advances in military technology had been during both the First and Second World Wars. He complained, for example, that the only remedy which had been found to the threat of the V-weapons had been to overrun the launch sites – 'the answer of infantry and not of science' – and warned that, as the range of long-distance weapons increased, such a solution would not always be available. From this, and coloured deeply by his personal prejudices, he surmised that 'in dealing with a nation that is periodically homicidal, I think no precaution is excessive'.[27]

Others shared these extreme sentiments – Lady Apsley, Conservative Member of Parliament for Bristol Central, suggested as early as September 1944 that the German people should be left only to 'the study of the higher humanities such as architecture and other peaceful pursuits' and that their scientists should be distributed across the countries of the United Nations to continue their research under supervision in laboratories there.[28] Even civil service assessments often reached similarly damning conclusions – in June, the Economic and Industrial Planning Staff (EIPS) produced a report which commented that Nazi Germany had 'succeeded in focussing every aspect of scientific activity, within the framework of a planned organisation, to waging war' and declared that it was the only nation which 'carried the prostitution of science to this extremity [*sic*]'.[29]

However, many others took a less extreme perspective. In April 1945, in an article in the *Daily Worker*, the eminent British geneticist, J.B.S. Haldane, wrote that he disagreed with strongly castigatory schemes on three grounds. First, because 'a great deal of German research, even in the last 12 years, has been of benefit to the whole of humanity'. Second, Haldane hoped that 'the Germans will ultimately take their place among the civilised peoples' and believed they could not do this 'without intellectual culture, which includes science' – as an example, he cited biological education as necessary to show the 'utter falsity of Hitler's racial theories'. Third, because of the length of time which it takes to put a discovery in fundamental science into practice, meaning that pure research itself posed no particular threat. Ultimately, Haldane felt that the banning of certain types of applied science, the requirement of having a license for any research and periodic inspections of facilities, without forewarning, would suffice to control German science.[30] Joseph Kenworthy, Baron Strabolgi, shared Haldane's view and dismissed more restrictive suggestions as impracticable and 'as Utopian as the Morgenthau plan for confining Germany to agriculture and pastoral pursuits', noting, however, that there was no reason 'why we should not keep an eye on them and control them'.[31]

Despite these more moderate and rehabilitative attitudes, it was the instinct to control and restrict which won out in the early part of the post-war period. Britain established a number of committees and agencies tasked with monitoring science in Germany and enforcing British measures – these included the German Science and Industry Committee (GSIC), the Scientific and Technical Intelligence Branch (STIB) and the Scientific Committee for Germany, as well as the Research Branch of CCG(BE) (Control Commission for Germany (British Element)), which included control of science as one element of its broader remit. Furthermore, in April 1946, the four-power Allied Control Council issued Law No. 25, entitled 'Control of Scientific Research', which forbade any applied or fundamental research 'of a wholly or primarily military nature', as well as any non-warlike research which would require the use of facilities which could also be used for military research.[32] This became the guiding principle for British policy on the control of science in Germany. Topics which were banned included research in the chemical, rubber, steel and synthetic fuel industries, as well as the manufacture of civilian aircraft out of concern that such work could conceal more sinister research on flying bombs, rocketry or dispersal methods for bacterial warfare.[33]

However, there were numerous problems with this approach which rapidly became evident. For a start, there was an argument that the technology which proved decisive in some future war might originate from an ostensibly peaceful field, such as light chemicals or electronics, which the Allies could not realistically justify banning. This served to highlight the ultimate futility of a distinctly restrictive approach and made its numerous other downsides seem far less palatable as a result.[34] One such downside was the question of what would happen to the numerous scientific experts who found themselves out of work as a result of the Allied prohibitions. E.E. Haddon, the Assistant Director of the Technical and Personnel Administration worried that 'scientists and technicians, particularly the first-class brains, are likely to accept unemployment less placidly than the others and may form or join subversive political groups of which, with their intelligence, they will probably become leaders.' He went on to advise that Intelligence Division among others should keep close watch on scientists, especially those who worked in fields which they identified as particularly dangerous: electronics, radar and biological warfare.[35] Research Branch issued a similarly cautionary report in December 1946 which espoused the view that it was possible, and even likely, that if some of the scientists and technicians who had spent the greater part of their working lives on research and development in fields now prohibited under Allied occupation laws, 'find that they can continue their work without detection, they may do so, partly in the hope of attracting the attention of foreign customers, and partly because of their intense interest in their subject'.[36]

As these comments suggest, the British were not only concerned with disgruntled German scientists forming the core of a dangerous German resurgence but also, as explored in the previous chapter, going to work for hostile foreign powers, especially the Soviet Union. Indeed, a concern voiced by many British officials was that strict implementation of Law No. 25 would leave many German specialists out of work, but who would 'find a ready market for their services with the Russians'.[37] This convergence of denial policy with control of science can be most clearly seen in the work of Research Branch, which was charged not only with monitoring any potentially dangerous German scientific research, but also with preventing too many German scientists leaving the British zone (especially for Soviet employment). This led them to advocate a 'conception of control' which was not 'merely the negative one of preventing the Germans from doing undesirable things' but also taking 'positive action to provide conditions in which German research can develop along the right line'.[38] This positive strategy, formulated loosely, meant giving 'as much encouragement as possible to peaceful research, and to all measures which increase the prestige of Western democratic ideals – in particular, interchange of scientific views, and increased facilities for scientific publications'.[39]

In time, this strategy became the order of the day for the British occupation authorities. Reconstruction helped to offset the costs of occupation and to build Germany up as a self-sufficient ally in the struggle against the Soviets and this was soon extended to science. In the words of David Cassidy, the Allies hoped that 'a rehabilitated science would serve as a democratising influence, foster educational reform, and serve as a foundation for long-range economic viability, hence political stability.'[40] German scientific institutions, such as the Kaiser Wilhelm Society, had originally been viewed mistrustfully by the occupiers, but were now to be rehabilitated as hubs of pro-Western, democratic thinking.[41] In fact, despite structural damage and material shortages, all six universities in the British zone had reopened by December 1945, which put the British ahead of the other occupiers in this regard.[42] It was also hoped that if scientists could be brought onside, they could form the vanguard of wider Allied efforts to 'democratise' Germany at large.[43] This formed part of a general re-education initiative, which proved, despite best intentions, to be little more than an optimistic ideal rather than a realistic and enforceable practical tactic.[44]

Another, more promising element of this approach was internationalism. For the twelve years that the Third Reich was in power and especially during the Second World War, German science had been isolated and cut off from the global community. Ironically, it was the circulation of reports by agencies such as BIOS (British Intelligence Objectives Sub-Committee) and FIAT (Field Information Agency, Technical) that initially served to reintroduce German science to international networks, though this later developed into more of a consensual, mutual exchange.[45] Indeed, the Austrian-born chemist, Frederick Paneth, who had fled Nazi Germany for Britain in 1933, wrote in a 1948 article for *Nature* that postwar investigations into German science 'could not fail to result in the wakening of the old spirit of international solidarity so sadly interrupted during the War'.[46] In addition, the Scientific Committee for Germany proposed placing research contracts from British government departments and private industry with German firms, which it was hoped would 'help us to keep German scientific activity out of subversive channels and will promote good morale and a healthy internationalism among German scientists'.[47] However, it was felt that while British industrialists wished to 'benefit by the fruits of past German research', through BIOS and other exploitation channels, it was 'doubted very much whether they wished German research to continue'.[48]

As this suggests, while the policies of exploitation and control of science often came into contact, they were separate initiatives with differing aims and methods. In July 1945, an EIPS memorandum stated: 'a distinction should be drawn between the control of German research pure and simple, and the positive exploitation of the results of German research for the benefit of this country and the United Nations generally.' This separation did not mean that the two programmes had no impact on one another. For example, many German scientists who had remained in Germany and were suffering under the restrictions imposed on their disciplines voiced resentment at what they considered to be the unjustly preferential treatment afforded to those scientists who had been recruited by the Allies. Their disenchantment was exacerbated by the belief that many of the men who had gone to Britain and the US were in fact lesser minds, who had only risen to prominence in the Third Reich on account of their unscrupulous political opportunism.[49] As a result, the recruitment of expert personnel also coincided closely, and often clashed, with the Allied denazification efforts.

Denazification and demilitarisation

All four occupying powers considered the removal of all traces of Nazism from public (and, to a lesser extent, private) life in Germany to be a top priority and employed various tactics to achieve this aim.[50] However, the process of denazification, often characterised as something of a moral crusade, soon faltered over issues of practicality. It became necessary to limit denazification in order to facilitate reconstruction in Germany and to allow the occupiers to build working relationships with the German people.[51] As time went on, it was also shaped by 'changes of view, occurring in high places, regarding the relative danger of Nazism and communism'.[52] The British approached denazification with particular pragmatism and soon gained a reputation for being a 'soft touch' in this respect. For example, the British prosecuted 22,296 people for war crimes, from a zonal population of 22,303,000 (representing about 0.1 per cent), whereas the Americans prosecuted 169,282, from a population of 17,174,000 (about 1 per cent, or 10 times as many).[53] The reasons for this more lenient approach are difficult to ascertain – on one hand, some have argued that Britain simply did not see the complete punishment and re-education of an entire country as a feasible aim, and did not have the resources to even try.[54]

On the other, some have suggested more prosaic considerations, such as the British belief that low-level Nazis would be more amenable to taking orders than clear anti-Nazis (many of whom were communists and socialists), especially in minor but necessary administrative positions, or that too firm a commitment to denazification could hamper vital German economic recovery.[55] Ultimately, most authorities were more concerned that too thorough a policy of denazification would impede the functioning of basic administration and services under their jurisdiction, than they were about ideals of retribution, revenge or justice.[56] Even the moral rectitude of the policy could be called into question – William Boulton, the head of the British Legal Division in Germany (the organisation ultimately responsible for denazification), described it as 'a temporary and evil necessity'[57] – and Winston Churchill also expressed his disapproval of the scheme, opining that 'retributive persecution is of all policies the most pernicious', showing that even Britain's bold and forthright wartime leader was essentially a post-war pragmatist.[58] In any case, denazification was a Sisyphean task, the conduct of which was likely to satisfy nobody.[59]

The denazification of science was arguably one of the more difficult elements of the policy as a whole. Large swathes of the German population believed they would be exempted from harsh retribution, either because of their (rarely convincing) anti-Nazi credentials or because their skills and experience would be essential for post-war administration and reconstruction.[60] Although, on the whole, this belief was quickly revealed to be a delusion once denazification measures took effect, the official history of the British occupation makes it clear that exceptions were made for various classes of 'indispensable' experts.[61] For example, the massive coal shortage in early 1946 led the denazification of mining engineers to be suspended, but not before a mine disaster, when no qualified engineer was on duty, led to the deaths of over 400 miners.[62] This approach also extended to German scientists, who found that details of their conduct during the Third Reich mattered less to the occupiers than proven loyalty to a profession of critical importance to a nation undergoing reconstruction.[63]

Furthermore, the occupying authorities bought into the notion that science was inherently apolitical and that German scientists, by the very merit of their profession, had generally not supported Nazism or had, in some cases, directly opposed it.[64] In September 1946, the Deputy Chiefs of Staff voiced the opinion that denazification of science would hardly be necessary as 'from a political point of view the records of scientists as a class were reasonably good'.[65] The US National Academy of the Sciences took this even further, expressing the belief that the scientific community had withdrawn into their ivory tower during the Third Reich and thus composed an 'island of non-conformity' within the regime.[66] Others preferred to view the situation in more practically beneficial but abstract terms – scientists, especially physicists, were to be seen as little more than tools, and tools could not be Nazified or denazified.[67] This view translated into practical action, or the distinct lack of it. The Kaiser Wilhelm Society, for instance, was left to largely denazify itself, which led to it further promoting the view that all German scientists had either resisted the Nazi regime or were victims of it.[68]

The exploitation initiative also provided ways to limit the severity of denazification. John Gimbel has noted that FIAT was sought out by many German specialists as a source of employment when their political records prevented them from finding work through more conventional channels.[69] In addition, many scientists felt that post-war service for one of the occupying powers would retrospectively justify their contentious wartime activities – after all, how could a researcher's work during the Third Reich be criticised, when the British or Americans wanted them to continue their work in Britain or the United States?[70] From the point of view of the Western Allies, the utilisation of German science was often seen as more important than a thorough process of denazification, and especially so when the Cold War spectre of Soviet recruitment loomed large.[71] In October 1946, Bertie Blount of Research Branch (who was a persistent critic of the failings of exploitation policy) complained acidly, and in no uncertain terms, about how successful Britain's 'denazification policy, as carried out by the clever young men of Intelligence Division, is being in driving ability and intelligence into the ranks of our enemies'.[72] Blount's sentiments were echoed by Herbert Cremer, a chemical engineer and member of the Scientific Committee for Germany, who considered it the 'height of folly' that by Britain's 'literal adherence to the [inter-Allied] denazification agreement, we should be helping to drive German scientists into the hands of the Russians, who themselves treated the same agreement with complete cynicism'.[73]

This was not a wholly accurate appraisal of the situation – as we have seen, this 'literal adherence' was not especially evident in the British zone and, as Mary Fulbrook has noted, retribution for Nazi-era crimes was often meted out far more harshly in the eastern zone[74] – but that the very idea of it was seen as inimical to successful exploitation is the salient point here. Certainly, the British did adjust their denazification policy to minimise its damaging effect on recruitment and they did not even deem it necessary to conceal this approach. In the House of Lords on 12 March 1946, Lord Chancellor William Jowitt responded to criticism by Lord Vansittart of the employment of politically questionable German individuals in Britain, by declaring 'I am willing to risk their being Nazis – and I think they probably are – so long as they are highly skilled technicians who will teach our people something which they did not previously know.'[75] Expediency trumped morality in no uncertain terms.

As with the vast majority of British occupation policies, both control of science and denazification were aimed primarily towards ensuring that Germany could never again wage an offensive war. The third aspect of this endeavour was a policy of widespread demilitarisation and disarmament, of industry as well as of the armed forces. A Gallup poll of January 1947 showed that 43 per cent of British respondents felt Germany would become an aggressor state again, though almost half of those could give no reasoning behind this judgement, while only 23 per cent believed Germany would become a democratic, peace-loving nation instead.[76] This was coloured in no small part by memories of the aftermath of the First World War, when the core of German militarism had been left intact. Furthermore, a shortage of relevant information had hindered effective demilitarisation and made it 'impossible to be sure that all the war material in question was surrendered'.[77] There was to be no repeat of the mistakes of 1918–19 and the first step in this process was to determine exactly what needed to be destroyed or confiscated. The given definition of 'war material' was: 'any material of whatever nature and wherever situated, intended for war on land, at sea, or in the air, or which is or may be or has been at any time in use by, or intended for use by, the armed forces, civil defence, or other formations or organisations.'[78] With the aid of this remarkably broad classification, handling conventional war material was fairly straightforward, certainly when compared to the more troublesome subject of German industry.

In September 1945, the Cabinet approved British policy on industrial disarmament which described it as 'of the greatest value to the United Nations, by lengthening the time between the start and the fruition of Germany's rearmament', though acknowledging that it did not 'in itself furnish security or avert the need for armed force'.[79] Any factories or plants directly associated with weaponry or war material had to be liquidated and there were three ways by which this could be achieved – they could be destroyed, dismantled and taken as reparations, or converted for use in the peacetime economy.[80] Of these, the middle option swiftly emerged as the most favourable for eliminating Germany's war potential. One clear reason for this was that dismantling represented something of a compromise between the conflicting aims of weakening Germany's military power and maintaining its economic viability.[81] So naturally obvious were the links between demilitarisation and reparations that the official British policy statement contained a clear distinction between the two and the assertion that disarmament measures 'should be carried out regardless of their effect on Germany's capacity to make reparation for the damage she has done'.[82] It should also be noted that the restrictions on German industry were partly motivated by a desire to allow Britain to expand its export capacity at Germany's expense, though this was consistently denied by the authorities.[83] There were serious risks to such a severe approach, however, most notably the concern that if industrial dismantling was carried out too thoroughly, it could cause economic crisis and incur greater expenditure for the occupiers.[84]

Once again, it was the nascent Cold War which really led to a reconfiguration of priorities and tactics. Initially, the four occupying powers had approached industrial disarmament on a quadripartite basis – indeed, the Allied Control Council Act of November 1945, which saw the dissolution of IG Farben and the confiscation of all its property, can be seen as a high point of post-war four-power co-operation.[85] However, the cracks soon began to show and British and American policymakers came to suspect that the Soviets were hoping that extensive dismantling in the western zones would provoke chaos and unrest and encourage the population to turn to communism.[86] Indeed, this was not a far-fetched idea – in 1947, the scheduled dismantling of a factory in Kiel which produced spare parts for diesel engines led to open confrontation between the German workers and British officials; only at gunpoint could the workers be induced to participate in the destruction of their own place of work.[87] From an exploitation point of view, there was an anxiety that expert German technicians and engineers would only stay in the British zone if 'congenial work' could be provided for them and this would be particularly problematic if factories and facilities concerned with forbidden topics like aerodynamics and ship design were forcibly dismantled.[88] For all of these reasons, most occupation officials therefore disagreed with punitive dismantling and preferred a humanitarian and practical reconstructive approach instead.[89] This attitude soon became the norm, especially as Germany was gradually welcomed into the broader western alliance, as an ally and not a vassal state.

In many ways, the central narrative of British occupation policy in Germany is that expediency consistently triumphed over morality. As a result of Britain's post-war economic limitations and the looming danger of the Cold War, rigid commitments to grand schemes, often formulated during the war, melted away and were replaced by a pervasive spirit of pragmatism and practicality, driven always by the need to preserve Britain's interests, ideally at the lowest possible cost. As such, denazification, demilitarisation, deindustrialisation and the strict control of science seemed counter-productive – they were motivated by a desire for retribution and to see Germany suffer for its actions, but in practice they served only to increase the costs of occupation, foment unrest among the German people, and threaten a hostile Germany entering into the orbit of the Soviet Union. As such, they were often shelved or wound down ahead of schedule, which has led to a common assertion that these initiatives failed. However, this is a difficult point to sustain – for example, describing denazification as a failure seems unfair when no clear goals were ever outlined against which to judge it.[90] On the whole, the primary stated aim of the British occupation was to guarantee that Germany would no longer pose a threat to world peace and, in this regard, it can be deemed a success, even if certain moral principles were sacrificed in order to achieve it.

The programme of exploitation formed a key part of the British occupation. It was able to range so widely and survive for so long precisely because it served the pragmatic and practical needs of the British occupation. It was a source of profit for Britain, both for the state and for private industry, which thus offset some of the considerable expenditure of occupation; it helped to keep German scientific and technological equipment and expertise out of the hands of the Soviets; and it ensured that, in any future war with Germany, Britain would be guaranteed the technological edge. However, while the dominance of expediency in British occupation policy initially sustained the exploitation scheme, its value diminished as priorities changed, to the point at which it threatened to jeopardise the full reconstruction of Germany, which, by the late 1940s, was judged to be the best strategy to secure British interests in Europe. Therefore, as this chapter has shown, exploitation can only be understood within the broader framework of the British occupation of Germany, and alongside the other, concurrent policies with which it came into contact.

Notes

1 Adam Tooze, 'Reassessing the Moral Economy of Post-War Reconstruction: The Terms of the West German Settlement in 1952', in Mark Mazower, Jessica Reinisch and David Feldman (eds.), *Post-War Reconstruction in Europe: International Perspectives, 1945–49* (Oxford: Oxford University Press, 2011), 47.
2 Barker, *Britain in a Divided Europe*, 56–7; Office of Population Research, 'The Demography of War: Germany', *Population Index*, 14 (1948), 299.
3 Meehan, *Strange Enemy People*, 13.
4 Lee, *Victory in Europe*, 17.

5   Philip Boobbyer, 'Lord Rennell, Chief of AMGOT: A Study of His Approach to Politics and Military Government (c.1940–43)', *War in History*, 25 (2018), 305.
6   Knowles, *Winning the Peace*, 35–6.
7   Ramsden, *Don't Mention the War*, 238–9.
8   Barbara Marshall, 'German Attitudes to British Military Government, 1945–47', *Journal of Contemporary History*, 15 (1980), 675.
9   Ramsden, *Don't Mention the War*, 237.
10 Knowles, *Winning the Peace*, 120.
11 TNA, FO 1039/671, 'CCG Script for "Germany under Control" exhibition in London', 1946.
12 Victor Rothwell, *Britain and the Cold War, 1941–1947* (London: Jonathan Cape, 1982), 59.
13 Barker, *Britain in a Divided Europe*, 16.
14 Anne Deighton, 'Cold War Diplomacy: British Policy Towards Germany's Role in Europe, 1945–49', in Turner (ed.), *Reconstruction in Post-War Germany*, 21. Price adjustments calculated on TNA Currency Converter [accessed online 18 July 2018, http://www.nationalarchives.gov.uk/currency-converter/].
15 Turner, 'British Policy towards German Industry', 67.
16 Aldrich, *Hidden Hand*, 191–2.
17 Lee, *Victory in Europe*, 11, 21.
18 Mawby, 'Revisiting Rapallo', 84–6.
19 TNA, FO 1039/671, 'CCG Script for "Germany under Control" exhibition in London', 1946.
20 Deighton, 'Cold War Diplomacy', 21.
21 Michael Balfour, *West Germany: A Contemporary History* (New York: St Martin's Press, 1983), 138.
22 Knowles, *Winning the Peace*, 121.
23 IWM, Documents.9531, 'Montgomery's Non-Fraternisation Instructions', March 1945.
24 John Farquharson, *The Western Allies and the Politics of Food: Agrarian Management in Post-War Germany* (Oxford: Berg, 1985), 254.
25 Bernstein, *Myth of Decline*, 77.
26 Charlie Hall, 'Pushed into Pragmatism: British Approaches to Science in Post-War Occupied Germany', *The International History Review*, 41 (2019).
27 Hansard, HL Deb, 29 May 1945, vol. cxxxvi, cc. 246–4.
28 Hansard, HC Deb, 9 September 1944, vol. cdiii, cc. 664–5.
29 TNA, CAB 124/544, 'Technical and Scientific Research in Germany After the War', 19 June 1945.
30 'Disarming German Science', *Daily Worker*, 2 April 1945, 2.
31 Hansard, HL Deb, 29 May 1945, vol. cxxxvi, cc. 251–2.
32 'ACC Law No. 25: Control of Scientific Research (29 April 1946)', in B. Ruhm v. Oppen (compiler), *Documents on Germany under Occupation, 1945–54*, (Oxford: Oxford University Press, 1955), 131–4.
33 TNA, CAB 124/544, 'Interim Report to the Secretary of State for Foreign Affairs', 20 April 1945.
34 Rothwell, *Britain and the Cold War*, 58.
35 TNA, AVIA 54/1403, 'Problem of Subversive Warlike Research in Germany', 17 February 1947.
36 TNA, AVIA 54/1403, 'The Problem of Subversive Warlike Scientific and Technical Research in Germany', 3 December 1946.
37 TNA, FO 1062/149, 'STRB policy report', July 1946.
38 TNA, FO 1062/149, 'The Control of Scientific Research in the British Zone of Germany', 23 April 1946.
39 TNA, AVIA 54/1403, 'The Problem of Subversive Warlike Scientific and Technical Research in Germany', 3 December 1946.

40 David Cassidy, 'Controlling German Science I: US and Allied Forces in Germany, 1945–1947', *HistoricalStudies in the Physical and Biological Sciences*, 24 (1994), 200.

41 Richard Beyler, Alexei Kojevnikov, and Jessica Wang, 'Purges in Comparative Perspective: Rules for Exclusion and inclusion in the Scientific Community under Political Pressure', in Carola Sachse and Mark Walker (eds.), *Politics and Science in Wartime: Comparative International Perspectives on the Kaiser Wilhelm Institute* (Chicago: University of Chicago Press, 2005), 44–5; David Cassidy, 'Controlling German Science II: Bizonal Occupation and the Struggle over West German Science Policy', *Historical Studies in the Physical and Biological Sciences*, 26 (1996), 197–239.

42 David Phillips, *Investigating Education in Germany: Historical Studies from a British Perspective* (Abingdon: Routledge, 2016), 58; Alan Beyerchen, 'German Scientists and Research Institutions in Allied Occupation Policy', *History of Education Quarterly*, 22 (1982), 297.

43 Richard H. Beyler and Morris F. Low, 'Science Policy in post-1945 West Germany and Japan: Between Ideology and Economics', in Mark Walker (ed.), *Science and Ideology: A Comparative History* (Abingdon: Routledge, 2005), 99.

44 David Welch, 'Priming the Pump of German Democracy: British Re-Education Policy in Germany after the Second World War', in Ian Turner (ed.), *Reconstruction in Post-War Germany*, 215–38; David Phillips, 'Aspects of Education for Democratic Citizenship in Post-War Germany', *Oxford Review of Education*, 38 (2012), 567–81; Glesni Euros, 'The Post-War British "Re-Education" Policy for German Universities and its Application at the Universities of Göttingen and Cologne (1945–1947)', *Research in Comparative and International Education*, 11 (2016), 247–66.

45 Beyler and Low, 'Science Policy', 100.

46 Frederick Paneth, 'Scientific Research in the British Zone of Germany', *Nature*, 161 (1948), 191–2.

47 TNA, CAB 124/1924, D.A. Johnston to H.L. Verry, November 1946.

48 TNA, CAB 124/1928, 'Minutes of 1st GSIC Meeting', 22 May 1946.

49 Krige, *American Hegemony*, 49.

50 Jill Jones, 'Eradicating Nazism from the British Zone of Germany: Early Policy and Practice', *German History*, 8 (1990), 145–62; Perry Biddiscombe, *The Denazification of Germany: A History, 1945–1950* (Stroud: Tempus, 2007); Ian Turner, 'Denazification in the British Zone', in Turner (ed.), *Reconstruction in Post-War Germany*, 239–67; Richard Bessel, *Nazism and War* (London: Weidenfeld & Nicolson, 2004), 173–82; Bessel, *Germany 1945*, 193–95.

51 Toby Thacker, *The End of the Third Reich: Defeat, Denazification and Nuremberg* (Stroud: Tempus, 2006), 153.

52 TNA, FO 1032/2555, 'Denazification', 6 March 1948.

53 Ramsden, *Don't Mention the War*, 222; OPR, 'Demography of War', 299.

54 Jeffery K. Olick, *In the House of the Hangman: The Agonies of German Defeat, 1943–1949* (Chicago: University of Chicago Press, 2005), 50.

55 Bessel, *Germany 1945*, 193.

56 Jones, 'Eradicating Nazism', 161.

57 Edith Raim, *Nazi Crimes against Jews and German Post-War Justice: The West German Judicial System during Allied Occupation, 1945–1949* (Oldenbourg: De Gruyter, 2015), 99–100.

58 Ramsden, *Don't Mention the War*, 219–25.

59 Raim, *Nazi Crimes against Jews*, 94.

60 Rebecca Boehling, *A Question of Priorities: Democratic Reform and Economic Recovery in Postwar Germany* (Oxford: Berghahn, 1996), 73.

61 F.S.V. Donnison, *Civil Affairs and Military Government North West Europe 1944–46* (London: HMSO, 1961), 361.

62 Ramsden, *Don't Mention the War*, 223.

63 Beyler, Kojevnikov and Wang, 'Purges in Comparative Perspective', 47.
64 Ibid., 44–5.
65 TNA, CAB 82/8, 'Minutes of 27th DCOS Meeting', 11 September 1946.
66 Simpson, *Blowback*, 34.
67 Mark Walker, 'The Nazification and Denazification of Physics' in Judt and Ciesla (eds.), *Technology Transfer*, 57.
68 Kristie Macrakis, *Surviving the Swastika: Scientific Research in Nazi Germany* (Oxford: Oxford University Press, 1993), 195.
69 John Gimbel, 'German Scientists, US Denazification Policy and the 'Paperclip Conspiracy'', *The International History Review*, 12 (1990), 446–7.
70 Walker, *Nazi Science*, 205–06.
71 Hoffmann, 'Germany is No More', 601.
72 TNA, FO 1032/170, Bertie Blount to Brig. Spedding, 19 October 1946.
73 TNA, CAB 124/1928, 'Minutes of 1st SCG Meeting', 7 January 1947.
74 Fulbrook, *Dissonant Lives*, 286–8.
75 Hansard, HL Deb, 12 March 1946, vol. cxl, c. 62.
76 Gallup, *Public Opinion Polls*, 148.
77 TNA, FO 1032/35, 'German War Material', 3 August 1944.
78 TNA, FO 1062/396, 'General Definition and Preliminary List', 24 October 1945.
79 TNA, FO 1032/169, 'UK Policy on Industrial Disarmament of Germany', 2 October 1945.
80 TNA, FO 1062/396, 'Liquidation of German War and Industrial Potential', 2 October 1945.
81 Jörg Fisch, 'Reparations and Intellectual Property' in Judt and Ciesla (eds.), *Technology Transfer*, 18.
82 TNA, FO 1032/169.
83 Alan Kramer, 'British Dismantling Politics, 1945–49: A Reassessment', in Turner (ed.), *Reconstruction in Post-War Germany*, 152–3.
84 Krige, *American Hegemony*, 22.
85 Peter Hayes, *Industry and Ideology: IG Farben in the Nazi Era* (Cambridge: Cambridge University Press, 1987), 378.
86 Douglas I. Bell, 'STEG and the occupation of Germany: Demilitarization as reconstruction, 1945–1953', *War in History*, 25 (2018), 116.
87 Marshall, 'German Attitudes', 667.
88 TNA, FO 1062/149, 'The Control of Scientific Research in the British Zone of Germany', 23 April 1946.
89 John Killick, *The United States and European Reconstruction, 1945–60* (New York: Routledge, 2013), 61.
90 Jones, 'Eradicating Nazism', 161.

10 Exploitation in context

The British programme of exploitation did not take place in a vacuum. As explored in the preceding chapter, it was one policy of many which comprised British occupation strategy in Germany after the Second World War. In addition, it was also a subject which had considerable potential to provoke controversy and debate. One aspect of this was the way in which it factored into lively discussions about reparations. Burdened by the enormous costs of waging the war and maintaining an occupation zone, and by the memories of the disastrous reparations policy pursued after the First World War, the British were keen to both extract recompense from the ruins of Germany and to do so in a way which did not increase the risks of a future European conflict. In some senses, exploitation presented a solution to this dilemma, but it is also clear that exploitation was much more than just a facet of a wider reparations scheme – its implications for intelligence, diplomacy and the arms race far exceed that of any mechanism for financial restitution. That said, exploitation cannot be understood without reference to reparations, nor can it be separated from the complex moral and legal frameworks within which it necessarily operated. British officials, eagerly seeking justification for their deeply acquisitive exploitation programme, often sought answers to essentially moral questions in the letter of the law, assuming that if they could find a legal mandate for exploitation, then any challenge to their right to exploit could be easily deflected or defended against.

Exploitation was not just contentious in terms of its relationship with reparations, the law or morality. Unsurprisingly, it also occupied a unique place within the public sphere, particularly in terms of the way in which it was reported in the press. This served to shape the British public's perception of the scheme and saw both positive and negative elements emphasised, sometimes simultaneously. While exploitation was largely a clandestine programme, it could not remain completely concealed – indeed, to be truly successful, it relied on public awareness – and the need to court, and respond to, popular opinion did play a role in the determination of policy. In examining exploitation within these various contexts – reparations, legality and morality, and the public domain – it is possible to establish a clearer understanding of the nature of this controversial programme and to grasp its numerous and far-reaching ramifications.

Reparations

In March 1944, Lieutenant-General Ronald Weeks, the Deputy Chief of the Imperial General Staff, predicted that German research and development information might be 'the only form of reparation which it will be possible to exact from Germany'.[1] To some extent, this proved a prescient observation – throughout its post-war lifespan, exploitation remained closely entwined with the pursuit of reparations from Germany. As with many other policies, the Allies' attitude towards reparations was shaped considerably by the experience at the end of the First World War. The approach adopted then had proved immensely unsuccessful for a number of reasons, not only failing to make Germany pay adequately for the war, but also generating much bitterness and unrest in Germany, which Hitler and the Nazis were able to turn into support for their programme of nationalist rejuvenation.[2] The main lesson learned was that it was ineffective to demand reparations in direct financial form – instead, payment in kind was to be encouraged.[3] This new approach was followed after the Second World War – Allied reparations from Germany would initially take the form of capital equipment, dismantled and shipped abroad, and would then be followed by annual deliveries of goods from current German production.[4]

The USA actively pressed for patents, secret processes and technical know-how to be included as part of these reparations in kind, as their post-war economy was enormous and at risk of overproduction and huge surpluses, so bringing in ordinary machinery or goods from Germany was not only uninteresting but actually undesirable. Partly as a result of these American pressures and partly due to a desire among all the Allies to legitimise their exploitation programmes, the four-power reparations agreements encompassed the removal of specialist equipment, documents and other material from Germany's laboratories, factories and research facilities, alongside more conventional reparations such as industrial plant or overseas financial holdings. However, the process of quantifying and attributing a specific economic value to the spoils of exploitation proved very difficult and often contentious. This was especially true with regard to the utilisation of expert personnel and their specialist knowledge, both of which were practically unquantifiable. As we have seen, the value of these so-called 'intellectual reparations' was perceived by the Allies to be extremely high but was also essentially impossible to calculate in strict financial terms. Moreover, it was very easy, in theory at least, for Germany to limit the benefit obtained from them by the creditor nations.[5]

As a result of the inter-Allied agreements, the exploitation programme was affected by the quest for reparations in two main ways. On the one hand, reparations provided a very useful panacea, justifying all physical removals from Germany and thus granting some legitimacy to exploitation; on the other, reparations were strictly governed, both domestically and internationally, often tying the hands of acquisitive exploitation agents. To examine the positive, complementary side of the relationship first, it was quickly established that equipment and documentation comprised an 'essential counterpart' to the industrial intelligence gathered by BIOS (British Intelligence Objectives Sub-Committee) and that the value of this intelligence would be 'seriously reduced' without having 'the physical material for purposes of experiment' in Britain.[6] Reparations channels offered a clear way to bring this material to Britain legally and directly. In addition, exploitation teams were permitted to visit sites earmarked for reparations right up until the point when they were handed over to the recipient power and sometimes even up until dismantling actually began, showing clear co-operation.[7] On occasion, incomplete exploitation could even lead to a particular facility being selected for reparations, as was the case with the Thyssen steel plant in Duisburg, which could produce 1,200 tons a month of 'special extra low-loss transformer steel' and which the Research Branch of CCG(BE) (Control Commission for Germany (British Element)) felt had been 'imperfectly exploited by BIOS'. It was suggested that either the plant continued to operate under British supervision or that it should be brought to Britain in its entirety, both of which constituted forms of reparations.[8]

On the other side of the coin, the relationship had great potential to be fractious. In October 1945, it was reported that the Deputy Chiefs of Staff (DCOS) committee was 'anxious to get as much equipment as possible out of Germany now, before items are frozen for reparations'.[9] Once reparations policy came into force, there were three permissible ways in which material could be removed from Germany: it could be taken as reparations, as long as it passed through all the necessary official international channels; it could be paid for in approved currency as a straightforward export; or, if regarded as 'booty', it could be 'removed outside the reparations procedure and without payment'. Naturally, for the exploitation agencies, who were keen to take as much as possible without having to pay for it or debit it against their internationally allocated reparations account, this last option was the most attractive. However, the material in question had to qualify as 'booty' for this route to be legitimate – the given definition of booty was: 'arms, munitions and implements of war, and all research and development facilities (including documents, material and training devices) relative thereto.' What this entailed in real terms was any equipment found within research establishments concerned solely with warlike subjects, as well as certain industrial items required as prototypes or for further examination in Britain.[10] Even with these fairly flexible criteria, exploitation officials constantly searched for new ways to remove equipment without having to go through the restrictive reparations channels.

For example, when the Board of Trade tried to push for scientific equipment, such as microscopes and chemical balances, to be removed as reparations, they came up against opposition from Research Branch, who insisted that there was too great a shortage of such equipment in Germany. The Board of Trade considered it probable that 'the Germans have successfully pulled the wool over the eyes of Research Branch' in this respect, but felt that any protest would be futile and decided that this should be written off as 'one more long drawn-out and losing battle ... on which no more effort need be expended'.[11] Instead, it was hoped that some of this equipment could be removed 'as a result of the limitation or prohibition of certain lines of research in Germany', thus exempting it from reparations restrictions.[12] Elsewhere, the Ministry of Aircraft Production argued that all material taken under the aegis of Operation Surgeon, from establishments such as LFA Volkenröde, should be considered 'booty', irrespective of its actual purpose.[13] Another tactic, which was used when reporting to the Inter-Allied Reparations Agency (established in Brussels in 1946 to handle the allocation of reparations to the claimant nations), was to point out that all information obtained from removed material was made internationally available in the published reports of BIOS and others and, therefore, these removals could 'not be regarded as constituting a unilateral acquisition of German reparation by the United Kingdom'.[14]

However, the overriding factor which influenced the relationship between exploitation and reparations was, as with so many other occupation policies, the international dimension. Reparations proved a fertile ground for disagreement between the Allies and a major subject of discussion at late wartime and early post-war conferences. At the Potsdam Conference in summer 1945, the Allies agreed that each of them could take what they felt was appropriate from their own zone of Germany, a decision which reflected the fact that the Soviet Union were already busily stripping their zone of anything of value anyway. In addition, to preserve harmonious East-West relations for as long as possible, and in acknowledgement of the Soviet Union's disproportionate sacrifice in the war, the Soviets were granted a substantial portion of industrial capital from the western zones, which was deemed 'unnecessary for the German peace economy' – 15 per cent of this was to be given in exchange for food and other raw materials, while an additional 10 per cent was provided free of charge.[15] Despite its good intentions, this arrangement soon gave rise to discontent among the Western Allies, particularly in Britain – both Ernest Bevin and Sholto Douglas, Montgomery's successor as commander of the British zone, criticised the reparations situation as 'a cow fed from the West and milked from the East'.[16]

This disparity was exacerbated by the increasingly obstructive reparations restrictions to which Britain adhered. During a discussion by the DCOS committee on this topic, Sidney Kirkman, Deputy Chief of the Imperial General Staff, voiced his fears 'that by too much red tape we should damage our own interests while the other Allies were helping themselves under the heading "booty"'. The remark was met with widespread agreement among other members of the committee.[17] With the passage of time, this debate diminished in importance as the focus of exploitation shifted overwhelmingly towards recruiting and utilising specialist German personnel, while reparations fell sharply out of favour, thus losing what little merit it had retained as a cover-all justification for equipment and material removals. One reason for this fall from grace was that the reparations scheme was engendering increasingly hostile foreign public opinion, particularly in Germany, and Britain had come to appreciate the importance of keeping the German people on side if they were to form a bulwark against Soviet expansion. Lieutenant-General Brian Robertson, the Deputy Military Governor of the British zone, wrote in February 1947 of his concerns that the discrepancy 'between our own economic requirements and our political objectives in Germany' may lead the German people 'to complain that we are treating them as the cat treats the mouse'.[18] This concern was heightened by the protests of German workers forced to participate in dismantling, as discussed in the previous chapter. Beyond Germany, there were those who expressed the opinion that the real purpose of the British reparations plan was to limit German competition in world markets for the sake of Britain's 'own selfish interests', an accusation which was firmly refuted, but which did contain a kernel of truth.[19]

Perhaps of greater importance was the fact that reparations were not serving their primary purpose of recouping some of Britain's expenses from the Second World War. Britain was spending approximately £80 million a year on its zone in Germany and claiming no more than £29 million in reparations – in the opinion of Hugh Dalton, the Chancellor of the Exchequer, Britain had ended up paying reparations *to* Germany.[20] In most cases, it was clear that the removal of an established plant was a far greater loss to Germany than it was a gain to the recipient nation.[21] This did not fit in with plans to rebuild a strong and prosperous Germany, as seen in the case of the Volkswagen plant in Wolfsburg, which the British chose not to dismantle on account of the potentially injurious effect it could have on any German economic revitalisation, as well as concerns about the disruption its introduction might cause to British automobile manufacturing.[22] Generally speaking, it was swiftly realised that the small benefit derived from dismantling factories for reparations was dwarfed by the benefit of reconstructing German industry, both for national self-sufficiency and general European security in the face of potential Soviet aggression. As John Farquharson has put it, despite pressure from the Treasury and the Board of Trade for extensive financial compensation to be extracted from Germany, those who pushed for greater German economic reconstruction carried the day and it is difficult to find a time when reparations were granted precedence in the British zone.[23]

Legality and morality

One of the great benefits afforded to the exploitation scheme by the push for reparations was that it provided it with some form of legal grounding. Furthermore, discussions of legality were entwined with, yet remained to some extent distinct from, parallel considerations of morality. To examine the legal perspective first – the concept of reparations has roots in international law, based on the principle that the victors and the vanquished enter into a contract by way of a peace treaty and this obliges the defeated nation to pay, in one form or another, for losing the war – a clear example of this is the Treaty of Versailles at the end of the First World War, despite the opprobrium which this agreement attracted both in Germany and elsewhere. At the end of the Second World War, Germany as a nation-state technically ceased to exist, so no peace treaty could be signed and '*de facto* reparations' became the norm, with the occupiers simply taking what they wanted from the territory they controlled. In this connection, intellectual reparations, such as the services of expert personnel, were especially problematic as they did not offer direct redress for actual losses suffered during the war by the victors.[24] In that sense, the term 'reparations', was hard to apply.

Booty, as we have seen, was far more flexible than reparations, but the Allies were still somewhat answerable to one another on how much they took from Germany. In October 1945, the Economic Division of the Control Commission persuaded the Treasury to help circumvent this accountability, by authorising a programme of 'advance deliveries', up to the value of around £1 million 'without consulting the Allies' as either 'the Allies have no claims or … if such claims exist they may safely be ignored'.[25] This whole issue was further complicated by the widespread and illegal looting of Germany in which troops of all nationalities indulged. According to some reports, Soviet soldiers perpetrated this with vengeful criminality, while British soldiers did it with the childlike mischief of a school bully, taking what they wanted but rarely resorting to any serious physical violence.[26] Nonetheless, it became a more complex problem when exploitation forces were involved, as their indiscriminate personal plunder threatened to shatter the already fragile legitimacy of the whole initiative. A directive issued to 30AU in late 1944 informed the men that 'anything belonging to the enemy can be taken provided that it can be utilised for the good of the unit … "Findings are keepings" but searching for something of personal value is looting.'[27]

Another legal issue which presented problems to the exploitation officials concerned German patents, with many believing that the British acquisition of these commercial secrets would have more profound consequences for the German economy than the destruction or dismantling of industrial material.[28] Indeed, many of the larger German firms, such as IG Farben and Siemens, began to demand payment from the British for the information which they had given up. These requests had little weight behind them as Article 12 of Control Council Proclamation No. 2 (issued as additional terms to the German surrender) unequivocally instructed German authorities to hand over 'all research, experiment, development and design' relating to war, 'whether in government or private establishments'.[29] As such, this element of the issue was rather easily handled – all CCG personnel were instructed that 'any such demands for payment should be answered with an immediate and firm refusal'.[30] This reflected an inherent hostility among the British authorities towards any attempts by the German people to restrict or challenge their ability as occupiers to exert their will over their zone of occupation. In March 1947, the Control Commission stated in no uncertain terms their opposition to 'the enactment of any legislation which might seem to cast doubt on our complete freedom to dispose of BIOS information as we wish in Germany or elsewhere'.[31]

A different side to this problem arose when German firms began to protest about their industrial secrets being made freely available to domestic rivals, after exploitation. This prompted considerable consternation among the British occupation authorities. On one hand, there was the opinion that Britain would be abusing its position as an occupying power if German firms were encouraged 'to utilise the secret processes (not necessarily patented, and therefore not actionable at law), divulged to Allied investigators without compensation to the owning firm'. On the other, bans against using this information could hinder Germany's economic recovery or limit the supplies of vital commodities, such as insulin.[32] Either way, the BIOS reports were publicly available from HM Stationery Office, so the information contained within was 'no longer a secret'.[33] If the technical material in question was patented, however, the German owners were in a slightly stronger position. Control Commission policy stated that any other German who used information in a BIOS report which was the subject of patent protection in Germany was 'liable for infringement' and the German patentee would 'be able to take action when the patent system is reactivated'.[34] This does show that the benefits of exploitation were not only derived by foreign occupiers but could also be of use to domestic rivals.[35] Furthermore, the availability of detailed technical reports also allowed other countries to advance within a particular market at Germany's expense, as shown in the rapid post-war development of Japan's dyestuffs industry, facilitated by the adoption of German techniques.[36]

Exploitation was not only challenged in legal terms but also came under criticism on moral grounds. Unsurprisingly, defeating a country as completely as Germany had been by May 1945, to the point of unconditional surrender and total occupation by foreign powers, and then proceeding to comprehensively remove not only large quantities of that country's specialised equipment and documents, but also many of its brightest minds and a wealth of scientific and technical know-how, prompted many questions about the moral conduct of the victorious Allies. The response which came from those in power in Britain, the USA, the Soviet Union and France was that these were fair and reasonable reparations; in this, they were relying on a concept which had a strong yet flexible legal basis and much historical precedent and thus minimised any further criticism. However, the moral questions surrounding reparations in general, and exploitation in particular, are more complex than this defence strategy suggests and often underpinned some of the legal issues already discussed. Understanding how Britain justified its exploitation programme in moral terms reveals a substantial amount about the context within which the scheme operated.

A good starting point is the term 'exploitation' itself. According to the *Oxford English Dictionary*, there are two definitions of the verb 'to exploit' – the first is 'to make use of (a resource etc.); derive benefit from', while the second is 'to utilise or take advantage of for one's own ends'.[37] The connotations traditionally associated with the post-war exploitation of German science and technology tend to reflect an emphasis on the second definition, but for those involved in the planning and execution of the programme, it is the first definition which would have seemed most apt. Clarence Lasby believes that the exploitation officials conceived of the word 'exploitation' in its military sense – 'to gain value from personnel' – a point which is reinforced by the subsequent decision to replace it with the far less provocative alternative, 'utilisation'.[38] Others argue that the notion that Germany was 'exploited' after the war does not hold up when the figures of economic gains made by the exploitation scheme are compared to the cost to Britain of supporting its zone of occupation and fending off disease, starvation and unrest – if all exploitation did was offset some of the costs of sustaining the German population, how immoral can it be judged to have been?[39] Certainly, minimising the burden of occupation on the long-suffering and war-weary British taxpayer was often rolled out as a rationale for the exploitation programme.[40]

The main justification offered up for exploitation by the officials concerned was that it was 'part of the price that the Germans were having to pay for losing the war'.[41] These were the words of Charles Ellis, the scientific advisor to the War Office, spoken to his colleagues on the Deputy Chiefs of Staff committee in November 1945. He felt that this approach would help to improve British public reception to the employment of German experts in Britain. These sentiments were echoed on the ground in Germany; in January 1947, the Control Commission reaffirmed that 'the right of the Allies to use information collected by Allied investigating agencies such as BIOS is one of the consequences of Germany losing the war.'[42] In short, the guiding principle was 'to the victor, the spoils' but the reality was slightly more complex than that. The Allies had the right to exploit Germany not simply because it had lost the war, but because it had lost a war which it had started. This is evidenced by the fact that materials and equipment taken by Germany from countries which it had occupied during the war, such as France, were restored to their original owners wherever possible. Germany's wartime removals were deemed wrong and in need of restitution, while the Allies' post-war removals were not, because Germany in 1945 was not an innocent victim, but the original perpetrator. All of this ties in with the notion that Britain promulgated an image of its contribution to the war as 'just' and 'good' in the latter half of the 1940s.[43]

Another area which forces us to look closely at the moral framework within which exploitation operated is the relationship with war crimes. For example, it can be argued that many of the rocket scientists recruited by the Allies (especially the USA) after the war, bore direct responsibility for the awful conditions of the Mittelwerk underground V-weapon factory, where slave labourers from the Dora concentration camp toiled in abysmal conditions, many being forced to stay beneath the ground for so long that they lost all sense of night and day.[44] In fact, Michael J. Neufeld has noted that more people died producing V-2 rockets than died from being hit by them.[45] Nonetheless, when American war crimes investigators sought to bring those responsible to justice they found that many rocket experts were shielded from prosecution by their newfound employment in the Allied countries.[46] Similarly, in 1944, the French-Romanian aviation engineer Henri Coanda, who had collaborated on certain designs with the Luftwaffe in France during the war, was freed from a Paris jail and likely saved from execution by CIOS (Combined Intelligence Objectives Sub-Committee) operatives who sought to interrogate him.[47] Incidents such as these reflect a wider lenience often shown by the Allies in cases where they felt the guilty party had something to offer them. For example, the senior executives of IG Farben (who were accused, among other things, of profiting from the use of slave labour) were let off with relatively mild sentences as it was hoped that their expertise (and capital) could help support the reconstruction of West Germany as a British and American ally.[48] As Mark Walker has put it, 'the armourers of the National Socialists were now judged by what they could do for their new employers, not for what they had done for Hitler.'[49]

Some German individuals even attempted to take advantage of the Allies' potentially contradictory aims – for instance, Albert Speer, Nazi Minister of

Armaments and War Production, bombarded his post-war interrogators with scientific details and industrial statistics, which he knew were of immediate interest to them, in order to divert them away from more troublesome topics, such as his culpability in cases of slave labour and concentration camp atrocities.[50] In addition, at the Doctors' Trial held in Nuremberg in 1946–47, the defendants' legal teams argued that the British and American use of data from Nazi human experimentation vindicated their argument that it is sometimes necessary to tolerate a lesser evil (the killing of some) to achieve a greater good (the saving of many).[51] This tactic was indicative of a wider moral quandary about whether the Allies could conscientiously use information gathered by the Nazis in the course of their inhumane, unethical, and often unscientific, medical experiments.[52] This dilemma did not prevent the data from being used altogether, however – a summary report prepared for the American military about the Nazi hypothermia experiments has been cited in peer-reviewed medical literature more than two dozen times since the end of the war and British air-sea rescue experts used the Nazi data to modify rescue techniques for those exposed to cold water.[53]

Overall, the story of exploitation rarely sits well within a conventional moral context. The concept of reparations provided some justification, as well as a much-needed legal grounding, for the scheme but also served to restrict it, in a way which the relevant officials often found stifling and ultimately harmful. However, under closer scrutiny, even reparations demands provided a flawed reasoning, not least because their execution was far more damaging to Germany than it was beneficial to the recipient nations. Also, if we take the Allied quest for reparations as a noble and idealistic cause, of which exploitation was merely one part, then its rapid abandonment in favour of German reconstruction in the latter half of the occupation period raises many other uncomfortable questions. In reality, it is perhaps more honest to see the pursuit of exploitation, while technically legal, as having no particular moral mandate. It can perhaps even be described as immoral, especially in the way in which recruitment spared certain individuals from prosecution as war criminals or impeded denazification proceedings. As highlighted in the previous chapter, exploitation was an inherently pragmatic endeavour, driven not by grand ideals but rather by direct strategic needs, not least protection against the perceived threat of the Soviet Union. That said, even without this oft-stated justification, unscrupulous exploitation would most likely have gone ahead anyway, because German technical expertise was deemed indispensable.[54] Therefore, where a language of morality is used to describe exploitation, this usually amounts to little more than a disingenuous attempt to present a more palatable version of this controversial programme, often for public consumption.

The public domain

Throughout this book, the narrative of exploitation has focused primarily on official policy and the work of military officials and civil servants, both in domestic and international terms. The British non-governmental actors have been few and far between, amounting to a limited selection of industrialists, privately employed scientists, frontline soldiers, and some others. These groups either had no influence over policy whatsoever, and were involved only through obedience to instructions from above, or were able to exert influence only because they had been inducted deliberately into the inner workings of the scheme by those in higher authority. What this discourse has therefore largely neglected to account for are any factors which lay beyond the control of the policymakers and their operatives – the clearest example of this is popular opinion, or how the notion of exploitation played out in the public domain in Britain. It is necessary to examine this, particularly through press response to the policy, in order to place exploitation in sufficient context and to understand its broader ramifications. Of course, while the press offers the most accessible window into the popular consciousness, it does not provide a perfect barometer of public opinion and should therefore be treated with caution. Nonetheless, in the immediate post-war years, British newspapers were still in a so-called 'golden age', with a wide readership, and can thus provide us with a decent insight into the way exploitation was perceived outside of Whitehall.[55]

The importance of public opinion was in the minds of exploitation policymakers from an early stage, which arguably presents a marked contrast with many other, more secretive intelligence operations. At a Department of Scientific and Industrial Research meeting in June 1945, 'it was recognised that public opinion might be offended by proposals to employ German scientists', but it was thought that if it was explained that the scientists were 'supernumerary to normal staffs' and were here for British benefit, 'there would be no serious outcry'.[56] When the DCOS committee discussed recruitment in August, the chair, First Sea Lord Charles Kennedy-Purvis, wished to know 'what difficulties, if any, [were] anticipated from local opposition within the establishments to which they were posted'. Similarly, Henry Hulme, the Director of Operational Research at the Admiralty, 'thought it was important to be prepared to answer the argument that these Germans would be taking other people's jobs and other people's houses'. He was reassured that there were plenty of vacant jobs and that German scientists would not be displacing British subjects.[57] Nonetheless, in the report which emerged from this meeting, one of the three potential objections to the recruitment scheme was 'public criticism of the employment, presumably with remuneration, of Germans who so recently directed the main German scientific effort against us'.[58]

Accordingly, the government departments responsible for exploitation followed a course of strategic publicity – a tactic devised primarily to pre-emptively defend the government from any potential criticism. On 19 December 1945, Stafford Cripps, the President of the Board of Trade, delivered a carefully drafted public statement to the House of Commons. Cripps announced that:

> It is the Government's policy to secure from Germany a knowledge of scientific and technical developments that will be of benefit to this country and to make such knowledge available to those who can use it. This step seems desirable since although we were generally ahead, there are certain fields in

which the Germans held a temporary lead. As part of this policy it is proposed to recruit ... a strictly limited number of German scientists and technicians of the highest grade for service in this country.

In order to strengthen his case, Cripps added that 'our American and Russian Allies are pursuing a similar policy'.[59] In choosing to characterise the aggressively acquisitive process of exploitation as an international contest, in which Britain's global standing and 'Great Power' status hung in the balance, Cripps hoped the scheme would be more palatable to Parliament and the general public alike, especially when considered in the light of widespread uncertainty about Britain's position in the new post-war world.[60] Pre-empting other potential criticism, he also reassured his audience that these men would be 'politically unobjectionable' and that 'in no case will a German be brought in to undertake work that could equally well be performed by a British subject'.[61] Despite this public statement, a degree of secrecy persisted; the following March, when Arthur Lewis, Member of Parliament (MP) for West Ham Upton, asked Cripps if he would release a list of the names, qualifications and political histories of the German experts being brought to Britain, Cripps refused on the grounds that it 'would not be practicable ... nor would it be desirable in the public interest'.[62] In this instance, it is clear that secrecy was employed not in the interests of security but to shield the project from moral opprobrium.[63]

Exploitation officials on the ground also took an interest in the role of the press; more specifically, they aimed to restrict their access to all elements of the programme. No press were admitted to the Dustbin internment camp, lest they ended up 'seriously prejudicing interrogations' being conducted there.[64] Similarly, EPES (Enemy Personnel Exploitation Section) officers were warned to be 'particularly careful in their dealings with the Press' and to 'check personally and thoroughly any authority which pressmen claim to have'.[65] The German scientists were warned along these lines, too. All the experts in residence at the BIOS Reception Centre in Hampstead were instructed not to 'give any statement to the Press or any other Person or write any article or grant any interview to any person concerning his service or otherwise give or facilitate any publicity in regard thereto'.[66] There were some examples of engagement with the press, but these were usually guarded and carefully managed interactions. For instance, some members of the press were permitted to visit the German chemical warfare experimental station at Raubkammer while Porton Group No. 1 was conducting its on-site investigations, but the journalists were deliberately given very few details about the nature of the new nerve agents which had been discovered there.[67]

Despite these measures, the British newspapers did indeed report on exploitation, though perhaps not with the alarmism or disapproval which the officials had anticipated and feared. On 29 June 1945, *The Times* printed a lengthy article from their military correspondent in Germany, headlined 'Germany's Secret Weapons', which contained information on seven categories of German military research which had been discovered at the end of the war, including chemical warfare, radio and optical equipment, and jet aircraft. It also detailed the actions of the British operatives responsible for uncovering this research, who had 'entered Germany with machinery organised to prevent the destruction or concealment of research work or plants of special kinds', adding that they had been 'more successful than they had dared to hope'.[68] Then, at the end of August, most of Britain's leading papers reported on the work of CIOS, following a statement made by President Truman on the subject. The *Manchester Guardian* wrote that British and American experts, following closely behind the Allied armies, had unearthed 'German war secrets which had value not only in relation to the war against Japan but also ... as a contributing factor in post-war scientific and industrial development'.[69] The *Daily Worker* even highlighted the potential of these developments to 'improve peacetime living conditions in all countries', specifically mentioning the possibility of a piloted rocket that could cross the Atlantic in 17 minutes and processes to make synthetic butter and alcohol from coal.[70] In all the articles, as with the two mentioned here, the emphasis was on the admirable boldness of the agents involved and the value and promise of the material which they had gathered.

Greater controversy arose when the recruitment of German experts to work in Britain was discussed. As discussed in Chapter Six, this often took on a distinctly local flavour, wherein a nimby mind-set among residents provoked opposition to German scientists working or living in a particular area, often reflected in local newspapers. At a more national level, there was far less criticism and the general spirit of the press reporting was neutral, if not positive. Indeed, in March, when *The Times* reported that 200 German scientists and technicians were being brought over by the Board of Trade to work in 'a purely advisory capacity for a limited period', in order to 'secure for British industry the best industrial intelligence from Germany', the focus was once again on the scheme's value and not on its questionable propriety.[71] In September 1947, the *Daily Express* noted certain success stories which had been brought about by the recruitment of German specialists, such as Otto Koenig's work on increasing sugar beet yields and an unnamed expert working for the Central Electricity Board on the production and installation of generators.[72] In some cases, the uprooted German scientists were even viewed with sympathy – when the first rocketeers arrived in Britain in November 1946, the *Daily Herald* described them as wearing 'threadbare overcoats and shabby hats' and it was noted that 'one of them only has one pair of socks; none of them has much more than he stands up in'. Much of the rest of the article focused on a breakdown of their salaries and tax obligations and their allocation of clothing coupons, all to highlight their poverty. Presuming a certain degree of insight, the correspondent even described how the experts were 'wondering whether it was going to be as good as they expected when they volunteered to work at Britain's experimental rocket plant'.[73]

This sympathetic, and even caring, attitude can also be seen in the sentiments of ordinary British citizens, often phrased in terms of criticism of the British treatment of the German recruits. William Proctor, the Labour MP for Eccles in Manchester, received a letter from a constituent in February 1946 which showed concern that 'the secrecy surrounding the fate of these scientists ... [is] not only fettering scientific progress, but bedevilling international relations'.[74] John Hynd, the Minister for Germany and Austria, instructed MPs to soothe their constituents' concerns on this matter by assuring them that the German detainees were not being mistreated, that they were returned to Germany and released as soon as the interrogation was complete and that 'the question of the detainee's political affiliations does not arise in connection with these interrogations.'[75] Comparisons with other countries also drove public critiques, but on different grounds: in March, the *Daily Express* printed a letter from W. Steed, of north London, which referred to the employment of Wernher von Braun by the United States, then asked, with evident reproach, 'are we in Britain doing nothing with German secret inventions?'[76]

This changing discourse can, in part, be ascribed to the growing fear of the Soviet Union and concerns about their recruitment of German scientists. On 29 October 1945, the *Manchester Guardian*, *Daily Mail*, and *Daily Express* all published a story, citing statements made by Günther Hillmann, acting director of the Kaiser Wilhelm Institute in Berlin, which noted that 'Soviet research on the atom bomb is being actively pursued with the assistance of German scientists and with German equipment and data.' Hillmann predicted that 'the Russians might develop their own atomic bomb within two years' and used this platform as an opportunity to condemn harsh restrictions on science in the western zones as the reason so many German experts were going over to the Soviet Union.[77] In March 1949, Brendan Bracken, the wartime Minister of Information, warned that the Soviet Union was 'building up the greatest submarine fleet in history', a process which was overseen by German scientists and with submarines helmed by German commanders.[78] Interestingly, when *The Times* covered Operation Osoaviakhim, their reporter made four distinctions between the Soviet and the British-American exploitation programmes – one, that the Western Allies favoured interrogation over recruitment; two, that the German experts in Britain and America were all there willingly; three, that the numbers in the West were far fewer than those taken by the Soviets; and four, that these latest deportations were 'new both in method and degree'. However, in an uncharacteristic example of strongly expressed opinion, *The Times* also described 'the whole business of competition between allies for German military secrets' as 'distasteful and disturbing'.[79]

The public domain was not just a potential source of criticism for the exploitation programme, it was also somewhat necessary for its ultimate success. The reports filed by CIOS and BIOS investigators were of no value if they were not adequately circulated to the firms and individuals who could best make use of them. Even with the mass distribution to libraries (outlined in Chapter Four), insufficient awareness of the scheme and its benefits was perceived as a significant issue throughout 1946, at least in Parliament. In February, Thomas Moore, MP for Ayr Burghs, accused the Board of Trade of disadvantaging British businesses by withholding reports from them, while there was wide circulation in the US; the President of the Board, Stafford Cripps, responded by rejecting the very premise of this complaint.[80] Then, in June, the Parliamentary Secretary to the Board of Trade, John Belcher, was challenged on the subject by Frederick Erroll, MP for Altrincham and Sale, and Leslie Solley, MP for Thurrock. Erroll stated that 'the fact remains that a large number of manufacturers are not receiving this information', and Solley asked whether there was 'any reason at all why the Government should not advertise to the people in the various industries that this information is available'. Belcher replied that the Board of Trade was doing exactly that, revealing that arguably the problem was not the ignorance of British business to the information on offer, but rather the ignorance of Members of Parliament to the true breadth of its dissemination.[81]

Nonetheless, new methods were sought to expand publicity and, on 10 December 1946, an exhibition of the work carried out by BIOS opened at the Board of Trade on Millbank, with the aim of allowing 'British industry to make the fullest use of the information now available about Germany's wartime advances in science and heavy industry'. In opening the exhibition, Stafford Cripps appealed to 'smaller firms without their own research departments' to allow the BIOS information 'to help them to introduce the latest manufacturing methods and processes'.[82] The main feature of the exhibition was a series of murals explaining the process by which British investigations into German industry were carried out, accompanied by photographs and relevant statistics. The exhibition also included a desk where visitors could lodge technical queries with the BIOS Information Section and another where they could place orders for specific reports (from a total available list of 1,039) with HMSO.[83] After its initial nine-day run in London, the exhibition moved on to visit 'the most important provincial industrial centres of Britain';[84] during its run, this touring exhibition attracted representatives from 20,000 firms.[85] Prior to this, in early 1946, the Science Museum in London also hosted an exhibition entitled 'German Aeronautical Developments', which included jet and rocket engines, bombsights, and both a V-1 and a V-2 – perhaps unsurprisingly, it was a highly popular showcase and drew many visitors.[86]

Similar exhibitions were mounted in the United States, such as one held by the Society of Automotive Engineers in Detroit, which highlighted German automobile technologies.[87] In Canada, a large press conference was held in December 1945 at the Chateau Laurier hotel in Ottawa and attended by over 100 press representatives. Here, exploitation officials who had travelled to Germany in the summer described what they had found, and examples of technology which had been brought back, from jet engines to a purportedly revolutionary butter-churn, were displayed as examples of how Canada, and Canadian private enterprises in particular, would profit from this scheme.[88] However, after this event, very little else was done to follow up and there were numerous complaints about the Canadian authorities' failure to advertise the scientific spoils of war to the relevant sectors of industry.[89]

In the USA, the short-term utilisation of a few German experts on military topics met only limited opposition, but this grew considerably when it was suggested that some German experts would be employed in the long-term and rose even further when the idea of these German specialists receiving US citizenship was mooted. A Gallup poll taken in December 1946 put the following question: 'It has been suggested that we bring over to America one thousand German scientists who used to work for the Nazis and have them work with our own scientists on scientific problems. Do you think this is a good or bad idea?' The respondents considered it a bad idea at a ratio of 10:7.[90] The press in the USA took an even more inflammatory approach than in Britain and, instead of emphasising the technical and financial benefits to the US of Project Paperclip, the newspapers preferred to scrutinise the nature of the immigrants themselves, describing them as 'the former pets of Hitler' and one as looking 'remarkably like a youthful Hermann Goering'. Rabbi Stephen S. Wise, the head of both the American Jewish Congress and the World Jewish Congress, wrote to the US Secretary of War, Robert Patterson, in April 1947 and described US recruitment of German scientists as 'deplorable', arguing that 'as long as we reward former servants of Hitler while leaving his victims in DP [displaced person] camps, we cannot even pretend that we are making any real effort to achieve the aims we fought for.'[91] As in Britain, however, much of this opposition melted away once the harsh realities of the Cold War and the possibilities of future conflict became increasingly apparent.[92]

Generally speaking, public reaction to the exploitation programme in Britain followed a fairly familiar trajectory. Initially, as early discussions by policymakers had predicted, there was a certain degree of uproar, fuelled in no small part by sensationalist reporting in the press. However, the primary source of the outrage was not that these men were scientists who may have played some part in the war effort against Britain, but rather that they presented added competition for jobs and homes and because they were Germans, who were often viewed with mistrust and hostility in Britain at this time.[93] As time went on, what little opposition there was dissipated swiftly and harmlessly on account of two main factors: firstly because of growing evidence of how much value could be derived from these men and their expertise (a cause advanced by the exploitation officials themselves); and secondly because of the growing threat of the Soviet Union. In this latter respect, the trend in public opinion mirrors very neatly those which can be observed in demilitarisation and denazification, reparations policy and Anglo-German relations as a whole.

Of course, the British exploitation of German science and technology did not only occupy a place in the popular consciousness of Britain, it was also pertinent in Germany. While the Allies' widespread utilisation of German expertise might have seemed of peripheral concern to many ordinary German citizens in an era of extreme deprivation, there were those who opposed it very actively. For instance, in January 1947, Erich Klabunde, an SPD (Social Democratic Party) member of the Hamburg *Bürgerschaft*, made a statement in which he described British exploitation thus:

> An English manufacturer would name his German counterpart and competitor and 'invite' him to England (whether the man comes voluntarily or not is questionable). They then discuss business and the German is gently persuaded to reveal secrets of his firm. When he refuses, he is kept in polite internment until he gets so tired of not being allowed to return to his family that he tells the Englishman what he wants to know regarding the state of German industry. Thus for about $6 a day the English business man gains the

deepest secrets of Germany's economic life. This is surely a priceless type of reparation.[94]

This prompted deeply unhappy responses from British officials – T-Force HQ felt it was 'most undesirable' that British authorities should have to 'justify their legal actions to a German' and deemed Klabunde's statement 'a direct attack on the authority of the Occupying Powers' which 'should be dealt with accordingly'.[95]

Similar remarks to those of Klabunde were made by Johannes Semler, a founder member of the Bavarian Christian Socialist Union party and a senior economic administrator within the Bizone, who was deeply critical of any Allied policies which he perceived to be harmful to his goal of German reconstruction. On 16 January 1948, he told the *Kölnische Rundschau* newspaper that the British had deliberately destroyed the German coal and steel industries, while less than two weeks earlier, in a speech to his party in Erlangen, he described American aid to Germany as 'chicken feed'. This comment eventually led to his dismissal from his post by the Allies, on the grounds that 'the general tone of your criticisms is not objective, but indicates an attitude of aggressive opposition towards the occupying powers.'[96] Generally speaking, the British were wary of any publicity afforded to exploitation in the German press, as when the Düsseldorf-based business newspaper, *Handelsblatt*, reported that the information gathered 'by the British Intelligence Service on the subject of German industrial technology has now reached the propositions of a monumental collection'.[97] Not all German views on exploitation originated from a commercial or economic point of view – in the New Year's Day sermon of Dr Otto Dibelius, the Protestant Bishop of Hamburg, in 1947, he lamented 'what connection with peace is there when German scientists and engineers are working in America, and probably elsewhere, too, on the production of armaments and instruments of destruction?'[98] As this suggests, public reactions to exploitation were mixed in both Britain and Germany – though were more commonly positive in the former and negative in the latter – but ultimately policymakers paid them little heed as the programme was driven by much larger and more urgent considerations.

To conclude, what this chapter has shown is that the British exploitation of German science and technology was a contentious and sometimes controversial programme throughout its lifespan. It formed a central but highly disputed part of broader Allied reparations demands which were, in themselves, extensively diplomatically charged and which contributed directly to the split between East and West which took place in the immediate post-war period. Even beyond the field of reparations, exploitation existed in a distinctly grey area, both legally and morally. It was motivated by multiple desires and objectives – to claim recompense for the cost of fighting the war and occupying Germany, to minimise the risk of a German military resurgence and to guarantee British security in the face of any future threat, be it of German, Soviet or other origin – but the authorities responsible were keen to justify it in less explicitly pragmatic terms. While a legal mandate of sorts was established, albeit one which smacked very much of 'victor's justice', a moral mandate proved far more elusive and often resembled little more than the age-old adage of 'to the victor, the spoils'. This was muddied even further when the prerogative of exploitation appeared to obstruct the course of more high-minded initiatives, such as the prosecution of war crimes. Many of these potent debates played out in the public domain, especially the press, but there was also public support for exploitation and sometimes even critiques that it was not being conducted vigorously or efficiently enough. Ultimately, the context within which exploitation existed was complex, subject to change and influenced by a wide range of factors and, in that sense, it largely reflected the nature of the programme itself.

Notes

1 TNA, FO 942/27, 'Post-Hostilities Equipment Policy', 29 March 1944.
2 Richard Bessel, *Germany after the First World War* (Oxford: Oxford University Press, 1993), 99–100.
3 Cairncross, *Price of War*, 10.
4 John Farquharson, 'Anglo-American Policy on German Reparations from Yalta to Potsdam', *English Historical Review*, 112 (1997), 906.
5 Fisch, 'Reparations and Intellectual Property', 14–7.
6 TNA, FO 1032/166, War Office to BERCOMB, 17 December 1945.
7 TNA, FO 1031/7, 'Advance Notes for Investigators', 2 September 1946.
8 TNA, 1032/167, Maj. D.E. Evans to Econ. Branch, 4 March 1946.
9 TNA, FO 1032/166, 'Report on Visit to London', 19 October 1945.
10 TNA, FO 1032/166, War Office to BERCOMB, 17 December 1945.
11 TNA, BT 211/117, S.A. Dakin to D. Wood, 15 August 1947.
12 TNA, FO 942/425, 'Draft Letter to be sent to industries', 14 November 1946.
13 TNA, FO 942/426, W.J. Deveen to G. Whitham, 31 August 1946.
14 TNA, BT 211/117, 'Reply from the Delegate of the United Kingdom', 9 September 1946.
15 Marc Trachtenberg, *A Constructed Peace: The Making of the European Settlement, 1945–63* (Princeton, NJ: Princeton University Press, 1999), 23–8.
16 Knowles, *Winning the Peace*, 50.
17 TNA, CAB 122/342, 'Minutes of 23rd Meeting of DCOS', 12 December 1945.
18 TNA, FO 943/42, Lieut-Gen. B. Robertson to COGA, 12 February 1947.
19 Hansard, HL Deb, 22 October 1947, vol. clii, c. 154.
20 Judt, *Postwar*, 123.
21 Backer, *Priming the German Economy*, 64.
22 Rieger, *People's Car*, 106–7.
23 Farquharson, 'Governed or Exploited?', 42.
24 Fisch, 'Reparations and Intellectual Property', 12–6.
25 TNA, FO 1062/114, E.A. Seal to E.W. Playfair, 23 October 1945.
26 Seth A. Givens, 'Liberating the Germans: The US Army and Looting in Germany during the Second World War', *War in History*, 21 (2013), 46.
27 TNA, ADM 223/500, December 1944.
28 Volker Koop, *Besetzt. Britische Besatzungspolitik in Deutschland* (Berlin: be.bra, 2007), 135.
29 'Control Council Proclamation No. 2 (20 September 1945)' in Ruhm v. Oppen, *Documents on Germany under Occupation*, 68–79.
30 TNA, FO 1032/1470A, CONFOLK to BERCOMB, 14 January 1947.
31 TNA, FO 1032/1470A, BERCOMB to CONFOLK, 1 March 1947.

32 TNA, FO 1032/1470A, CONFOLK to BERCOMB, 31 December 1946.
33 TNA, FO 1032/1470A, BERCOMB to CONFOLK, 14 January 1947.
34 TNA, FO 1032/1470A, 1 March 1947.
35 Farquharson, 'Governed or Exploited?', 39.
36 Nakajima, 'Allied Forces', 208.
37 Della Thompson (ed.), *Oxford English Dictionary,* 9th ed. (Oxford: Oxford University Press, 1998), 475.
38 Lasby, *Project Paperclip,* 4.
39 Farquharson, 'Governed or Exploited?', 42.
40 Koop, *Besetzt,* 132.
41 TNA, CAB 82/3, 'Minutes of 20th Meeting of DCOS', 21 November 1945.
42 TNA, FO 1032/1470A, BERCOMB to CONFOLK, 14 January 1947.
43 Toby Haggith, 'Great Britain: Remembering a Just War (1945–1950)', in Lothar Kettenacker and Torsten Riotte (eds.), *The Legacies of Two World Wars: European Societies in the Twentieth Century* (Oxford: Berghahn, 2011), 225–56; Welch and Fox, *Justifying War,* 1–20.
44 Nikolaus Wachsmann, *KL: A History of the Nazi Concentration Camps* (London: Abacus, 2016), 443–4.
45 Neufeld, *Rocket and the Reich,* 296.
46 Ibid., 270–1.
47 Koerner, 'Technology Transfer from Germany to Canada', 102.
48 Hayes, *Industry and Ideology,* 378–9.
49 Walker, *Nazi Science,* 204.
50 Sereny, *Albert Speer,* 552.
51 Michael Grodin, 'Historical Origins of the Nuremberg Code', in George Annas and Michael Grodin (eds.), *The Nazi Doctors and the Nuremberg Code: Human Rights in Human Experimentation* (Oxford: Oxford University Press, 1992), 133.
52 Stephen G. Post, 'The Echo of Nuremberg: Nazi Data and Ethics', *Journal of Medical Ethics,* 17 (1991), 42–4.
53 Arthur L. Caplan, 'The Ethics of Evil: The Challenge and the Lessons of Nazi Medical Experiments', in William R. Lafleur, Gernot Böhme, and Susumu Shimazono (eds.), *Dark Medicine: Rationalising Unethical Medical Research* (Bloomington, IN: Indiana University Press, 2007), 67.
54 Neufeld, *Rocket and the Reich,* 270.
55 Mark Connelly, 'The British People, the Press and the Strategic Air Campaign against Germany, 1939–45', *Contemporary British History,* 16 (2002), 40.
56 TNA, FO 1032/176, 'Minutes of Meeting of DSIR', 25 June 1945.
57 TNA, CAB 82/3, 'Minutes of 8th Meeting of DCOS', 1 August 1945.
58 TNA, CAB 122/343, 'Policy for the Exploitation of German Science and Technology', 1 August 1945.
59 Hansard, HC, 19 December 1945, vol. cdxvii, cc. 1504–5W.
60 Reynolds, *Britannia Overruled,* 169.
61 Hansard, HC, 19 December 1945, vol. cdxvii, cc. 1504–5W.
62 Hansard, HC, 18 March 1946, vol. cdxx, c. 295W.
63 Balmer, *Secrecy and Science,* 20.
64 TNA, FO 1031/75, 8 October 1946.
65 TNA, FO 1031/68, 'EPES Standing Instructions', 24 October 1945.
66 TNA, FO 1031/19, 'Sample Contract for German Visitors', 14 March 1947.
67 Schmidt, *Secret Science,* 168–9.
68 'Germany's Secret Weapons', *The Times,* 29 June 1945, 5.
69 'Nazi War Secrets Useful in Peace', *Manchester Guardian,* 27 August 1945, 2.
70 'To America in 17 Minutes by 3,000 Mile Rocket?', *Daily Worker,* 27 August 1945, 4.
71 'Research Workers from Germany', *The Times,* 9 March 1946, 4.
72 Chapman Pincher, 'Germans called in to help us out', *Daily Express,* 4 September 1947.

73 'Rocket Men May Get £600 a Year', *Daily Herald*, 8 November 1946, 5.
74 TNA, PREM 8/373, Dr J.W. Jeffery to Mr W.J. Proctor, MP, 1 February 1946.
75 TNA, AVIA 54/1294, John Hynd, MP to A.M.F. Palmer, MP, 13 February 1946.
76 'Letters to the Editor', *Daily Express*, 8 March 1946, 2.
77 'Germans Working for Russians', *Manchester Guardian*, 29 October 1945, 5.
78 Hansard, HC, 8 March 1949, vol. cdlxii, c. 1118.
79 'Deported German Workers', *The Times*, 29 October 1946, 5.
80 Hansard, HC Deb, 25 February 1946, vol. cdxix, c. 1554.
81 Hansard, HC Deb, 3 June 1946, vol. cdxxiii, c. 1589–90.
82 'German Advances in Science', *The Times*, 10 December 1946, 2.
83 TNA, BT 211/541, 'BIOS Exhibition', 25 November 1946.
84 'Reports on German Industrial and Scientific Progress', *Nature*, 158 (1946), 867–8.
85 Farquharson, 'Governed or Exploited?', 37.
86 Nahum, 'I believe the Americans have not yet taken them all!', 111–2.
87 Koerner, 'Technology Transfer from Germany to Canada', 110.
88 Ibid., 99–100.
89 Ibid., 110.
90 Lasby, *Project Paperclip*, 191–2.
91 Laney, *German Rocketeers in the Heart of Dixie*, 40–1.
92 Lasby, *Project Paperclip*, 213.
93 Weber-Newth and Steinert, *German Migrants in Post-War Britain*, 152.
94 TNA, FO 1031/19, 'German Scientists interrogated in UK', 30 January 1947.
95 TNA, FO 1031/19, 'Disclosure of Information by German Scientists', 6 February 1947.
96 James C. Van Hook, *Rebuilding Germany: The Creation of the Social Market Economy, 1945–1957* (Cambridge: Cambridge University Press, 2004), 155–6.
97 TNA, BT 211/541, 'Information Section: Publicity in German Press', 15 October 1946.
98 'German Bishop on Lack of Peace Spirit', *The Times*, 2 January 1947, 3.

Conclusion

The British exploitation of German science and technology was only made possible by the unique circumstances of the occupation of Germany. A modern, advanced nation, with a sizeable research and development sector, lay at the mercy of other, similarly advanced nations, all of whom sought recompense for the previous conflict and military advantages for any future one. As a result, exploitation was able to grow to a relatively enormous scale during the occupation period, but it also became increasingly untenable once the occupation was brought to an end. As we have seen, by this point, it was considered in Britain's best interests to restore Germany to a position of self-sufficiency and relative strength in Europe. The central objectives of exploitation ran counter to this vision and therefore it had to be set aside. Nonetheless, the outcomes and implications of exploitation had a lifespan which extended far beyond 1949. In part, this is because the scientific and technological spoils of war, whether in material or personnel form, continued to benefit British armouries and industries. In addition, exploitation contributed to the start of a major period of transnational technology transfer, scientific intelligence-gathering and international arms races which ran throughout the latter half of the twentieth century. Furthermore, the cultural legacies of exploitation are still encountered remarkably frequently today, though their origins in this poorly understood post-war programme are not always fully acknowledged. Ultimately, this conclusion offers an opportunity to reflect on the ending of exploitation, its direct and indirect aftermaths, and its broader significance in history.

Epilogue I: The end of exploitation

Throughout the relatively short period during which exploitation was a key policy aim of the British occupation authorities, it was regularly driven into conflict with other concurrent initiatives, many of which were concerned with the rebuilding and rehabilitation of western Germany in the face of changing international and domestic pressures. Both John Gimbel and John Farquharson have characterised this as a dispute between governors and exploiters, though Farquharson convincingly challenges the more adversarial connotations of this when applied to the British zone.[1] Ultimately, it was this dispute, and the somewhat inexorable triumph of the governors, which led to the eventual demise of the exploitation programme. In Britain, the governors were led by the Foreign Office, with support from the British Element of the Control Commission for Germany (CCG(BE)) in Berlin, while the exploiters were able to count on support from the Board of Trade (which was ostensibly representing the interests of large swathes of British industry) and the Treasury, led by the staunchly anti-German Chancellor, Hugh Dalton.[2]

Unsurprisingly, these debates were not confined to the domestic politics of Whitehall and the international dimension played an important role, too. The departments which sought an end to exploitation in both Britain and the US looked to each other to advance their own cause, in the hope that if one country decided to curtail or terminate exploitation missions then the other would have to follow suit, especially as the movement towards an economic merger of the British and American zones gained momentum.[3] As bizonal fusion began in 1947, pressure to reduce the burdensome costs of the occupation mounted and the most promising solution was to facilitate the economic reconstruction of the western portion of Germany.[4] In February, E.G. Lewin of Research Branch wrote to the Economic Sub-Commission of the CCG to inform them that the Control Office in London was 'anxious that we should try to reach agreement with the Americans and French that technical investigations such as BIOS (British Intelligence Objectives Sub-Committee) and FIAT (Field Information Agency, Technical) teams should be wound up simultaneously in all three western zones'. The given date for this conclusion was 31 March 1947.[5]

One month later, the three Western Allies issued a joint proclamation which confirmed that technical investigations had been taking place in Germany since June 1945 (though, as we have seen, they actually began long before that) and acknowledged that 'many Allied Governments have sent in teams of investigators who have profited from facilities offered them by zone authorities'. They also pointed out that the results of the investigations were 'public and available to all'. The main message of this proclamation was about the future, however, and ran thus:

> British and US and French authorities, having regard to the current German economic situation in the western zones and to increasing difficulties of providing accommodation etc., have decided to bring all technical investigations in field under BIOS and FIAT auspices to a close after 15 May 1947. No industrial technical investigators of the above organisations will be permitted to enter British, US and French zones of Germany and all these industrial technical investigations will be terminated by 30 June 1947.[6]

The dissolution of BIOS followed this proclamation fairly swiftly. By the beginning of November, it had been merged with the Technical Intelligence Section and the Board of Trade Documents Unit, all subsumed into the Technical Information and Documents Unit.[7] However, the reality did not always match so closely the lines laid down in the inter-Allied statement, which had been offered mainly for public consumption, in Germany and elsewhere.

T-Force, the logistics arm of the exploitation programme, continued to operate long after June 1947. This continuation was justified in a number of ways.

One main line of argument was that T-Force was simply collecting documents from German firms which had been included on lists of requirements before the deadline for the end of technical investigations. In fact, they published further such lists on 16 October, 22 October and 8 December 1947 and frankly admitted that some of the firms concerned may not know they still had materials to deliver but asserted that this would not be accepted as a satisfactory excuse for non-compliance.[8] T-Force also argued that it had numerous other responsibilities to attend to in Germany – these included the removal of equipment earmarked as reparations or booty by BIOS investigators, the chaperoning of reparations teams throughout the British zone and the facilitation of visits of property owners wishing to inspect their interests in Germany.[9] Additionally, T-Force felt it had a part to play in denial policy, acting as a 'co-ordinating agency ... in Germany for identifying, locating, security clearing, and movement' of selected German scientists and warning that failure to sustain these efforts would entail 'probable loss to the UK of Germans that we can ill afford to spare'.[10] Despite all these excuses, T-Force was disbanded on 1 August 1948 and its remaining responsibilities handed over to Regional Administrative Offices and the Joint Export Import Agency.[11]

As suggested by the mention made of it by T-Force when trying to justify their continued existence, denial policy had the greatest lasting power of any element of the exploitation initiative. In fact, it ended up outlasting the programme from which it had originally emerged – the Deputy Chiefs of Staff (DCOS) scheme was not terminated until July 1949,[12] and Matchbox continued to operate until February 1951, after the establishment of the Federal Republic of Germany and the German Democratic Republic as nation states.[13] The unnaturally long life of denial policy is attributable, almost entirely, to the continuing deterioration in relations between the Soviet Union and the West. The perceived likelihood of war in Europe, prompted by Soviet aggression, increased exponentially during this period. In March 1949, it was considered important to 'guard against the possibility of German scientists having to be left in Germany and therefore assisting an invading power' and so a 'mobilisation plan' was drawn up, which essentially consisted of a continually updated list of 'German scientists who would have a real value to a hostile power', with the idea that as many as possible of these scientists should be evacuated to Western countries 'in the event of an emergency threatening'.[14] This 'Critical List of German Scientists' contained some 30 atomic specialists, as well as approximately 20 experts in other key subjects, such as aerodynamics and biological warfare.[15] Despite the intention to keep this list up-to-date, this did not always translate into reality. Bertie Blount, the Director of Research Branch, highlighted the inclusion of one aeronautical expert who had been 'working on his own farm for the last three years and presumably has become less valuable as an aerodynamicist in the process'. Blount questioned whether men such as this were really 'worthy of special treatment in an emergency'.[16]

Denial policy also eventually drew to a close, in part because of the changing attitudes of German scientists. The benefits of the Anglo-American reconstruction of science in Germany, coupled with fear of the USSR and another war, drove the German experts to actively seek closer ties with the West – the Soviet Union had become a dangerous threat, not a desirable alternative, and therefore there was no longer any need for Britain and the USA to forcibly deny these men to the Soviets.[17] Nonetheless, the aforementioned plans to extricate valuable German experts from Germany in the face of a Soviet invasion remained in place. On the whole, exploitation was terminated at the end of the 1940s because it no longer served Britain's broader interests in Germany and beyond. During the height of the occupation period, when exploitation was seen as a panacea which could provide reparations, punish and disarm the Germans and help defend against the Soviets, it attained a momentum and level of support which meant that it frequently took priority over other, more principled initiatives, such as denazification or reconstruction. It prevailed over these parallel policies because it offered real and immediate benefits, which outweighed any possible costs or risks. However, the issue with a programme being justified entirely on grounds of expediency is that it can quickly fall out of favour when circumstances change and less encouraging cost-benefit calculations are made. This is what happened with exploitation, primarily because British policymakers came to realise that Britain's European and global interests were best served, not by appropriating Germany's scientific and technological developments for themselves, but by rebuilding a strong, pro-Western German nation which would be resistant to both domestic communism and external Soviet aggression. In this way, the British exploitation scheme was very much a product of its time and was essentially unable to survive outside the unique political environment of the late wartime and early post-war period.

Epilogue II: Impacts and legacies

The story of British exploitation does not end with the programme's formal termination. Its impacts and legacies were manifold and far-reaching and pose a number of interesting questions about the scheme itself. Perhaps the biggest among these is: how much did Britain benefit from the scientific and technological spoils of war? Unfortunately, it is almost impossible to give a definitive answer to this question. John Gimbel dedicated over a quarter of his 1990 work on American exploitation, *Science, Technology and Reparations*, to the question of 'evaluating the take', but could not provide a definitive answer. He did, however, quote a US Commerce Department official who described the accumulation of information by the Combined Intelligence Objectives Sub-Committee (CIOS), BIOS and similar agencies as 'the greatest transfer of mass intelligence ever made from one country to another'.[18] Part of the problem with forming estimates is that many of the scientific spoils of war came in the form of expertise or know-how which are, by their very definition, unquantifiable. Another issue is that different contemporary groups and organisations produced different estimates, each somewhat skewed by their various motives and agendas. Assessing the basic facts, such as the number of reports filed or the number of experts recruited, is difficult enough because efforts were spread across various agencies and overall tallies were not always kept. Some, including Michael Howard who served with T-Force after the war, suspect that the files containing official government valuations have either been destroyed or remain closed to the public.[19] Even where this information can be obtained, appraisals of how much benefit these spoils actually rendered to Britain are so open to interpretation as to be essentially futile.

All of this does not mean that attempting to assess the value of exploitation is entirely without merit and the information available does help us to further our understanding of the programme, even if only in the broadest possible strokes. Turning first to the material spoils, one of the most useful resources which emerged from the British exploitation programme was the collection of BIOS, CIOS and FIAT Final Reports, which estimates suggest numbered around 3,000 in total and covered all manner of topics.[20] There were 1,900 BIOS reports on non-military subjects alone.[21] As we have seen, a government publicity drive brought these reports to the attention of business owners throughout Britain who almost certainly implemented some of the techniques described to increase the efficiency or output of their enterprises – calculating the total value of these improvements would be utterly impossible, but we can be confident that they existed and were fairly widespread. Alongside these reports, BIOS and its related organisations were responsible for the removal of equipment and material too. Under reparations arrangements, Britain received just under 380,000 tons of dismantled German capital equipment, of which only around a quarter would have been potentially productive. A much larger amount of material, most of which would have been far more useful, was taken outside of the official reparations channels and thus remains unaccounted for.[22]

Some material within these removals would have been more valuable than others. For example, under the auspices of Operation Surgeon, some 14,000 tons of aeronautical equipment was removed, including highly specialised wind tunnels and equipment for investigating high-velocity flight at stratospheric altitudes.[23] Much of this went to furnish the new Royal Aircraft Establishment at Farnborough and would, therefore, have played a direct role in British aeronautical developments moving forward. Elsewhere, however, this process was less successful, particularly where the recipients were private industry and not the state. This was often because technology transfer cannot operate as a one-way street – firms set to receive capital equipment from abroad would need to make a financial investment, bring in new staff or retrain existing staff and adjust to a new way of doing business. In many cases, they decided that the new machinery was not worth the cost and hassle, thus diminishing the potential benefits of exploitation.[24]

The financial value of these material spoils is therefore even harder to ascertain. In 1952, the Inter-Allied Reparations Agency (IARA) reported that Britain had taken approximately 204 million Reichsmarks (RM; at 1938 value) worth of reparations from Germany, in the form of industrial equipment and shipping. This equated to around £15 million in 1950, or roughly £470 million today.[25] By contrast, the United States took only RM50 million and France RM168 million; the Soviet Union meanwhile were not included in IARA estimates.[26] All of these statistics are deeply flawed, however, partly due to currency conversion issues but also because much of what was taken by the Allied nations, including Britain, was not reported to the IARA. The main category of removals left out of reparations reports was 'booty', which included much that was taken under the auspices of exploitation, especially with regard to military material. 'Booty' removals were not recorded at all before 1 January 1945 (a period when much was taken) so the given total for this – RM1.2 million, or approximately £85,000 (£2.6 million today) – is far from accurate.[27] It is also worth noting that, from the perspective of British industrial concerns, the greatest benefit of exploitation was not what they gained, but rather what the Germans lost. In short, the damage that exploitation and dismantling wrought to their German rivals allowed them to increase their export capacity at the expense of these German firms.

This, in turn, feeds into the other side of the story which needs to be assessed here – that is, the German perspective. While the Allies were quick to assert that Germany had lost no more than her leadership in some industries and techniques, many Germans did not agree.[28] As mentioned briefly in the previous chapter, some German politicians did begin to question and challenge the exploitation scheme in public. One of the most prominent critics was Gustav W. Harmssen, a Bremen Senator with responsibility for economic affairs – in 1948, he produced a report which estimated the total value of the patents, industrial secrets and similar assets removed from Germany by the occupation forces to be about $5 billion.[29] This enormously inflated figure was no doubt politically motivated. Harmssen was vocally opposed to Allied dismantling and, while Bremen fell under US control, he believed the American occupiers were being pressured by the British into pursuing a particularly harsh and punitive policy, in order to minimise future German economic competition.[30] Indeed, many of these complaints were directed more at the Allied programme of industrial dismantling than at scientific and technological exploitation, though the two schemes ultimately fell under the same broad umbrella. In particular, just as dismantling threatened to severely damage German industry's productive capacity, exploitation was accused of being little more than a 'conveyor belt' for commercial espionage.[31] Both tactics sought to give British business a competitive edge over their renascent German rivals.

Major German firms also took action during the valuation stage, which makes reaching a convincing conclusion about the worth of removed equipment even more difficult. In some cases, they sought to make machinery seem much older than it was, partly to make it less attractive and partly to conceal the fact that in many cases it had been purloined from another nation, such as France or Czechoslovakia, during Nazi occupation. The equipment they were trying to cling on to was itself a result of an earlier exploitation scheme, of which German companies were the beneficiaries rather than the targets. In other cases, representatives of German industrial concerns 'talked up' the value of certain items in the hope that they would therefore receive inflated compensation for their losses, though these attempts were only very infrequently successful.[32] One other element to take into account when considering the German perspective on the cost of exploitation and dismantling is the possible positive effect which such action had on the German economy. As Werner Abelshauser has argued, the process was actually a virtual prerequisite for the West German 'economic miracle' of the 1950s because the material removals opened the way for rapid renovation and streamlining of the industrial capital stock.[33] On the whole, while it is certainly true that British industry did encourage exploitation and reparations authorities to pursue courses of action which would profit them at the expense of their German competitors, the reality is that the process was perhaps no more beneficial to Britain's economy than it was to Germany's.

As has been shown, attempting to ascertain the value of the material spoils taken from Germany at the end of the war is extremely difficult. However, it pales in comparison to the challenge of estimating the value of expertise and know-how acquired through personnel exploitation, which remains, for all intents and purposes, essentially impossible. Even the total number of specialists recruited by Britain remains elusive – Glatt's estimate of somewhere in the region of 800–1,000 seems likely, but this figure certainly includes a large quantity of lower-skilled technicians and engineers whose impact would have been minimal.[34] However, as explored in Chapter Six, within this number were also several high-grade experts who exerted a very real influence on areas of British industry, such as brick-making or sugar refinement. In the defence sphere, the areas which ben-efitted most were aeronautics and guided projectile development, many of the German experts in which had been recruited through Operation Surgeon. German expertise contributed to the design of the English Electric Lightning supersonic jet fighter, the Black Knight missile, and, much later, Concorde – all of which remain peaks of post-war British technological production.[35] It really is little won-der that, in August 1948, the Defence Research Policy Committee reported that 'many of our German scientists are settling down and becoming of real value to our long term research programmes'.[36] This assessment was certainly borne out by the facts – Blue Steel, the principal British nuclear missile between 1963 and 1968, was based on earlier German designs, while Concorde did not come into service until 1976. One of the main reasons why the monetary value of German contributions is so incalculable is because there is no way of knowing how long it would have taken, nor how much it would have cost, for Britain to achieve these developments without German involvement.[37]

Another approach is to compare the fruits of British exploitation with those of other nation's efforts. The obvious point of comparison is the American scheme which, as mentioned in the introduction of this book, is often seen as dwarfing the British programme, almost to the point of irrelevance. In fact, numerous sources provide similar figures for both schemes, of somewhere in the region of 1,000 experts, though these feature the same sizeable margins for error.[38] Famous indi-viduals such as Wernher von Braun, and especially his connection with the US space programme and the Moon landings, make it seem as though the USA was the overall winner in the struggle for the spoils. However, the wealth of scien-tific and technological expertise on offer in Germany means that American suc-cess did not necessarily mean British failure. While Britain struggled to compete with the USA in the recruitment of some of the really high-profile names, Britain was still able to recruit many influential individuals, such as Johannes Schmidt (on rocketry), Hellmuth Walter (on submarines and rocketry) and Martin Winter,

Hans Multhopp, Dietrich Küchemann, Johanna Weber, Karl Doetsch and many others on aeronautics.[39] In addition, it is fair to suggest that British recruitment was deliberately sparing and targeted, only bringing in the experts which government and industrial research establishments could successfully absorb without major disruption. In some areas, such as turbojet design, British officials chose not to pursue German expertise because they felt it could offer little to the already advanced British research and development work in that field.[40] It would also be fair to argue that British recruitment of German specialists to work on civil-industrial topics, rather than for the military, was more extensive than that of the USA. In short, the conventional narrative of Britain being left behind by the United States in the employment of expert German personnel does not hold up to genuine scrutiny.

Ultimately, all discussion about the financial value of the British exploitation of German science and technology has to be conducted in light of the fact that Britain had spent approximately £140 million in financial aid to Germany by April 1947, and spent around £43 million on food supplies to the British zone alone in the first eight months of that year.[41] Any gains made through exploitation would have gone only a very small way to offsetting these costs of occupation. The introduction of bread rationing in Britain in July 1946 (a measure which was avoided during the war), so that grain could be diverted to Germany, just goes to show that naked profiteering was hardly the dominant British motive. Furthermore, as German reconstruction became the order of the day, Britain created an environment in their zone of Germany which allowed the German people to rebuild and eventually prosper.[42] So, to revisit the question of 'evaluating the take' of exploitation, and whether this is, in essence, a futile endeavour, it is perhaps more useful to reframe the debate. Indeed, what matters more than the actual, quantifiable worth of scientific and technological exploitation (which is elusive), is the perception of its value held by British officials and policymakers, which remained high throughout the occupation period and which therefore continued to drive the scheme forward, even in the face of various obstacles.

The longer-term impacts of exploitation are not only detectable in terms of its influence on the subsequent research and development of the recipient nations and of Germany. There were also ramifications for scientific intelligence, arms races and transnational technology transfer throughout the Cold War era. Turning first to scientific intelligence, a clear example of exploitation's legacy here grew out of denial policy. When the criteria for which experts warranted inclusion in Matchbox were drawn up, the third category of individual to be included was: 'scientists and technicians who are valuable, not for their professional competence, but because they can give intelligence of value to us about Russian sponsored research and development'.[43] In time, this evolved into a major part of British scientific and technical intelligence-gathering on the USSR. The main endeavour in this regard was Operation Dragon Return, run by the Scientific and Technical Intelligence Branch, which questioned exploited scientists and technicians, alongside defectors, refugees and ex-POWs, who returned to the western half of Germany from the Soviet Union in the late 1940s and 1950s. In total, some 2,500 specialists returned from the Soviet Union before 1958 and, of these, around 25 per cent ended up in West Germany (the others preferring to stay in the more politically amenable East).[44]

The information these individuals could provide allowed the British to fill in several 'black holes' in intelligence coverage of the Soviet weapons programmes.[45] As Paul Maddrell has written, 'the first post-war penetration of Soviet military capability by British intelligence was a by-product of its effort to complete the victory over Germany.'[46] However, one major shortcoming of this approach was that the British and American intelligence services became overly reliant on German experts as sources of information on major Soviet military-scientific projects. In many cases, the paranoid Soviet authorities had not allowed German workers access to the more sensitive projects and those that did get access were rarely allowed to return to Germany, except under excessive supervision. This skewed the intelligence that the British and Americans gathered and led them to believe, for instance, that the Soviets would not be able to successfully detonate an atomic bomb before 1955 at the earliest. When they actually managed to do so in August 1949, it came as a huge shock to the West.[47]

Meanwhile, the principle of denial gradually involved into the broader policy of scientific non-proliferation, which was pursued actively by the United States and its allies, including Britain, against the Soviet Union during the Cold War. When Britain developed the first commercial jet airliner, the de Havilland Comet, in 1952, it was feared that if the Soviets were to lay their hands on one, they could use the Rolls-Royce Avon engine technology to power a long-range bomber force. In order to minimise this risk, the British government insisted that all foreign buyers who acquired Comets abided by a set of rules. As such, no Comet was permitted to fly to, or over, communist-held territory, all maintenance staff working on the Comet were to be screened for security risk and all spare Avon engines outside Britain were to be held either on British overseas territory or in a British-owned building under British supervision.[48] These almost laughably restrictive guidelines were all imposed with fear of the Soviet Union in mind. Elsewhere, exploitation tactics were revived to try and gain scientific intelligence on the USSR – for instance, the American Central Intelligence Agency (CIA) established dedicated committees such as the Joint Soviet Materials Intelligence Committee and the Joint Materiel Intelligence Agency. Their mission was to co-ordinate the collection and analysis of captured equipment and other foreign materials by the armed services and then derive and disseminate useful intelligence from this material.[49] The use of captured technology remained a powerful tool in the arsenal of Cold War scientific intelligence agencies.

Another important legacy of exploitation can be seen in the continuing transnational transfer of German technology and expertise, especially in the key fields of aeronautics and rocketry. As explored in Chapter Seven, German specialists worked in countries as diverse as Spain, Egypt, Brazil and India in the decades following the end of the Second World War. These nations often chose to bring in German experts to develop domestic aircraft industries or missile armouries, rather than simply buying from major powers such as the USA, Britain or the

Soviet Union, because they wished to maintain a 'non-aligned' status in the Cold War. Curiously, these native industries very rarely produced anything of genuine quality, even with German expert input, and the governments involved often did turn to the United States or the USSR for assistance in due course.[50] Nonetheless, in the words of Michael J. Neufeld, this ongoing movement of German scientists and technicians became part of 'a complex pattern of superpower relations and postcolonial nation-building'.[51]

An interesting postscript to this phenomenon can be seen in Egypt's recruitment of German rocket scientists to create a domestic missile programme. In Israel, the prospect of a hostile Arab state arming itself with long-range weapons, coupled with the involvement of former Nazis, gave rise to enormous fear and anger.[52] Israel responded by deploying Mossad agents on an assassination campaign against the German scientists (codenamed Operation Damocles) which resulted in, among other things, the death of five Egyptian workers at a rocket factory in Heliopolis and injuries to one German scientist's secretary when she unwittingly opened a letter bomb.[53] This hugely violent reaction to Egypt's rocket programme was made all the more severe because the scientists involved had links to the Third Reich, which was abhorrent to a nation whose very identity was tightly entangled with the Holocaust. Echoes of this were seen again when the West German rocketry company OTRAG test-fired some rockets in Zaire in the late 1970s – communist regimes and other African nations accused OTRAG of 'building military bombardment vehicles for either West German neo-Nazis or for white South Africans'.[54] These incidents refute the notion that post-war employment overseas absolved former Nazi scientists of any association with the crimes of the Third Reich. Moreover, the debates surrounding personnel exploitation, especially in terms of balancing strategic pragmatism with moral scruples, did not end in 1949. The legacy of this scheme was a long and often poisonous one.

A similar effect can be seen in exploitation's cultural legacy, which takes many forms. For instance, V-2 rockets can be found in museums across the world, including the Science Museum in London, the Australian War Memorial in Canberra, and the National Air and Space Museum in Washington, DC. Meanwhile in Germany, the Deutsche Museum in Munich exhibits one test section from a supersonic wind tunnel that spent years in Britain after the war.[55] These objects rarely reveal much about the movements which they underwent in the post-war period or the politics involved therein. The information screen which accompanies the V-2 on display at London's Imperial War Museum does, however, hint at it – a paragraph about Wernher von Braun ends with this line: 'His legacy, and those of the rockets he designed, is deeply ambiguous.'[56] Indeed, von Braun presents a very interesting case when considering legacies of exploitation. He has become the subject of a form of hero worship in the United States, especially around Huntsville, Alabama, where he and his team were based from 1950 onwards and where a local civic centre and various other landmarks now bear his name.[57] Nonetheless, his past was not entirely forgotten. In 1965, the musician and satirist Tom Lehrer penned a song about von Braun, in which he was described as 'a man whose allegiance / is ruled by expedience' and mockingly quoted as saying 'once the rockets are up / who cares where they come down?!'[58] In addition, when a 1960 biopic of von Braun was released in 1960, entitled *I Aim at the Stars*, comedian Mort Sahl suggested it should be subtitled 'but sometimes I hit London'.[59] While these comments suggest that the Nazi past of Wernher von Braun, and others like him, had not been completely erased, the fact that it became the subject of humour rather than of public outcry, shows how swiftly such individuals had been rehabilitated, at least in the countries which took the lead in post-war recruitment. In this way, the outcomes of exploitation have become a somewhat unremarkable part of our cultural fabric, while the process behind them remains largely obscured.

Concluding remarks

To conclude, I would like to draw attention to five key points which I believe are central to understanding the story of British exploitation of German science and technology after the Second World War. The first two are in the form of challenges to, and attempts to debunk, certain myths which continue to surround the scheme almost seventy years after its conclusion. To begin with, the erroneous notion that exploitation was some clandestine government conspiracy or cover-up, either hidden from the nation's most senior political decision-makers or from the general public, or both. While this angle might help to sell books and newspapers, the reality is that exploitation was known about and directed from the highest levels of power in Whitehall and Westminster. Key ministries were involved in its planning and execution, including the Ministry of Supply, the Foreign Office and the Board of Trade. Personnel recruitment, in the military sphere at least, was overseen by the Deputy Chiefs of Staff committee, which was, in turn, answerable to the War Cabinet. There are clear examples of Prime Minister Clement Attlee participating in discussions on exploitation and indeed giving his assent to certain developments therein. In terms of public awareness, while this may have been carefully managed by the responsible authorities, exploitation of both material and expert personnel was widely reported in the press and many such reports were based on official statements made by those very authorities or by senior political figures in the House of Commons. While there may have been some public criticism of certain aspects of the scheme, the general reaction was one of acceptance or even disinterest, especially outside of localities where German scientists were actually stationed. In short, exploitation was a conventional government policy in terms of management and oversight and, while some more sensitive elements were handled by intelligence agencies behind closed doors, it would not be fair to characterise it as a conspiracy or cover-up, or even as especially secretive.

Secondly, as has been explored both in Chapter Seven and elsewhere in this conclusion, but which bears repeating, the British scheme was not considerably smaller nor considerably less successful than its American, or indeed Soviet, counterparts. To be sure, the resources available to Britain in all its post-war endeavours were somewhat less than those of their two larger wartime allies, but that is not to say that they did not pursue a similarly active and acquisitive exploitation policy. The surfeit of equipment, material and documents on offer amidst the ruins of the Third Reich meant that they did not struggle to acquire the material spoils which they sought and BIOS Final Reports were considered generally superior to foreign equivalents and enjoyed a fairly global readership. In terms of recruitment, while the British may not have secured the internationally renowned figures which the USA was able to entice, nor did they take a quantity of workers to match that deported in the Soviet programme, they were still able to bring in a decent number of first-rate experts who had a genuinely notable effect on various fields of research and development in Britain, both in defence and civil industry, from the 1940s onwards. Certainly, some German specialists declined British employment offers as they saw Britain as a country of the past in a world which looked sure to be dominated by the two superpowers but, broadly speaking, Britain was able to recruit exactly the right number of scientists and technicians to render genuine benefits without disrupting existing structures and workforces. In this sense, British exploitation should not be dismissed as small and uninteresting when considered alongside the American and Soviet programmes, but rather should be held up against them as an equal and as a point of comparison.

The third point which I think goes some way to encapsulating the nature of British exploitation involves the motives behind it, the way in which it related to other policies and initiatives being pursued contemporaneously and the trajectory which it followed. Ultimately, exploitation was a pragmatic endeavour. Any discussion of it being inspired by a spirit of restitution, of Germany the perpetrator paying recompense to Britain the victim, can only really be seen as an act of retrospective justification, aimed at giving a controversial programme an unassailable moral grounding to defend it from criticism, both domestic and international. In reality, exploitation was motivated by prosaic and entirely practical considerations and it was so popular with policymakers because it served many purposes at once. It helped minimise the risk of a German military resurgence, it ensured that Britain could not be attacked by any weapon which it did not possess itself, it limited Soviet offensive capacity and it offered some economic benefit, both for the British state and for private industry. This granted a certain degree of power to the organisations responsible for exploitation and ensured that they often prevailed over those running parallel initiatives or policies – exploitation always came first. However, as the Soviet threat grew and British strategy hardened, with the Soviet Union as an enemy and Germany as an ally, exploitation fell from favour as quickly as it had risen. It certainly helped that by this time, at the end of the 1940s, most of the really valuable scientific spoils of war had already been gathered, in terms of both material and personnel, so there was less to lose by terminating the programme. In short, exploitation was very much a product of its time, arising from a very specific set of circumstances and then fading away again once those circumstances changed.

This, in turn, informs my fourth point, about the way in which exploitation fits into the relevant chronology. When exploitation was first conceived, and when its earliest precursors first started operating, the Second World War was still in full flow. The earliest advocates of an exploitation scheme suggested that it would help guarantee the victory over Germany and Japan and only after that might it be useful in securing Britain's broader interests. In addition, it was only because the war ended as it did, with Germany's unconditional surrender followed by a period of total occupation, that exploitation could be enacted in a meaningful way at all. It was also the more advanced weapons developments of the Second World War – the rocket, the jet engine, even the atomic bomb – that made investigating German armouries an enticing prospect in the first place. However, despite all of that, exploitation was also a distinctly Cold War phenomenon. Even before the war in Europe was over, let alone in the Pacific, it was being seen in terms of a struggle for the best spoils between East and West. Exploitation was the first chapter of the Cold War arms race and set the scene for much of what followed. It was an integral part of early Cold War strategy and should be seen as such – denial policy was, in essence, scientific containment, to be seen alongside the political containment of the Truman Doctrine and the economic containment of the Marshall Plan. Moreover, exploitation belongs completely in the critical transitional period which Britain, and much of the world, underwent between 1943 and 1949. It was a vehicle which saw Britain move from fighting the last war to fighting the next. It was shaped by the dynamic of changing enemies in this period, but it also served to advance that dynamic – the contest over German science and technology was one of many factors which ensured the East-West polarisation of the Cold War.

My final point is twofold and, to some degree, self-contradictory. On the one hand, the post-war British exploitation of German science and technology is a unique and standalone phenomenon and should be studied as such, at least in part. As this book has shown, it had a distinct and self-contained narrative – a story with a beginning, a middle and an end – which is interesting in its own right. This wholesale removal of a defeated nation's scientific and technological resources, including recruitment of expert personnel, by its military conquerors, has no parallel in modern history and deserves greater scrutiny. There is, without doubt, more to be said about the exploitation schemes of all four occupying powers, as well as about the involvement of other nations, including those whose participation in the Second World War was peripheral at best. On the other hand, what makes the study of exploitation so fascinating and so important is how it relates to broader themes. It would be excellent to see exploitation feature more prominently in histories of the military-industrial complex, of scientific intelligence, of post-1945 arms races, of transnational transfers of technology and expertise and of the Second World War and Cold War more generally. This would hopefully lead to a stronger presence for the programme in the public consciousness and would ensure that never again could it be described as a 'forgotten history'. Indeed, the British exploitation of German science and technology after the Second World War, with all its implications for geopolitics, military strategy and the relationship between science and power, deserves to be remembered.

Notes

1 Gimbel, *Science, Technology, and Reparations*, 133; Farquharson, 'Governed or Exploited?', 24.
2 Farquharson, 'Governed or Exploited?', 40.

3 Gimbel, *Science, Technology, and Reparations*, 130.
4 Robert W. Carden, 'Before Bizonia: Britain's Economic Dilemma in Germany, 1945–46', *Journal of Contemporary History*, 14 (1979), 538.
5 TNA, FO 1031/7, E.G. Lewin to ECOSC, 28 February 1947.
6 TNA, FO 1031/7, 'Joint British-US-French Statement', 27 March 1947.
7 TNA, FO 1031/9, 'BIOS Papers: General', 22 October 1947.
8 Gimbel, *Science, Technology, and Reparations*, 132.
9 TNA, FO 1031/4, 'The Future of T-Force/Organisation', 2 April 1947.
10 TNA, FO 1065/12, Derek Wood to T. Mackay, 28 July 1947.
11 TNA, FO 1031/4, 'T-Force Disbandment', 2 June 1948.
12 TNA, AVIA 54/1295, 'DCOS Scheme', July 1949.
13 Maddrell, *Spying on Science*, 34.
14 TNA, AVIA 54/1295, E.E. Haddon to DRP, 4 March 1949.
15 TNA, AVIA 54/1403, I. Worsfold to B. Lockspeiser, 11 April 1949.
16 TNA, AVIA 54/1403, B.K. Blount to B. Lockspeiser, 5 April 1949.
17 Krige, *American Hegemony*, 53–4.
18 Gimbel, *Science, Technology and Reparations*, 147.
19 Howard, *Otherwise Occupied*, 334.
20 Simons, *Operation Lusty*, 201.
21 Churchill Archives Centre, Metals Society, METL.1/METL.2, CIOS and BIOS reports.
22 Glatt, 'Reparations', 984–5.
23 Uttley, 'Operation Surgeon', 10.
24 Koerner, 'Technology Transfer from Germany to Canada', 108–10.
25 Farquharson, 'Governed or Exploited?', 41. Price adjustments calculated on TNA Currency Converter [accessed online 5 September 2018, http://www.nationalarchives. gov.uk/currency-converter/].
26 TNA, T 294/15, 'IARA Assembly report 1951', 22 January 1951.
27 Farquharson, 'Governed or Exploited?', 40–41. Price adjustments calculated on TNA Currency Converter [accessed online 5 September 2018, http://www.nationalarchives. gov.uk/currency-converter/].
28 Gimbel, *Science, Technology and Reparations*, 102.
29 TNA, FO 1036/13, 'Reparations: the Harmssen report', 1948.
30 Henry Burke Wend, *Recovery and Restoration: U.S. Foreign Policy and the Politics of Reconstruction of West Germany's Shipbuilding Industry, 1945–1955* (Westport, CT: Greenwood, 2001), 100.
31 Gimbel, *Science, Technology and Reparations*, 108.
32 Glatt, 'Reparations', 978–80.
33 Werner Abelshauser, 'Immaterial Reparations and the Reintegration of West Germany into the World Market', in Judt and Ciesla (eds.), *Technology Transfer*, 107.
34 Glatt, 'Reparations', 937.
35 Uttley, 'Operation Surgeon', 11–3.
36 TNA, AVIA 54/1826, W.F. Barnett, MoS to W.S. Polley, HM Treasury, 24 August 1948.
37 Nahum, 'I believe the Americans have not yet taken them all!', 124–6.
38 Neufeld, 'Nazi Aerospace Exodus', 53; Ciesla and Trischler, 'Legitimation through Use', 174.
39 Uttley, 'Operation Surgeon', 12.
40 Nahum, 'I believe the Americans have not yet taken them all!', 124.
41 Farquharson, 'Governed or Exploited?', 42.
42 Knowles, *Winning the Peace*, 146–7.
43 TNA, FO 1032/1231A, Lt-Col. W.H.A. Bishop to COGA, 12 April 1947.
44 Andre Steiner, 'The Return of German "Specialists" from the Soviet Union to the German Democratic Republic: Integration and Impact', in Judt and Ciesla (eds.), *Technology Transfer*, 126.

45 Goodman, *Official History of the JIC*, 279-80; Dylan, *Defence Intelligence and the Cold War*, 112.
46 Maddrell, *Spying on Science*, 17.
47 Aldrich, *Hidden Hand*, 224–6.
48 Engel, '"We are not concerned who the buyer is", 43–7.
49 David, 'Scavenging for Intelligence'.
50 Edgerton, *Shock of the Old*, 123–4.
51 Neufeld, 'Nazi Aerospace Exodus', 49.
52 Sirrs, *Nasser in the Missile Age*, 55–67.
53 Robert Howard, *Operation Damocles: Israel's Secret War against Hitler's Scientists, 1951–67* (London: Thistle, 2013).
54 Pirard, 'German Rockets in Africa', 891.
55 Neufeld, 'Nazi Aerospace Exodus', 60.
56 V-2 display, Imperial War Museum, London, 2018.
57 Laney, *German Rocketeers*, 71–2.
58 Tom Lehrer, 'Wernher von Braun', 1965 [accessed online 6 September 2018, https://www.youtube.com/watch?v=QEJ9HrZq7Ro].
59 Lance Morror, 'The Moon and the Clones', *Time*, 30 August 1998.

Glossary

All abbreviations, acronyms and codenames are as they appear in the original source material.

30 Assault Unit *Also known as: 30 Commando, 30 Advanced Unit, 30AU* – Admiralty-sponsored intelligence commando unit. Brainchild of Ian Fleming, assistant to the Director of Naval Intelligence. Active in North Africa, the Mediterranean, Operation Overlord and the invasion of Germany. Precursor to exploitation by way of technique and objectives. Had a reputation as piratical and careless – ironically nicknamed '30 Indecent Assault Unit'.

Abwehrkommando Advance intelligence commando unit of the German *Abwehr* (military intelligence). Used often during the early stages of the war, a component of *Blitzkrieg* tactics. Served as an inspiration for 30 Assault Unit and other exploitation operations.

Alsos *from the classical Greek word for 'sacred grove', a play on the name of its initiator, Leslie Groves* – Anglo-American (but US-led) scientific intelligence mission, with a particular focus on nuclear physics and the German atomic bomb project. Brainchild of Brigadier-General Leslie Groves, head of the Manhattan Project, and led by Professor Samuel Goudsmit and Colonel Boris T. Pash. Active in Italy, France and Germany; disbanded in late 1945. One of the first iterations of the exploitation programme and a main inspiration for later, expanded efforts.

British Intelligence Objectives Sub-Committee (BIOS) Whitehall committee responsible for co-ordinating the British scientific and exploitation efforts after the war. Emerged from the disbanded Anglo-American CIOS (see CIOS) in July 1945. Comprised of representatives from the Admiralty, the War Office, the Air Ministry, the Foreign Office, the Ministry of Supply, the Ministry of Aircraft Production, the Board of Trade, the Ministry of Fuel and Power, the Department for Scientific and Industrial Research and the Government of the Dominion of Canada. Chaired initially by Professor R.P. Linstead. Did not have its own pool of investigators but was tasked with developing lists of targets, making arrangements for investigators to visit the sites in Germany and for collating and making available their final reports.

Combined Intelligence Objectives Sub-Committee (CIOS)   *Briefly initially known as the Combined Intelligence Priorities Committee (CIPC)* – Anglo-American committee responsible for co-ordinating the British and US scientific and exploitation efforts during the latter part of the war. Comprised of representatives from seven British and seven American departments: (British) Foreign Office, Ministries of Economic Warfare, Supply and Aircraft Production, and the Intelligence sections of all three Armed Services; (US) State Department, Foreign Economic Administration, Office of Strategic Services, Office of Scientific Research and Development (OSRD), and the three Forces' Intelligence divisions. Chaired by Brigadier T.J. Betts (US Army), with Professor R.P. Linstead (British civilian) as vice-chair. Did not have its own pool of investigators but was tasked with developing lists of targets, making arrangements for investigators to visit the sites in liberated Europe and Germany and for collating and making available their final reports. Disbanded, with SHAEF, in July 1945.

Control Commission for Germany (British Element) (CCG(BE))   British component of the Allied Control Commission; responsible for administering the British zone of Occupied Germany. Headquartered in Bad Oeynhausen, near Hannover. Worked in concert with the administrations of the other main Allies (USA, France, Soviet Union) through the Control Council in Berlin.

Control Office for Germany and Austria (COGA)   British government office responsible for the British occupation of Germany (and, briefly, Austria), based in Whitehall. Enacted policy through the CCG(BE) (see above) and the British Army of the Rhine (BAOR).

Darwin Panel   British committee tasked with facilitating the recruitment of German experts in civilian fields in Britain and later responsible for 'exclusive exploitation' – the employment of German specialists directly by private firms. Comprised of representatives from the Department of Scientific and Industrial Research, Board of Trade, Control Commission for Germany, Home Office, Treasury, German Economic Division, Admiralty, Security Services and the Ministries of Supply, Labour, Health, Agriculture and Fisheries, Aircraft Production, and Fuel and Power. Chaired by Sir Charles Darwin, director of the National Physical Laboratory.

Denial policy   Efforts by both the British and the Americans to minimise the benefits which the Soviet Union was able to derive through exploitation, often by preventing them from securing the services of German experts who were deemed valuable. This approach came to define much of the Anglo-American exploitation programme in its later years of operation.

Deputy Chiefs of Staff (DCOS)   British committee comprised of the deputy chiefs of staff from the armed forces. Responsible for many matters but, in terms of exploitation, their largest contribution was the so-called 'DCOS scheme' which provided for the British recruitment of German experts in military fields.

Dustbin Detention centre located at Schloß Kransberg, near Frankfurt-am-Main. Initially operated on an Anglo-American basis, but later migrated to exclusive American control. Detainees were primarily German scientific and technical experts who the Allies wished to interrogate and potentially recruit; among the most eminent was Albert Speer.

Enemy Personnel Exploitation Section (EPES) Component of FIAT (see below) specifically tasked with the detention, interrogation and recruitment of German scientists and technicians. Its forward section was particularly active in Berlin and the Soviet zone and responsible for securing the services of German specialists located in these areas.

Field Information Agency, Technical (FIAT) Anglo-American organisation responsible for many of the logistical demands of exploitation on the ground in Germany. With the dissolution of SHAEF, it was split into separate but co-operative British and American elements. Crucially, it facilitated the visits of British investigators to the American zone and vice versa. The British element was headed up by Brigadier R.J. Maunsell.

Inkpot British detention centre for German scientists and technicians, based at the Beltane School in Wimbledon, London.

Operation Backfire Anglo-American, but overwhelmingly British-led, project to assemble and test-fire V-2 rockets off the coast of northern Germany at Cuxhaven. Conducted primarily by German personnel with British supervision and observation. Achieved three launchings, two of which were successful. Considered a great achievement within the British military.

Operation Matchbox Key element of British denial policy (see above), based around a transit hotel where German scientists and technicians could stay in order to prevent their recruitment by the Soviets.

Operation Osoaviakhim Major Soviet operation which saw approximately 2,300 German scientists and technicians (along with their families) deported, often forcefully, from the Soviet zones of Germany and Berlin to the USSR. It took place in the early hours of 22 October 1946 and was conducted by the Soviet security service, the NKVD. It had largely positive implications for the British (and Americans) as it scared many German experts and encouraged them to actively seek employment in Britain or the USA.

Operation Surgeon British scheme to exploit German aeronautical expertise after the war, both through the examination and evacuation of equipment and facilities and through the interrogation and recruitment of scientists and technicians.

Project Paperclip The United States' major policy of recruitment of German scientists and technicians. Co-ordinated primarily by the Joint Intelligence Objectives Agency (JIOA). Estimates suggest some 1,500 German experts were recruited under this scheme, dwarfing the parallel British efforts.

Research Branch Component of the CCG(BE) (see above) which was responsible both for controlling German science after the war as well as enacting elements of denial policy. Had a close relationship with exploitation, which could be both complementary and conflicting.

T-Force   The British military element responsible for most of the logistical workload of exploitation. They travelled with the Allied advance across Europe after D-Day and were tasked with seizing and securing key facilities so that they could subsequently be visited by CIOS or BIOS (see above) investigators. They also played some part in the detention of key German individuals, the evacuation of German equipment and the provision of transport and accommodation for investigators.

Bibliography

Archival Sources

The National Archives, Kew

UK Atomic Energy Authority papers
AB 1/110	Investigation of nuclear physics developments in Germany	1945

Admiralty papers
ADM 178/392	Take–over of German naval war factories	1945–1946
ADM 202/308	30 Assault Unit	1944–1945
ADM 223/349	No 30 Assault Unit: target lists for operations in Germany	1944
ADM 223/500	30 Assault Unit and 30 Commando: papers	1942–1945
ADM 223/501	30 Assault Unit: targets	1944–1945

Air Ministry papers
AIR 19/434	British Bombing Research Mission	1945–1947
AIR 20/1715	Future of German Scientists: paper	1945
AIR 20/1722	Exercise "Post Mortem"	1945
AIR 20/4818	British Bombing Survey Unit	1945–1953
AIR 40/1178	A.D.I.(K) periodical progress reports 1–5	1945
AIR 40/1779	Joint "Crossbow" Working Committee and Joint "Crossbow" Committee	1944

Ministry of Aviation papers
AVIA 9/83	Sir Roy Fedden's mission to Germany	1945
AVIA 10/70	Alsos Mission: report	1943–1944
AVIA 12/82	Operation Surgeon: memorandum	1946
AVIA 12/191	Reparations from Germany	1943–1947
AVIA 15/2209	PERSONNEL: Scientific and Technical Staff (Code 35/2): Consideration of use of captured enemy technicians and scientists	1945–1946
AVIA 22/940	German reparations: Canadian requirements	1945–1947
AVIA 54/1294	Employment of Germans in UK under DCOS scheme: general	1946–1949
AVIA 54/1295	Employment of Germans in UK under DCOS scheme: general	1946–1953

AVIA 54/1403	Employment of German scientists and technicians: denial policy	1946–1950
AVIA 54/1404	Halstead Exploiting Centre: review of activities and eventual closure	1946–1950
AVIA 54/1826	Long–term employment of German scientists	1947–1950

Board of Trade papers

BT 211/23	Lists of BIOS, CIOS, FIAT and JIOA reports	1946
BT 211/62	'Operation Bottleneck': policy and arrangements	1946–1947
BT 211/116	Reparations assessment teams: policy	1945–1946
BT 211/117	Disposal of British Intelligence Objectives Sub-Committee equipment	1946–1947
BT 211/541	Meeting papers	1945–1946

Cabinet Office papers

CAB 21/1421	Co–ordination of scientific intelligence	1940
CAB 69/7	Papers DO (45) 7 – DO (45) 12	1945
CAB 79/37	Minutes of Meetings nos. 182–202	1945
CAB 79/68	Minutes of Meetings (O) nos. 291–323	1943
CAB 81/24	Papers 1–10 (1943); Papers 1–45 (1944)	1943–1944
CAB 81/47	Meetings 1–8, Papers 1–27 (1944); Meetings 1–3, Papers 1–18 (1945)	1944–1945
CAB 81/92	Meetings 1(0)–75(0)	1944
CAB 81/93	Meetings 1(0)–83(0)	1945
CAB 81/133	Papers: 41–75(0)	1946
CAB 81/134	Papers: 76(0)–110(0)	1946
CAB 82/3	Deputy Chiefs of Staff Committee: minutes of meetings 1–24	1945
CAB 82/6	Deputy Chiefs of Staff Committee: papers 1–66	1945
CAB 82/8	Deputy Chiefs of Staff Committee: minutes of meetings 1–39	1946
CAB 121/429	Policy towards German research and development establishment; employment of German scientists by the Allies vol. IV	1946–1947
CAB 122/342	Handling of German science and technology	1945–1946
CAB 122/343	Exploitation of German scientists and technicians	1945
CAB 122/349	Exploitation of German scientists and technicians	1946–1947
CAB 122/357	German scientists (Civil)	1945–1946
CAB 122/360	German scientists (Civil)	1947–1948
CAB 122/363	Allocation policy on samples of secret weapons	1945
CAB 124/544	Proposed Advisory Committee on the Control of Scientific Research and Development in Germany	1945–1946
CAB 124/1924	Scientific Committee for Germany	1946–1951
CAB 124/1928	Scientific Committee for Germany: minutes of meetings	1946–1949
CAB 126/333	Operation "Epsilon": discussions between Professor Blackett and German scientists detained at Farm Hall Godmanchester	1945
CAB 131/1	Meetings: 1–35	1946
CAB 131/5	Meetings: 1–27	1947

| CAB 158/2 | Joint Intelligence Sub-Committee: memoranda 51–85 | 1947 |
| CAB 176/8 | Secretariat Minutes (1945) 1457–1910 | 1945 |

Ministry of Defence papers

| DEFE 2/1107 | 30 Assault Unit: mobilisation, control, disbandment, Honours and Awards | 1943–1945 |
| DEFE 10/66 | Inter-Departmental Committee on German Scientists: minutes of meetings and memoranda | 1947 |

Foreign Office papers

FO 371/71038	Disbandment of Unterluss Works Centre	1948
FO 935/1	Research and Development Centres in Germany	1944–1945
FO 935/25	General correspondence	1944–1945
FO 935/140	British Bombing Survey Unit	1944–1946
FO 936/39	Field Information Agency (Technical)	1945
FO 942/8	Draft Armistice Terms: 1 Article 21(a)	1944
FO 942/27	Enemy Research and Development Sub-Committee	1944
FO 942/79	Combined Intelligence Sub-Committee	1944
FO 942/425	Reparation: urgent requirements of scientific apparatus	1946
FO 942/426	Operation "Surgeon"	1946–1947
FO 943/42	Reparations: dismantling policy	1947
FO 1010/20	Reparations: deliveries and restitutions	1945–1949
FO 1012/420	Reparations Operation "Trademark": vol. I	1945–1946
FO 1012/421	Reparations Operation "Trademark": vol. II	1946–1947
FO 1013/373	Operation "Matchbox": accommodation for German "consultants"	1948–1950
FO 1031/2	Reactivation of the hotel industry: vol. I	1947–1948
FO 1031/4	Termination and transfer of 'T' Force commitments	1947–1948
FO 1031/5	Liaison with Russians, French and other Allies: policy	1945–1946
FO 1031/6	Liaison with Russians, French and other Allies: individual cases	1945–1947
FO 1031/7	BIOS investigators; policy: vol. III	1946–1947
FO 1031/9	BIOS papers: general	1947
FO 1031/10	Visits, business interests and policy: vol. I	1945–1946
FO 1031/12	Operation "Abstract": interrogation of Prof. Dr Wernher von Braun, guided missiles, etc.	1947
FO 1031/19	Exploitation of German scientists and technicians: Policy	1946–1947
FO 1031/20	Exploitation of German scientists and technicians: Policy	1945–1949
FO 1031/22	Export of technicians	1947
FO 1031/25	Reports on EPES/FIAT activities	1947
FO 1031/49	History of 'T' Force	1945–1946
FO 1031/50	BIOS – Minutes of Meetings	1945–1946
FO 1031/51	British Intelligence Objectives Sub-Committee – general	1945
FO 1031/53	Quadripartite policy	1945–1946
FO 1031/59	Intelligence reports from Berlin: vol. I	1946

FO 1031/65	Personnel to be denied to Russians and establishments operated by them	1945–1946
FO 1031/67	German scientists and technicians from Soviet zones	1945–1947
FO 1031/68	EPES	1946–1948
FO 1031/69	DUSTBIN: policy	1946
FO 1031/74	Scientific and Technological Branch policy on unethical medicine and medical war crimes	1945–1946
FO 1031/75	EPES: policy	1945–1946
FO 1031/83	DUSTBIN: bacteriological warfare	1945
FO 1031/85	V Weapon personnel agreement to share	1945
FO 1031/86	Poison gas: interrogation and reports	1945
FO 1031/138	Russian affairs: recruitment of scientists	1946
FO 1032/35	Disarmament: directive on information required on German war material	1944–1945
FO 1032/164	Employment of scientists in UK on Darwin Panel scheme	1945–1947
FO 1032/166	Evacuation of equipment to UK; reparations: general	1945–1947
FO 1032/167	Evacuation of equipment to UK; reparations: clearances by RB	1945–1946
FO 1032/169	Liquidation, industrial disarmament: policy and procedure	1945–1947
FO 1032/170	Denazification: policy	1945–1947
FO 1032/176	British Intelligence Objectives Sub-Committee policy	1945–1947
FO 1032/177	British Intelligence Objectives Sub-Committee: organisation	1945–1947
FO 1032/179	British Intelligence Objectives Sub-Committee instructions and information on captured documents	1945–1947
FO 1032/247	Bacteriological research	1945–1947
FO 1032/297	Employment of German scientists in UK on Darwin Panel scheme: interrogation and detention policy	1945–1946
FO 1032/300	Employment of German scientists in UK on Darwin Panel scheme: defence research scheme; policy applicable to both schemes	1945–1947
FO 1032/302	Employment of German scientists in UK on Darwin Panel scheme: defence research scheme; policy for families	1946–1947
FO 1032/470	Organisation and functions of Combined Intelligence Objectives Sub-Committee teams	1945
FO 1032/475	Organisation and functions of Combined Intelligence Objectives Sub-Committee	1945
FO 1032/1231A	Intelligence: Operation "Matchbox"	1947
FO 1032/1231B	Intelligence: Operation "Matchbox"	1948
FO 1032/1459	Field Information Agency Technical	1945–1948
FO 1032/1470A	Field Information Agency Technical/T Force policy regarding access to technical targets in Allied zones: vol. I	1945–1947
FO 1032/2555	Denazification	1948
FO 1036/13	Reparations: the Harmssen report	1948

FO 1039/671	CCG script for "Germany under Control" exhibition in London	1946
FO 1039/672	Reports from Scientific and Technical Intelligence Branch	1946
FO 1050/67	Intelligence Division: formation of Scientific and Technical Intelligence Branch	1946
FO 1050/1419	Combined Intelligence Objectives Sub-Committee black list: geographically arranged. Processed.	1944
FO 1050/1421	Combined Intelligence Objectives Sub-Committee grey list	1944–1945
FO 1062/114	Operation "Surgeon"	1945–1948
FO 1062/149	Scientific and Technical Research Board minutes and correspondence: vol. I	1946
FO 1062/396	Industrial disarmament: policy	1945–1947
FO 1065/12	Organisation and future of Field Information Agency Technical and 'T' Force	1945–1948

GCHQ papers

HW 8/104	History of 30 Commando	1942–1946

Ministry of Labour papers

LAB 8/1198	Darwin Panel: employment of German Scientists, specialists and technicians for civil purposes in the United Kingdom	1945–1947
LAB 8/1450	General policy and correspondence	1946–1948

Office of the Prime Minister papers

PREM 3/21/3	British Bombing Research Mission	1944–1945
PREM 8/373	Interrogation of German Scientists in United Kingdom and their subsequent return to Germany	1945–1946

Treasury papers

T 294/15	Assembly reports 1951 and 1961	1951–1958

War Office papers

WO 193/432	Combined Intelligence Priorities Committee	1944–1945
WO 204/12455	Marine Einsatz Kommando 80	1944–1945
WO 204/12911	Abwehrkommandos: activities, staffing, accommodation etc.	1945
WO 208/2174	Field technical assessment by Porton group	1944–1945
WO 208/2183	Reports on phosphorus–nitrogen compounds Tabun and Sarin	1945
WO 208/3974	Interrogation of Dr K Blome Director of German Biological Warfare Activities	1945
WO 208/4280	Reports on potential bacteriological targets visited	1945
WO 208/4969	ASHCAN: interrogation etc. reports	1945
WO 219/1630A	T Force Planning: intelligence targets to be captured by a special force during and advance	1944–1945
WO 219/1668	30 Advanced Unit: supply of information to Combined Intelligence Objectives Sub-Committee of Joint Intelligence Committee	1945

WO 219/1669	Collection of economic intelligence: policy and organisation	1944–1945
WO 219/1986	Field Information Agency, Technical: T Force organisation	1944–1945
WO 219/1987	Field Information Agency, Technical: reports on T Force operations and activities, special technical investigations, lessons learned, etc.	1944–1945
WO 219/2165	Operation Backfire	1945

Imperial War Museum, London/Duxford

Private Papers

05/48/1	Private Papers of Edward C. Aspden	1945
06/27/1	Private Papers of Major P.A. Chittenden	1945
09/21/1	Private Papers of Mr Gilbert A. Hunter	1946
99/76/1	Private Papers of Lieutenant-Commander John Bradley, RNVR	1945
99/36/1	Private Papers of Brigadier W.P.T. Roberts CBE	1945
10/7/1	Private Papers of Monica Maurice	1947

CIOS Reports

Overall Report.	'The Intelligence Exploitation of Germany'	1945
XXVI–37	'The Treatment of Shock from Prolonged Exposure to Cold, Especially in Water'	1945
XXVIII–56	'Rockets and Guided Missiles'	1945
XXIX–21	'Miscellaneous Aviation Medical Matters'	1945
XXXI–86	'Chemical Warfare Installations in the Munsterlager Area'	1945
XXXII–125	'German Guided Missile Research'	1945

Miscellaneous Items

Documents. 9531	Montgomery's Non-Fraternisation Instructions	1945
Misc. 21/382	'Report on Operation Backfire'	1945
Official Report	'German Organisation and Personalities Engaged in Research and Development of Armaments during the Second World War'	1948
Official Report	'Reports on German and Japanese Industry – Classified List No. 18'	1948

Churchill College, Cambridge

Metals Society (METL.1/METL.2) – CIOS and BIOS Reports

| I–1 | 'Radar and Controlled Missiles' | 1944 |
| X–13 | 'Visit to Eindhoven, Holland' | 1944 |

Published Primary Sources

Press

Hansard

House of Commons debates
House of Lords debates
Written Questions and Responses

Newspapers and Magazines

Bucks Herald
Daily Express
Daily Herald
Daily Mail
Daily Telegraph
Daily Worker
Dundee Courier
Lancashire Evening Post
Manchester Guardian
Nature
New York Times
The Times
Time
World's Press News

Journal Articles

Landauer, Carl, 'The German Reparations Problem', *Journal of Political Economy*, 56 (1948), 344–347.
Morrison, Philip, 'Alsos: The Story of German Science', *Bulletin of the Atomic Scientists*, 3:12 (1947), 354–365.
Office of Population Research, 'The Demography of War: Germany', *Population Index*, 14 (1948), 291–308.
Von Laue, Max, 'The Wartime Activities of German Scientists', *Bulletin of the Atomic Scientists*, 4:4 (1948), 103.

Films/TV/Media

Kaufman, Philip (director), *The Right Stuff* (USA: Warner Bros., 1983).
Kubrick, Stanley (director), *Dr Strangelove or: How I Learned to Stop Worrying and Love the Bomb* (USA: Columbia Pictures, 1964).
Lehrer, Tom. 'Wernher von Braun'. Song recorded in 1965 [accessed online 6 September 2018, https://www.youtube.com/watch?v=QEJ9HrZq7Ro].
Poliakoff, Stephen (screenwriter), *Close to the Enemy* (UK: BBC, 2016).
Russo, Joe, and Anthony Russo (directors), *Captain America: The Winter Soldier* (Disney Studios, 2014).

Websites

Churchill, Winston. 'MIT Mid-Century Convocation'. Speech delivered on 31 March 1949 [accessed online 22 January 2016, http://libraries.mit.edu/archives/exhibits/midcentury/mid-cent-churchill.html].

Published Primary Sources, Memoirs and Autobiography

Baxter III, James Phinney, *Scientists against Time* (New York: Little Brown, 1946).
Bernstein, Jeremy (ed.), *Hitler's Uranium Club: The Secret Recordings at Farm Hall* (New York: Springer, 2001).
British Bombing Survey Unit, *The Strategic Air War against Germany* (London: Frank Cass, 1988).
Bush, Vannevar, *Pieces of the Action* (New York: Morrow, 1970).
Churchill, Winston, *The Second World War*, vol. VI (London: Cassell, 1953).
Danchev, Alex, and Daniel Todman (eds.), *Field Marshal Lord Alanbrooke: War Diaries 1939–1945* (London: Weidenfeld & Nicolson, 2001).
Eisenhower, Dwight D., *The Papers of Dwight David Eisenhower*, vol. VI (Baltimore: Johns Hopkins University Press, 1978).
Fedden, Roy, *Britain's Air Survival: An Appraisement and Strategy for Success* (London: Cassell, 1957).
Foreign Office, *Instructions for British Servicemen in Germany 1944* (London: Foreign Office, 1944).
Foreign Relations of the United States (FRUS) [accessed online 1 August 2015: http://digital.library.wisc.edu/1711.dl/FRUS].
Frank, Charles (ed.), *Operation Epsilon: The Farm Hall Transcripts* (Bristol: Institute of Physics Publishing, 1993).
Gallup, George H., *The Gallup International Public Opinion Polls: Great Britain, 1937–1975* (New York: Random House, 1976).
Goudsmit, Samuel A., *Alsos* (New York: American Institute of Physics Press, 1996).
Groves, Leslie R., *Now It Can Be Told: The Story of the Manhattan Project* (London: Andre Deutsch, 1963).
Howard, Michael, *Otherwise Occupied: Letters Home from the Ruins of Nazi Germany* (Tiverton, Devon: Old Street, 2010).
Huzel, Dieter K., *From Peenemünde to Canaveral* (Westport: Greenwood, 1962).
Jones, R.V., 'Scientific Intelligence', *RUSI Journal*, 92 (1947), 352–369.
Jones, R.V., *Most Secret War: British Scientific Intelligence, 1939–1945* (London: Coronet, 1978).
Jones, R.V., *Reflections on Intelligence* (London: Heinemann, 1989).
Kennan, George F., *Memoirs 1925–1950* (London: Hutchinson, 1968).
Neitzel, Sönke (ed.), *Tapping Hitler's Generals: Transcripts of Secret Conversations, 1942–45*, Translated by Geoffrey Brooks (London: Frontline Books, 2013).
Nutting, David, (ed.), *Attain by Surprise: The Story of 30 Assault Unit* (Chichester: David Colver, 1997).
Pash, Boris T., *The Alsos Mission* (New York: Charter Books, 1980).
Ruhm von Oppen, Beate (compiler), *Documents on Germany under Occupation, 1945–54* (Oxford: Oxford University Press, 1955).
Speer, Albert, *Inside the Third Reich* (London: Sphere, 1971).

Secondary Literature

Abelshauser, Werner, 'Immaterial Reparations and the Reintegration of West Germany into the World Market', in Matthias Judt and Burghard Ciesla (eds.), *Technology Transfer out of Germany after 1945* (Amsterdam, Harwood, 1996), 107–118.

Adler, K.H., 'Selling France to the French: The French Zone of Occupation in Western Germany, 1945–c.1955', *Contemporary European History*, 21 (2012), 575–595.

Agar, Jon and Brian Balmer, 'British Scientists and the Cold War: The Defence Research Policy Committee and Information Networks, 1947–1963', *Historical Studies in the Physical and Biological Sciences*, 28 (1998), 209–252.

Ahrens, Michael, *Die Briten in Hamburg: Besatzerleben, 1945–1958* (Munich: Dölling und Galitz, 2011).

Albrecht, Ulrich, 'Military Technology and National Socialist Ideology', in Monika Renneberg and Mark Walker (eds.), *Science, Technology and National Socialism* (Cambridge: Cambridge University Press, 2003), 88–125.

Aldrich, Richard J. (ed.), *British Intelligence, Strategy and the Cold War, 1945–1951* (Abingdon: Routledge, 1992).

Aldrich, Richard J. (ed.), *Intelligence, Defence, and Diplomacy: British Policy in the Post War World* (Ilford: Frank Cass, 1994).

Aldrich, Richard J. 'British Intelligence and the Anglo-American "Special Relationship" During the Cold War', *Review of International Studies*, 24 (1998), 331–351.

Aldrich, Richard J. *The Hidden Hand: Britain, America and Cold War Secret Intelligence* (London: John Murray, 2001).

Allport, Alan, *Demobbed: Coming Home after the Second World War* (New Haven, CT: Yale University Press, 2010).

Andrew, Christopher, 'Intelligence in the Cold War', in Melvyn P. Leffler and Odd Arne Westad (eds.), *The Cambridge History of the Cold War*, vol. II (Cambridge: Cambridge University Press, 2010), 417–437.

Annan, Noel, *Changing Enemies: The Defeat and Regeneration of Germany* (London: Harper Collins, 1995).

Annas, George J. and Michael A. Grodin (eds.), *The Nazi Doctors and the Nuremberg Code: Human Rights in Human Experimentation* (Oxford: Oxford University Press, 1992).

Augustine, Dolores L., *Red Prometheus: Engineering and Dictatorship in East Germany, 1945–1990* (Cambridge, MA: MIT Press, 2007).

Backer, John H., *Priming the German Economy: American Occupational Policies, 1945–8* (Durham, NC: Duke University Press, 1971).

Balfour, Michael, *West Germany: A Contemporary History* (New York: St Martin's Press, 1983).

Balmer, Brian, *Britain and Biological Warfare: Expert Advice and Science Policy*, 1930–65 (Basingstoke: Palgrave, 2001).

Balmer, Brian, 'The UK Biological Weapons Program', in Mark Wheelis, Lajos Rozsa and Malcolm Dando (eds.), *Deadly Cultures: Biological Weapons since 1945* (Cambridge, MA: Harvard University Press, 2006).

Balmer, Brian, *Secrecy and Science: A Historical Sociology of Biological and Chemical Warfare* (Abingdon: Routledge, 2012).

Bar-Zohar, Michel, *The Hunt for the German Scientists* (London: Arthur Barker, 1967).

Barker, Elisabeth, *Britain in a Divided Europe, 1945–1970* (London: Weidenfeld and Nicolson, 1972).

Barrett, A.G.M. and D.H.R. Barton, 'Darwin, Sir Charles Galton (1887–1962)', in *Oxford Dictionary of National Biography* (Oxford: Oxford University Press, 2004) [accessed online 6 August 2015, http://www.oxforddnb.com/view/article/32716].

Barrett, A.G.M. and D.H.R. Barton, 'Linstead, Sir (Reginald) Patrick (1902–1966)', in *Oxford Dictionary of National Biography*, (Oxford: Oxford University Press, 2004) [accessed online 21 July 2015, http://www.oxforddnb.com/view/article/34549].

Bath, Alan, *Tracking the Axis Enemy: The Triumph of Anglo-American Naval Intelligence* (Lawrence, KS: University Press of Kansas, 1998).

Bauman, Zygmunt, *Modernity and the Holocaust* (Cambridge: Polity, 2000).

Baylis, John, *Anglo-American Relations since 1939: The Enduring Alliance* (Manchester: Manchester University Press, 1997).

Becklake, John, 'German Rocket Engineers in Britain—their influence revisited', *Acta Astronautica*, 59 (2006), 510–515.

Beevor, Antony, *Berlin: The Downfall 1945* (London: Penguin, 2003).

Bell, Douglas I., 'STEG and the Occupation of Germany: Demilitarization as Reconstruction, 1945–1953', *War in History*, 25 (2018), 103–125.

Benbow, Tim, 'The Royal Navy and Sea Power in British Strategy, 1945–1955', *Historical Research*, 91 (2018), 375–398.

Berghahn, Volker, 'Technology, Reparations, and the Export of Industrial Culture. Problems of the German-American Relationship, 1900–1950', in Matthias Judt and Burghard Ciesla (eds.), *Technology Transfer out of Germany after 1945* (Amsterdam: Harwood, 1996), 1–10.

Bernstein, George L., *The Myth of Decline: The Rise of Britain since 1945* (London: Pimlico, 2004).

Bessel, Richard, *Germany after the First World War* (Oxford: Oxford University Press, 1993).

Bessel, Richard, *Nazism and War* (London: Weidenfeld & Nicolson, 2004).

Bessel, Richard, *Germany 1945: From War to Peace* (London: Pocket Books, 2010).

Bessel, Richard and Dirk Schumann, 'Introduction: Violence, Normality, and the Construction of Postwar Europe', in Richard Bessel and Dirk Schumann (eds.), *Life after Death: Approaches to a Cultural and Social History of Europe during the 1940s and 1950s* (Cambridge: Cambridge University Press, 2003), 1–14.

Beyerchen, Alan D., *Scientists under Hitler: Politics and the Physics Community in the Third Reich* (New Haven, CT: Yale University Press, 1977).

Beyerchen, Alan D., 'German Scientists and Research Institutions in Allied Occupation Policy', *History of Education Quarterly*, 22 (1982), 289–299.

Beyler, Richard, Alexei Kojevnikov, and Jessica Wang, 'Purges in Comparative Perspective: Rules for Exclusion and inclusion in the Scientific Community under Political Pressure', in Carola Sachse and Mark Walker (eds.), *Politics and Science in Wartime: Comparative International Perspectives on the Kaiser Wilhelm Institute* (Chicago: University of Chicago Press, 2005), 23–48.

Beyler, Richard, Alexei Kojevnikov, Jessica Wang and Morris F. Low, 'Science Policy in post-1945 West Germany and Japan: Between Ideology and Economics', in Mark Walker (ed.), *Science and Ideology: A Comparative History* (Abingdon: Routledge, 2005), 97–123.

Biddiscombe, Perry, *Werwolf! The History of the National Socialist Guerrilla Movement, 1944–46* (Toronto: University of Toronto Press, 1998).

Biddiscombe, Perry, *The Denazification of Germany: A History, 1945–1950* (Stroud, UK: Tempus, 2007).

Biddle, Wayne, *Dark Side of the Moon: Wernher von Braun, the Third Reich and the Space Race* (London: W.W. Norton, 2009).

Bird, Kai and Martin J. Sherwin, *American Prometheus: The Triumph and Tragedy of J. Robert Oppenheimer* (New York: Vintage, 2006).

Bloxham, Donald, 'British War Crimes Policy in Germany, 1945–1959: Implementation and Collapse', *Journal of British Studies*, 42 (2003), 91–118.

Boehling, Rebecca, *A Question of Priorities: Democratic Reform and Economic Recovery in Postwar Germany* (Oxford: Berghahn, 1996).

Boobbyer, Philip, 'Lord Rennell, Chief of AMGOT: A Study of His Approach to Politics and Military Government (c.1940–43)', *War in History*, 25 (2018), 304–327.

Bower, Tom, *Blind Eye to Murder: Britain, America and the Purging of Nazi Germany – A Pledge Betrayed* (London: Andre Deutsch, 1981).

Bower, Tom, *The Paperclip Conspiracy: The Battle for the Spoils and Secrets of Nazi Germany* (London: Grafton, 1988).

Breitman, Richard, *Official Secrets: What the Nazis Planned, What the British and Americans Knew* (New York: Hill & Wang, 1998).

Broadbery, Stephen and Peter Howlett, 'Blood, Sweat and Tears: British Mobilisation for World War II', in Roger Chickering, Stig Förster and Bernd Greiner, *A World at Total War: Global Conflict and the Politics of Destruction, 1937–1945* (Cambridge: Cambridge University Press, 2005), 157–176.

Bud, Robert and Philip Gummett (eds.), *Cold War, Hot Science: Applied Research in Britain's Defence Laboratories, 1945–1990* (London: Science Museum, 2002).

Bud, Robert and Philip Gummett (eds.), 'Introduction: Don't You Know There's a War On?', in Robert Bud and Philip Gummett (eds.), *Cold War, Hot Science: Applied Research in Britain's Defence Laboratories, 1945–1990* (London: Science Museum, 2002), 1–28.

Budiansky, Stephen, *Battle of Wits: The Complete Story of Codebreaking in World War II* (New York: Touchstone, 2000).

Buruma, Ian, *Year Zero: 1945 – A History* (London: Atlantic, 2013).

Cairncross, Alec, *The Price of War: British Policy on German Reparations, 1941–1949* (Oxford: Blackwell, 1986).

Caplan, Arthur L., 'The Ethics of Evil: The Challenge and the Lessons of Nazi Medical Experiments', in William R. Lafleur, Gernot Böhme, and Susumu Shimazono (eds.), *Dark Medicine: Rationalising Unethical Medical Research* (Bloomington, IN: Indiana University Press, 2007), 63–72.

Carden, Robert W., 'Before Bizonia: Britain's Economic Dilemma in Germany, 1945–46', *Journal of Contemporary History*, 14 (1979), 535–555.

Carruthers, Susan L., *The Good Occupation: American Soldiers and the Hazards of Peace* (Cambridge, MA: Harvard University Press, 2016).

Carson, Cathryn, 'Knowledge Economies: Toward a New Technological Age', in Michael Geyer and Adam Tooze (eds.), *The Cambridge History of the Second World War*, vol. III (Cambridge: Cambridge University Press, 2015), 196–219.

Carter, Gradon P., *Porton Down: 75 Years of Chemical & Biological Research* (London: HMSO, 1992).

Carter, Gradon P. and Graham S. Pearson, 'British biological warfare and biological defence, 1925–1945', in Erhard Giessler and John Ellis van Courtland Moon (eds.), *Biological and Toxin Weapons: Research, Development and Use from the Middle Ages to 1945* (Oxford: Oxford University Press, 2009), 168–189.

Cassidy, David, *Uncertainty: The Life and Science of Werner Heisenberg* (New York: W.H. Freeman, 1991).

Cassidy, David, 'Controlling German Science I: US and Allied Forces in Germany, 1945–1947', *Historical Studies in the Physical and Biological Sciences*, 24 (1994), 197–235.

Cassidy, David, 'Controlling German Science II: Bizonal Occupation and the Struggle over West German Science Policy, 1946–1949', *Historical Studies in the Physical and Biological Sciences*, 26 (1996), 197–239.

Chandler, Adam, 'Eichmann's Best Man Lived and Died in Syria', *The Atlantic*, 1 December 2014 [accessed online 27 June 2018, https://www.theatlantic.com/international/archive/2014/12/eichmanns–best–man–lived–and–died–in–syria/383296/].

Chen, Chern, 'Former Nazi Officers in the Near East: German Military Advisors in Syria, 1949–56', *The International History Review*, 40 (2018), 732–751.

Christopher, John, *The Race for Hitler's X-Planes* (Stroud: The History Press, 2012).

Ciesla, Burghard, and Helmuth Trischler, 'Legitimation through Use: Rocket and Aeronautical Research in the Third Reich and the USA', in Mark Walker (ed.), *Science and Ideology: A Comparative History* (Abingdon: Routledge, 2003), 156–185.

Clark, Ronald W., *Tizard* (London: Methuen, 1965).

Coast, David and Jo Fox, 'Rumour and Politics', *History Compass*, 13 (2015), 222–234.

Collier, Basil, *The Battle of the V-Weapons, 1944–5* (London: Hodder & Stoughton, 1964).

Confino, Alon, 'Dissonance, Normality, and the Historical Method: Why did Some Germans Think of Tourism after May 8, 1945?', in Richard Bessel and Dirk Schumann (eds.), *Life after Death: Approaches to a Cultural and Social History of Europe during the 1940s and 1950s* (Cambridge: Cambridge University Press, 2003), 323–348.

Connelly, Mark, 'The British People, the Press and the Strategic Air Campaign against Germany, 1939–45', *Contemporary British History*, 16 (2002), 39–48.

Connelly, Mark, *Reaching for the Stars: A History of Bomber Command* (London: I.B. Tauris, 2014).

Cooter, Roger, Mark Harrison and Steve Sturdy (eds.), *War, Medicine and Modernity* (Stroud: Sutton, 1998).

Cornwell, John, *Hitler's Scientists: Science, War and the Devil's Pact* (London: Penguin, 2013).

Cox, Sebastian, 'Introduction: An Unwanted Child – The Struggle to Establish a British Bombing Survey', in British Bombing Survey Unit, *The Strategic Air War against Germany* (London: Frank Cass, 1988), xvii–xxii.

Craig, Campbell, and Sergey Radchenko, *The Atomic Bomb and the Origins of the Cold War* (New Haven, CT: Yale University Press, 2008).

Creswell, Michael and Marc Trachtenberg, 'France and the German Question, 1945–55', *Journal of Cold War Studies*, 5:3 (2003), 5–28.

Crim, Brian E., *Our Germans: Project Paperclip and the National Security State* (Baltimore, MD: Johns Hopkins University Press, 2017).

Daniels, Mario, 'Brain Drain, innerwestliche Weltmarktkonkurrenz und nationale Sicherheit. Die Kampagne der westdeutschen Chemieindustrie gegen Wissenstransfers in die USA in den 1950er Jahren', *Vierteljahrshefte für Zeitgeschichte*, 64 (2016), 491–516.

David, James E., 'Scavenging for Intelligence: The U.S. Government's Secret Search for Foreign Objects during the Cold War', *National Security Archive Briefing Books*, 616 (2018) [accessed online 4 September 2018, https://nsarchive.gwu.edu/briefing–book/intelligence/2018–01–31/scavenging–intelligence–us–governments–secret–search–foreign].

Defrance, Corine, 'La mission du CNRS en Allemagne (1945–1950): Entre exploitation et contrôle du potentiel scientifique allemand', *Revue pour l'Histoire du CNRS*, 5 (2001), 54–65.

Deighton, Anne, 'Cold War Diplomacy: British Policy Towards Germany's Role in Europe, 1945–9', in Ian Turner (ed.), *Reconstruction in Post-War Germany: British Occupation Policy and the Western Zones, 1945–55* (Oxford: Berg, 1989), 15–34.

Deighton, Anne, *The Impossible Peace: Britain, the Division of Germany and the Origins of the Cold War* (Oxford: Clarendon, 1990).

Deighton, Anne, 'Britain and the Cold War, 1945–1955', in Mervyn P. Leffler and Odd Arne Westad (eds.), *The Cambridge History of the Cold War*, vol. I (Cambridge: Cambridge University Press, 2012), 112–132.

Dockrill, Saki, 'Britain's Strategy for Europe: must West Germany be Rearmed? 1949–51', in Richard Aldrich (ed.), *British Intelligence, Strategy and the Cold War, 1945–51* (Abingdon: Routledge, 1992), 193–214.

Donnison, F.S.V., *Civil Affairs and Military Government North West Europe 1944–6* (London: HMSO, 1961).

Dovey, H.O., 'Maunsell and Mure', *Intelligence and National Security*, 8 (1993), 60–77.

Duffy, James, *Target America: Hitler's Plan to Attack the United States* (Westport, CT: Praeger, 2004).

Dumbrell, John, *A Special Relationship: Anglo-American Relations in the Cold War and After* (Basingstoke, UK: Macmillan, 2001).

Dylan, Huw, *Defence Intelligence and the Cold War: Britain's Joint Intelligence Bureau, 1945–64* (Oxford: Oxford University Press, 2014).

Ebert, Hans J., Johann B. Kaiser and Klaus Peters, *Willy Messerschmitt: Pioneer of Aviation Design* (Atglen, PA: Schiffer, 1999).

Eckert, Astrid M., *The Struggle for the Files: The Western Allies and the Return of German Archives after the Second World War* (Cambridge: Cambridge University Press, 2012).

Edgerton, David, 'British Scientific Intellectuals and the Relations of Science, Technology, and War', in Paul Forman and Jose M. Sanchez–Ron (eds.), *National Military Establishments and the Advancement of Science and Technology* (Dordrecht: Kluwer, 1996), 1–35.

Edgerton, David, *Warfare State: Britain, 1920–1970* (Cambridge: Cambridge University Press, 2005).

Edgerton, David, *Britain's War Machine: Weapons, Resources and Experts in the Second World War* (London: Allen Lane, 2011).

Edgerton, David, *The Shock of the Old: Technology and Global History since 1900* (Oxford: Oxford University Press, 2011).

Engel, Jeffrey A., '"We are not concerned who the buyer is": Engine Sales and Anglo-American Security at the Dawn of the Jet Age', *History and Technology*, 17 (2000), 43–67.

Epstein, Katherine, *Torpedo: Inventing the Military-Industrial Complex in the United States and Great Britain* (Cambridge, MA: Harvard University Press, 2014).

Erlichman, Camilo, and Christopher Knowles (eds.), *Transforming Occupation in the Western Zones of Germany: Politics, Everyday Life and Social Interactions, 1945–55* (London: Bloomsbury, 2018).

Euros, Glesni, 'The Post-War British "ReEducation" Policy for German Universities and its Application at the Universities of Göttingen and Cologne (1945–1947)', *Research in Comparative and International Education*, 11 (2016), 247–266.

Evans, Rob, *Gassed: British Chemical Warfare Experiments on Humans at Porton Down* (London: Stratus, 2000).

Farquharson, John, *The Western Allies and the Politics of Food: Agrarian Management in Post-War Germany* (Oxford: Berg, 1985).

Farquharson, John, 'The British Occupation of Germany, 1945–6: A Badly Managed Disaster Area?', *German History*, 11 (1993), 316–338.

Farquharson, John, 'Anglo-American Policy on German Reparations from Yalta to Potsdam', *English Historical Review*, 112 (1997), 904–926.

Farquharson, John, 'Governed or Exploited? The British Acquisition of German Technology, 1945–48', *Journal of Contemporary History*, 32 (1997), 23–42.

Feigel, Lara, *The Bitter Taste of Victory: In the Ruins of the Reich* (London: Bloomsbury, 2016).

Feiveson, Harold A., Alexander Glaser, Zia Mian and Frank N. von Hippel, *Unmaking the Bomb: A Fissile Material Approach to Nuclear Disarmament and Non–Proliferation* (Cambridge, MA: MIT Press, 2014).

Felton, Mark, 'Adlerhorst – The Führer's Secret Castle', 20 March 2016 [accessed online 28 June 2018, http://markfelton.co.uk/publishedbooks/adlerhorst–hitlers–forgotten–headquarters/].

Fisch, Jörg, 'Reparations and Intellectual Property', in Matthias Judt and Burghard Ciesla (eds.), *Technology Transfer out of Germany after 1945* (Amsterdam: Harwood, 1996), 11–26.

Folly, Martin, '"The impression is growing … that the United States is hard when dealing with us": Ernest Bevin and Anglo-American Relations at the Dawn of the Cold War', *Journal of Transatlantic Studies*, 10 (2012), 150–166.

Ford, Roger, *Germany's Secret Weapons in World War II* (Staplehurst UK: Spellmount, 2000).

Fort, Adrian, *Prof: The Life and Times of Frederick Lindemann* (London: Jonathan Cape, 2003).

Foschepoth, Josef and Rolf Steininger (eds.), *Die britische Deutschland– und Besatzungspolitik, 1945–1949* (Paderborn, Germany: Ferdinand Schöningh, 1985).

French, David, *Army, Empire and Cold War: The British Army and Military Policy, 1945–1971* (Oxford: Oxford University Press, 2012).

Fry, Helen, *The M Room: Secret Listeners Who Bugged the Nazis in World War 2* (London: Marranos Press, 2012).

Fulbrook, Mary, *Dissonant Lives: Generations and Violence through the German Dictatorships* (Oxford: Oxford University Press, 2011).

Gareau, Frederick H., 'Morgenthau's Plan for Industrial Disarmament in Germany', *The Western Political Quarterly*, 14 (1961), 517–534.

Giessler, Erhard, 'Biological warfare activities in Germany, 1934–1945', in Erhard Giessler and John Ellis van Courtland Moon (eds.), *Biological and Toxin Weapons: Research, Development and Use from the Middle Ages to 1945* (Oxford: Oxford University Press, 2009), 91–126.

Giffard, Hermione, 'Engines of Desperation: Jet Engines, Production and New Weapons in the Third Reich', *Journal of Contemporary History*, 48 (2013), 821–844.

Giffard, Hermione, *Making Jet Engines in World War II: Britain, Germany, and the United States* (Chicago: University of Chicago Press, 2016).

Gimbel, John, *The American Occupation of Germany: Politics and the Military, 1945–1949* (Stanford, CA: Stanford University Press, 1968).

Gimbel, John, 'US Policy and German Scientists: The Early Cold War', *Political Science Quarterly*, 101 (1986), 433–451.

Gimbel, John, 'The American Exploitation of German Technical Know–How after World War II', *Political Science Quarterly*, 105 (1990), 295–309.

Gimbel, John, 'German Scientists, US Denazification Policy and the "Paperclip Conspiracy"', *The International History Review*, 12 (1990), 441–465.

Gimbel, John, *Science, Technology, and Reparations: Exploitation and Plunder in Postwar Germany* (Stanford: Stanford University Press, 1990).

Gimbel, John, 'Science, Technology & Reparations in Postwar Germany', in Jeffrey Diefendorf, Axel Frohn and Hermann-Josef Rupieper (eds.), *American Policy and the Reconstruction of West Germany, 1945–55* (Cambridge: Cambridge University Press, 1993), 175–196.

Givens, Seth A., 'Liberating the Germans: The US Army and Looting in Germany during the Second World War', *War in History*, 21 (2013), 33–54.

Glatt, Carl, 'Reparations and the Transfer of Scientific and Industrial Technology from Germany: A Case Study of the Roots of British Industrial Policy and of Aspects of British Occupation Policy in Germany between Post–World War II Reconstruction and the Korean War', Ph.D. dissertation, European University Institute, Florence, (1994).

Glees, Anthony, 'The Making of British Policy on War Crimes: History as Politics in the UK', *Contemporary European History*, 1 (1992), 171–197.

Goda, Norman J.W., 'The Gehlen Organisation and the Heinz Felfe Case', in David A. Messenger and Katrin Paehler (eds.), *The Nazi Past: Recasting German Identity in Postwar Europe* (Lexington, KY: University Press of Kentucky, 2015), 271–294.

Goodchild, James, 'R.V. Jones and the Birth of Scientific Intelligence', Ph.D. dissertation, University of Exeter, (2013).

Goodchild, James, 'Exploitation of Displaced European Refugees and Axis Prisoners of War in Britain, 1939–49', in Sandra Barkhof and Angela K. Smith (eds.), *War and Displacement in the Twentieth Century: Global Conflicts* (Abingdon: Routledge, 2014), 103–133.

Goodchild, James, *A Most Enigmatic War: R.V. Jones and the Genesis of British Scientific Intelligence, 1939–1945* (Warwick: Helion, 2017).

Goodman, Michael S., 'Jones' Paradigm: The How, Why, and Wherefore of Scientific Intelligence', *Intelligence and National Security*, 24 (2009), 236–256.

Goodman, Michael S., *The Official History of the JIC*, vol. I (Abingdon: Routledge, 2014).

Goschler, Constantin and Michael Wala, *Keine Neue Gestapo: Das Bundesamt für Verfassungsschutz und die NS-Vergangenheit* (Berlin: Rowohlt, 2015).

Gossel, Daniel, *Briten, Deutsche und Europa. Die Deutsche Frage in der britischen Außenpolitik, 1945–1962* (Stuttgart: Franz Steiner, 1999).

Graebner, Norman A., and Edward M. Bennett, *The Versailles Treaty and its Legacy: The Failure of the Wilsonian Vision* (Cambridge: Cambridge University Press, 2011).

Graham-Dixon, Francis, *The Allied Occupation of Germany: The Refugee Crisis, Denazification and the Path to Reconstruction* (London: IB Tauris, 2013).

Green, John, 'Obituary – Dr Johanna Weber', Royal Aeronautical Society [accessed online 20 July 2018, https://www.aerosociety.com/news/obituary–dr–johanna–weber/].

Greenwood, Sean, 'Coal and the Origins of the Cold War: the British Dilemma over Coal Supplies from the Ruhr', in Michael F. Hopkins, Michael Kandiah, and Gillian Staerck (eds.), *Cold War Britain, 1945–1964: New Perspectives* (Basingstoke, UK: Palgrave Macmillan, 2003), 143–154.

Grodin, Michael, 'Historical Origins of the Nuremberg Code', in George Annas and Michael Grodin (eds.), *The Nazi Doctors and the Nuremberg Code: Human Rights in Human Experimentation* (Oxford: Oxford University Press, 1992), 121–144.

Grossmann, Atina, *Jews, Germans and Allies: Close Encounters in Occupied Germany* (Princeton, NJ: Princeton University Press, 2009).

Gummett, Philip, *Scientists in Whitehall* (Manchester: Manchester University Press, 1980).

Haddon, Catherine, 'Union Jacks and Red Stars on Them: UK Intelligence, the Soviet Nuclear Threat and British Nuclear Weapons Policy, 1945–1970', Ph.D. dissertation, Queen Mary, University of London, (2008).

Haggith, Toby, 'Great Britain: Remembering a Just War (1945–1950)', in Lothar Kettenacker and Torsten Riotte (eds.), *The Legacies of Two World Wars: European Societies in the Twentieth Century* (Oxford: Berghahn, 2011), 225–256.

Hall, Charlie, 'A Conflict of Interests? British Efforts in the Pursuit of Post-War Justice and Technical Intelligence in Occupied Germany', M.A. dissertation, University of Kent, 2013.

Hall, Charlie, 'British Exploitation of German Science and Technology from War to Post-War, 1943–1948', Ph.D. dissertation, University of Kent, (2017).

Hall, Charlie, 'Pushed into Pragmatism: British Approaches to Science in Post-War Occupied Germany', *The International History Review*, 41 (forthcoming, 2019).

Hampshire, A. Cecil, *The Secret Navies* (London: William Kimber, 1978).

Harris, Robert and Jeremy Paxman, *A Higher Form of Killing: The Secret History of Chemical and Biological Warfare* (London: Arrow, 2002).

Harris, Sheldon H., *Factories of Death: Japanese Biological Warfare, 1932–1945, and the American Cover-Up* (Abingdon: Routledge, 2002).

Hart, John D., 'The Alsos Mission, 1943–1945: A Secret U.S. Scientific Intelligence Unit', *International Journal of Intelligence and Counter-Intelligence*, 18 (2005), 508–537.

Hartcup, Guy, *The Effect of Science on the Second World War* (Basingstoke: Macmillan, 2000).

Hastings, Max, *The Secret War* (London: William Collins, 2016).

Hathaway, Robert, *Ambiguous Partnership: Britain and America, 1944–47* (New York: Columbia University Press, 1981).

Hathaway, Robert, *Great Britain and the United States: Special Relations since World War II* (Boston, MA: Twayne's, 1990).

Hayes, Peter, *Industry and Ideology: IG Farben in the Nazi Era* (Cambridge: Cambridge University Press, 1987).

Henke, Klaus-Dietmar, *Die amerikanische Besetzung Deutschlands* (Munich: Oldenbourg, 1995).

Hennessy, Peter, *The Secret State: Whitehall and the Cold War* (London: Penguin, 2003).

Hinsley, F.H., *British Intelligence in the Second World War* (London: HMSO, 1984).

Hitchcock, William I., *France Restored: Cold War Diplomacy and the Quest for Leadership in Europe, 1944–1954* (Chapel Hill, NC: UNC Press, 1998).

Hoffmann, Stefan-Ludwig, 'Germany is No More: Defeat, Occupation and the Postwar Order', in Helmut Walser Smith (ed.), *The Oxford Handbook to Modern German History* (Oxford: Oxford University Press, 2011), 593–614.

Hogg, Ian V., *German Secret Weapons of the Second World War: The Missiles, Rockets, Weapons and New Technology of the Third Reich* (London: Greenhill Books, 1999).

Hohne, Heinrich, *Canaris*, translated by J. Maxwell Brownjohn (London: Secker & Warburg, 1979).

Holloway, David, *Stalin and the Bomb: The Soviet Union and Atomic Energy, 1939–1956* (New Haven, CT: Yale University Press, 1994).

Home, R.W. and Morris F. Low, 'Postwar Scientific Intelligence Missions to Japan', *Isis*, 84 (1993), 527–537.

Hooper, H.O., 'Weeks, Ronald Morce, Baron Weeks (1890–1960)', *Oxford Dictionary of National Biography* (Oxford: Oxford University Press, 2004) [accessed online 28 April 2016, http://www.oxforddnb.com/view/article/36814].

Howard, Robert, *Operation Damocles: Israel's Secret War against Hitler's Scientists, 1951–67* (London: Thistle, 2013).

Hunt, Linda, *Project Paperclip: The United States Government, Nazi Scientists and Project Paperclip, 1945–1990* (New York: St Martin's Press, 1991).

Huxford, Grace, *The Korean War in Britain: Citizenship, Selfhood and Forgetting* (Manchester: Manchester University Press, 2018).

Jacobsen, Annie, *Operation Paperclip: The Secret Intelligence Program that Brought Nazi Scientists to America* (Paris: Hachette, 2014).

Jeffery, Keith, *MI6: The History of the Secret Intelligence Service, 1909–1949* (London: Bloomsbury, 2011).

Jeffreys, Diarmuid, *Hell's Cartel: IG Farben and the Making of Hitler's War Machine* (London: Bloomsbury, 2008).

Johnson, Brian, *The Secret War* (London: Pen & Sword, 2004).

Jones, Evan, 'The Employment of German Scientists in Australia after World War II', *Prometheus: Critical Studies in Innovation*, 20 (2002), 305–321.

Jones, Jill, 'Eradicating Nazism from the British Zone of Germany: Early Policy and Practice', *German History*, 8 (1990), 145–162.

Joubert de la Ferté, Philip, *Rocket* (London: Hutchinson, 1957).

Judt, Matthias and Burghard Ciesla (eds.), *Technology Transfer out of Germany after 1945* (Amsterdam, Harwood, 1996).

Judt, Tony, *Postwar: A History of Europe since 1945* (London: Pimlico, 2007).

Junker, Detlef (ed.), *The United States and Germany in the Era of the Cold War*, vol. I (Cambridge: Cambridge University Press, 2004).

Kershaw, Ian, *Hitler, 1889–1936: Hubris* (London: Penguin, 1998).

Kershaw, Ian, *Hitler, 1936–1945: Nemesis* (London: Penguin, 2000).

Kershaw, Ian, *The End: Germany, 1944–1945* (London: Penguin, 2012).

Kettenacker, Lothar, *Germany since 1945* (Oxford: Oxford University Press, 1997).

Kevles, Dan, 'Cold War and Hot Physics: Science, Security, and the American State, 1945–56', *Historical Studies in the Physical and Biological Sciences*, 20 (1990), 239–264.

Killick, John, *The United States and European Reconstruction, 1945–60* (New York: Routledge, 2013).

Knowles, Christopher, *Winning the Peace: The British in Occupied Germany, 1945–1948* (London: Bloomsbury, 2017).

Koerner, Steven T, 'Technology Transfer from Germany to Canada after 1945: A Study in Failure?', *Comparative Technology Transfer and Society*, 2 (2004), 99–124.

Koop, Volker, *Besetzt. Britische Besatzungspolitik in Deutschland* (Berlin: be.bra, 2007).

Kramer, Alan, 'Demontagepolitik in Hamburg', in Josef Foschepoth and Rolf Steininger (eds.), *Die britische Deutschland- und Besatzungspolitik, 1945–1949* (Paderborn, Germany: Ferdinand Schöningh, 1985), 165–180.

Kramer, Alan, 'British Dismantling Politics, 1945–9: A Reassessment', in Ian Turner (ed.), *Reconstruction in Post-War Germany: British Occupation Policy and the Western Zones, 1945–55* (Oxford: Berg, 1989), 125–154.

Krammer, Arnold, 'Technology Transfer as War Booty: The US Technical Oil Mission to Germany, 1945', *Technology and Culture*, 22 (1981), 68–103.

Krige, John, *American Hegemony and the Postwar Reconstruction of Science in Europe* (Cambridge, MA: MIT Press, 2008).

Kruger, Lee, *Logistics Matters and the US Army in Occupied Germany, 1945–1949* (Basingstoke, UK: Palgrave, 2016).

Kurowski, Franz, *The Brandenburger Commandos: Germany's Elite Warrior Spies in World War II* (Mechanicsburg, PA: Stackpole Books, 2005).

Kurtz, Michael J., *America and the Return of Nazi Contraband: The Recovery of Europe's Cultural Treasures* (Cambridge: Cambridge University Press, 2006).

Laney, Monique, *German Rocketeers in the Heart of Dixie: Making Sense of the Nazi Past in the Civil Rights Era* (New Haven, CT: Yale University Press, 2015).

Lasby, Clarence G., *Project Paperclip: German Scientists and the Cold War* (New York: Atheneum, 1971).

Le Tissier, Tony, *Race for the Reichstag: The 1945 Battle for Berlin* (London: Frank Cass, 1999).

Lee, Sabine, *Victory in Europe: Britain and Germany since 1945* (London: Pearson, 2001).

Leggett, Don and Charlotte Sleigh (eds.), *Scientific Governance in Britain, 1914–1979* (Manchester: Manchester University Press, 2016).

Lewis, Julian, *Changing Direction: British Military Planning for Post–War Strategic Defence, 1942–47* (London: Sherwood, 1988).

Liberman, Peter, 'The Spoils of Conquest', *International Security*, 18 (1993), 125–153.

Liberman, Peter, *Does Conquest Pay? The Exploitation of Occupied Industrial Societies* (Princeton, NJ: Princeton University Press, 1996).

Lichtblau, Eric, *The Nazis Next Door: How America Became a Safe Haven for Hitler's Men* (Charlottesville, VA: Mariner, 2014).

Lindee, Susan, 'The Repatriation of Atomic Bomb Victim Body Parts to Japan: Natural Objects and Diplomacy', *Osiris*, 13 (1999), 379–409.

Locke, Ian, 'Post-War Germany – Britain's Lost Opportunity?', *History Today*, 47:8 (1997), 11–17.

Lomas, Daniel W.B., *Intelligence, Security and the Attlee Governments, 1945–51* (Manchester: Manchester University Press, 2016).

Lomas, Daniel W.B., 'The Drugs Don't Work: Intelligence, Torture and the London Cage, 1940–8', *Intelligence and National Security*, 33 (2018), 918–929.

Long, Bronson, *No Easy Occupation: French Control of the German Saar, 1944–1957* (Rochester, NY: Camden House, 2015).

Longden, Martin A.L., 'From "Hot War" to "Cold War": Western Europe in British Grand Strategy, 1945–1948', in Michael F. Hopkins, Michael Kandiah, and Gillian Staerck (eds.), *Cold War Britain, 1945–1964: New Perspectives* (Basingstoke: Palgrave Macmillan, 2003), 111–126.

Longden, Sean, *T-Force: The Forgotten Heroes of 1945* (London: Constable, 2010).

Longmate, Norman, *Hitler's Rockets: The Story of the V-2s* (London: Hutchinson, 1985).

Lowenhaupt, Henry S., 'On the Soviet Nuclear Scent', *Studies in Intelligence*, 11 (1967), 13–29.

Ludvigsen, Karl, *Battle for the Beetle* (Cambridge, MA: Robert Bentley, 2000).

Maas, Ad and Hans Hooijmaijers (eds.), *Scientific Research in World War II: What Scientists did in the War* (Abingdon: Routledge, 2009).

MacDonogh, Giles, *After the Reich: From the Liberation of Vienna to the Berlin Airlift* (London: John Murray, 2007).

MacIsaac, David, *Strategic Bombing in World War Two: The Story of the USSBS* (New York: Garland, 1976).

MacLeod, Roy, '"All for Each and Each for All": Reflections on Anglo-American and Commonwealth Scientific Cooperation, 1940–1945', *Albion*, 26 (1994), 79–112.

MacLeod, Roy, (ed.), *Science and the Pacific War: Science and Survival in the Pacific, 1939–45* (Dordrecht, Holland: Kluwer, 2000).

MacLeod, Roy, 'Scientists', in Jay Winter (ed.), *The Cambridge History of the First World War*, vol. II (Cambridge: Cambridge University Press, 2014), 434–459.

Macrakis, Kristie, *Surviving the Swastika: Scientific Research in Nazi Germany* (Oxford: Oxford University Press, 1993).

Maddrell, Paul, 'Operation Matchbox and the Scientific Containment of the USSR', in Peter Jackson and Jennifer Siegel (eds.), *Intelligence and Statecraft: The Use and Limits of Intelligence in International Society* (Westport: Greenwood, 2005), 173–206.

Maddrell, Paul, *Spying on Science: Western Intelligence in Divided Germany, 1945–61* (Oxford: Oxford University Press, 2006).

Maddrell, Paul, 'British-American Scientific Intelligence Collaboration during the Occupation of Germany', *Intelligence and National Security*, 15 (2008), 74–94.

Maguire, G., *Anglo-American Policy towards the Free French* (Basingstoke: Macmillan, 1995).

Mahnken, Thomas, Joseph Maiolo and David Stevenson (eds.), *Arms Races in International Politics: From the Nineteenth to the Twenty-First Century* (Oxford: Oxford University Press, 2016).

Mahoney, Leo J., 'A History of the War Department Scientific Intelligence Mission (Alsos), 1943–1945', Ph.D. dissertation, Kent State University, (1981).

Margolian, Howard, *Unauthorized Entry: The Truth about Nazi War Criminals in Canada, 1946–1956* (Toronto: University of Toronto Press, 2000).

Marshall, Barbara, 'German Attitudes to British Military Government, 1945–47', *Journal of Contemporary History*, 15 (1980), 655–684.

Maulucci Jr., Thomas W. and Detlef Junker (eds.), *G.I.s in Germany* (Cambridge: Cambridge University Press, 2013).

Mawby, Spencer, 'Revisiting Rapallo: Britain, Germany and the Cold War, 1945–1955', in Michael F. Hopkins, Michael Kandiah, and Gillian Staerck (eds.), *Cold War Britain, 1945–1964: New Perspectives* (Basingstoke, UK: Palgrave Macmillan, 2003), 81–94.

May, Alex, *Britain and Europe since 1945* (Abingdon: Routledge, 2014).

McGovern, James, *Crossbow and Overcast* (London: Arrow, 1968).

McKercher, B.J.C., *Transition of Power: Britain's Loss of Global Pre–eminence to the United States, 1930–1945* (Cambridge: Cambridge University Press, 1999).

McKitrick, Frederick L., *From Craftsmen to Capitalists: German Artisans from the Third Reich to the Federal Republic, 1939–1953* (Oxford: Berghahn, 2016).

McNeill, William H., *The Pursuit of Power: Technology, Armed Force and Society Since AD1000* (Oxford: Blackwell, 1983).

McPartland, Mary A., 'The Farm Hall Scientists: The United States, Britain and Germany in the New Atomic Age, 1945–6', Ph.D. dissertation, George Washington University, (2013).

Meehan, Patricia, *A Strange Enemy People: Germans Under the British, 1945–50* (London: Peter Owen, 2001).

Mellor, David, *The Role of Science and Industry: Australia in the War of 1939–1945* (Canberra: Australian War Memorial, 1958).

Messenger, David A. and Katrin Paehler (eds.), *The Nazi Past: Recasting German Identity in Postwar Europe* (Lexington, KY: University Press of Kentucky, 2015).

Michail, Evgenios, 'After the War and After the Wall: British Perceptions of Germany following 1945 and 1989', *University of Sussex Journal of Contemporary History*, 3 (2001), 1–12.

Millar, George, *The Bruneval Raid: Stealing Hitler's Radar* (London: Cassell, 2002).

Miscamble, Wilson D., *The Most Controversial Decision: Truman, the Atomic Bombs, and the Defeat of Japan* (Cambridge: Cambridge University Press, 2011).

Moreno, Jonathan D., *Undue Risk: Secret State Experiments on Humans* (New York: Routledge, 2001).

Mueller, Gordon H., 'Rapallo Reexamined: A New Look at Germany's Secret Military Collaboration with Russia in 1922', *Military Affairs*, 40 (1976), 109–117.

Naimark, Norman, *The Russians in Germany: A History of the Soviet Zone of Occupation, 1945–1949* (Cambridge, MA: Harvard University Press, 1995).

Nakajima, Yuki, 'The Allied Forces and the Spread of German Industrial Technology in Post-War Japan', in Pierre-Yves Donzé & Shigehiro Nishimura (eds.), *Organizing Global Technology Flows: Institutions, Actors, and Processes* (Abingdon: Routledge, 2014), 197–212.

Nettl, J.P., *The Eastern Zone and Soviet Policy in Germany, 1945–50* (New York: Octagon, 1977).

Nahum, Andrew, '"I believe the Americans have not yet taken them all!": The Exploitation of German Aeronautical Science in Post-War Britain', in Helmuth Trischler and Stefan Zeilinger (eds.), *Tackling Transport* (London: Science Museum, 2003), 99–138.

Neufeld, Michael J., *The Rocket and the Reich: Peenemunde and the Coming of the Ballistic Missile Era* (Cambridge, MA: Harvard University Press, 1995).

Neufeld, Michael J., *Von Braun: Dreamer of Space, Engineer of War* (New York: A.A. Knopf, 2007).

Neufeld, Michael J., 'The Nazi Aerospace Exodus: Towards a Global, Transnational History', *History and Technology*, 28 (2012), 49–67.

O'Reagan, Douglas, 'Science, Technology, and Know-How: Exploitation of German Science and the Challenges of Technology Transfer in the Postwar World', Ph.D. dissertation, University of California, Berkeley, (2014).

O'Reagan, Douglas, 'French Scientific Exploitation and Technology Transfer from Germany in the Diplomatic Context of the Early Cold War', *The International History Review*, 37 (2014), 366–385.

O'Reagan, Douglas, *Taking Nazi Technology: Allied Scientific Espionage and Exploitation of German Technology after the Second World War* (Baltimore, MD: Johns Hopkins University Press, forthcoming).

Oleynikov, Pavel V., 'German Scientists in the Soviet Atomic Project', *The Non-Proliferation Review*, 7:2 (2000), 1–30.

Olick, Jeffrey K., *In the House of the Hangman: The Agonies of German Defeat, 1943–1949* (Chicago: University of Chicago Press, 2005).

Orde, Anne, *The Eclipse of Great Britain: The United States and British Imperial Decline, 1895–1956* (Basingstoke, UK: Macmillan, 1996).

Overy, Richard, *Why the Allies Won* (London: Jonathan Cape, 1995).

Overy, Richard, *Russia's War* (London: Penguin, 1998).

Overy, Richard, *The Bombing War: Europe, 1939–1945* (London: Penguin, 2013).

Overy, Richard, '"Instructive for the future": the Interrogation of the Major War Criminals in Germany, 1945', in Christopher Andrew and Simona Tobia (eds.), *Interrogation in War and Conflict: A Comparative and Interdisciplinary Analysis* (Abingdon: Routledge, 2014), 93–109.

Pearson, John, *The Life of Ian Fleming* (London: Jonathan Cape, 1966).

Peden, G.C., *Arms, Economics and British Strategy: From Dreadnoughts to Hydrogen Bombs* (Cambridge: Cambridge University Press, 2007).

Peltzer, Lilli, *Die Demontage deutscher naturwissenschaftlicher Intelligenz nach dem Zweiten Weltkrieg: Die Physikalisch-Technische Reichsanstalt, 1945–1948* (Berlin: ERS, 1995).

Perlman, Susan McCall, 'US Intelligence and Communist Plots in Post-War France', *Intelligence and National Security*, 33 (2018), 376–390.

Phillips, David, 'Aspects of Education for Democratic Citizenship in Post-War Germany', *Oxford Review of Education*, 38 (2012), 567–581.

Phillips, David, *Investigating Education in Germany: Historical Studies from a British Perspective* (Abingdon: Routledge, 2016).

Pirard, Theo, 'German Rockets in Africa: The Explosive Heritage of Peenemünde', *Acta Astronautica*, 40 (1997), 885–898.

Pope, Rex, 'British Demobilization after the Second World War', *Journal of Contemporary History*, 30 (1995), 65–81.

Post, Stephen G., 'The Echo of Nuremberg: Nazi Data and Ethics', *Journal of Medical Ethics*, 17 (1991), 42–44.

Presas i Puig, Albert, 'Technological Transfer as a Political Weapon: Technological Relations between Germany and Spain from 1918 to the early 1950s', *Journal of Modern European History*, 6 (2008), 218–236.

Priemel, Kim C. and Alexa Stiller (eds.), *Reassessing the Nuremberg Military Tribunals: Transitional Justice, Trial Narratives, and Historiography* (Oxford: Berghahn, 2012).

Proctor, Robert, 'Nazi Doctors, Racial Medicine, and Human Experimentation', in George J. Annas and Michael A. Grodin (eds.), *The Nazi Doctors and the Nuremberg Code: Human Rights in Human Experimentation* (Oxford: Oxford University Press, 1992), 17–31.

Raim, Edith, *Nazi Crimes against Jews and German Post-War Justice: The West German Judicial System during Allied Occupation, 1945–1949* (Oldenbourg: De Gruyter, 2015).

Ramsden, John, *Don't Mention the War: The British and the Germans since 1890* (London: Little, Brown, 2006).

Rankin, Nicholas, *Ian Fleming's Commandos: The Story of 30 Assault Unit in WWII* (London: Faber & Faber, 2011).

Reed, Bruce Cameron, *The History and Science of the Manhattan Project* (New York: Springer, 2014).

Reinisch, Jessica, *The Perils of Peace: The Public Health Crisis in Occupied Germany* (Oxford: Oxford University Press, 2013).

Renneberg, Monika and Mark Walker (eds.), *Science, Technology and National Socialism* (Cambridge: Cambridge University Press, 2003).

Reynolds, David, *Britannia Overruled: British Policy and World Power in the 20th Century* (Harlow: Longman, 1991).

Reynolds, David, 'Great Britain', in David Reynolds (ed.), *The Origins of the Cold War in Europe: International Perspectives* (New Haven, CT: Yale University Press, 2004), 77–95.

Richards, Pamela, *Scientific Information in Wartime: The Allied-German Rivalry, 1939–45* (Westport, CT: Greenwood, 1994).

Richelson, Jeffrey T., *Spying on the Bomb: American Nuclear Intelligence from Nazi Germany to Iran and North Korea* (New York: W.W. Norton, 2006).

Rieger, Bernhard, *The People's Car: A Global History of the Volkswagen Beetle* (Cambridge, MA: Harvard University Press, 2013).

Roberts, Geoffrey, *Stalin's Wars: From World War to Cold War, 1939–53* (New Haven, CT: Yale University Press, 2006).

Rothwell, Victor, *Britain and the Cold War, 1941–1947* (London: Jonathan Cape, 1982).

Rotter, Andrew J., *Hiroshima: The World's Bomb* (Oxford: Oxford University Press, 2008).

Rüger, Jan, 'OXO: Or, the Challenges of Transnational History', *European History Quarterly*, 40 (2010), 656–668.

Rüger, Jan, 'Revisiting the Anglo-German Antagonism', *The Journal of Modern History*, 83 (2011), 579–617.

Schake, Kori, *Safe Passage: The Transition from British to American Hegemony* (Cambridge, MA: Harvard University Press, 2017).

Schaller, Michael, *The American Occupation of Japan: The Origins of the Cold War in Asia* (Oxford: Oxford University Press, 1985).

Schmaltz, Florian, *Kampfstoff-Forschung im Nationalsozialismus: zur Kooperation von Kaiser-Wilhelm-Instituten, Militär und Industrie* (Göttingen: Wallstein, 2005).

Schmidt, Ulf, *Justice at Nuremberg: Leo Alexander and the Nazi Doctors' Trial* (Basingstoke: Palgrave Macmillan, 2004).

Schmidt, Ulf, 'Scars of Ravensbrück: Medical Experiments and British War Crimes Policy, 1945–1950', *German History*, 23 (2005), 20–49.

Schmidt, Ulf, *Karl Brandt, the Nazi Doctor: Medicine and Power in the Third Reich* (London: Continuum, 2007).

Schmidt, Ulf, 'Accidents and Experiments: Nazi Chemical Warfare Research and Medical Ethics during the Second World War', in Don Carrick and Michael Gross (eds.), *Military Medical Ethics for the 21st Century* (Farnham: Ashgate, 2013), 225–244.

Schmidt, Ulf, *Secret Science: A Century of Poison Warfare and Human Experiments* (Oxford: Oxford University Press, 2015).

Schwarz, Hans–Peter, 'The Division of Germany, 1945–1949', in Mervyn P. Leffler and Odd Arne Westad (eds.), *The Cambridge History of the Cold War*, vol. I (Cambridge: Cambridge University Press, 2012), 133–153.

Sears, William R., 'Project Paperclip' book review, *Bulletin of the Atomic Scientists*, 28:6, (1972), 55–6.

Sereny, Gitta, *Albert Speer: His Battle with Truth* (London: Picador, 1996).

Siddiqi, Asif, 'Germans in Russia: Cold War, Technology Transfer, and National Identity', *Osiris*, 24 (2009), 120–143.

Simons, Graham, *Operation Lusty: The Race for Hitler's Secret Technology* (Barnsley: Pen & Sword, 2016).

Simpson, Christopher, *Blowback: America's Recruitment of Nazis and its Impact on the Cold War* (London: Weidenfeld & Nicolson, 1988).

Šimůnek, Michal and Miloš Hořejš, 'Exploatace (okupovaného) spojence. Aktivity tzv. Sdruženého podvýboru pro zpravodajské úkoly (C.I.O.S.) a Britského podvýboru pro zpravodajské úkoly (B.I.O.S.) v Československu na příkladě Škodových závodů v Plzni, 1945–47' [Exploiting an (Occupied) Ally: The Activities of CIOS and BIOS in Post–war Czechoslovakia as Exemplified by the Škoda Pilsen Company, 1945–1947], in I. Janovský, J. Kleinová, H. Stříteský (eds), *Věda a technika v Československu v letech 1945–1960* (Prague: NTM, 2010), 385–406.

Sirrs, Owen L., *Nasser and the Missile Age in the Middle East* (Abingdon: Routledge, 2006).

Slaveski, Filip, *The Soviet Occupation of Germany: Hunger, Mass Violence and the Struggle for Peace* (Cambridge: Cambridge University Press, 2013).

Smith, Frederick, 2nd Earl of Birkenhead, *The Prof in Two Worlds: The Official Life of Professor F.A. Lindemann, Viscount Cherwell* (London: Collins, 1961).

Stanley, Ruth, 'German-speaking Armaments Engineers in Argentina and Brazil, 1947–1963', in Oliver Rathkolb (ed.), *Revisiting the National Socialist Legacy: Coming to Terms with Forced Labor, Expropriation, Compensation, and Restitution* (New Brunswick, NJ: Transaction, 2004), 205–225.

Steff, Reuben, *Strategic Thinking, Deterrence and the US Ballistic Missile Defence Project: From Truman to Obama* (Farnham: Ashgate, 2013).

Steiner, Andre, 'The Return of German "Specialists" from the Soviet Union to the German Democratic Republic: Integration and Impact', in Matthias Judt and Burghard Ciesla (eds.), *Technology Transfer out of Germany after 1945* (Amsterdam, Harwood, 1996), 119–130.

Stewart, Irvin, *Organising Scientific Research for War: The Administrative History of the Office of Scientific Research and Development* (Boston, MA: Little Brown, 1948).

Stocker, Jeremy, *Britain and Ballistic Missile Defence, 1942–2002* (London: Frank Cass, 2004).

Stokes, Raymond, *Divide and Prosper: The Heirs of I.G. Farben under Allied Authority, 1945–1951* (Berkeley, CA: University of California Press, 1988).

Stokes, Raymond, 'Intellectual Spoils', *Science*, 248 (1990), 1241.

Süss, Dietmar, *Death from the Skies: How the British and Germans Survived Bombing in World War Two* (Oxford: Oxford University Press, 2014).

Thacker, Toby, *The End of the Third Reich: Defeat, Denazification and Nuremberg* (Stroud: Tempus, 2006).

Thompson, Della (ed.), *Oxford English Dictionary*, 9th ed. (Oxford: Oxford University Press, 1998).

Thorpe, Charles, *Oppenheimer: The Tragic Intellect* (Chicago: University of Chicago Press, 2006).

Todman, Daniel, *Britain's War: Into Battle, 1937–1941* (London: Penguin, 2017).

Tooze, Adam, *The Wages of Destruction: The Making and Breaking of the Nazi Economy* (London: Penguin, 2008).

Tooze, Adam, 'Reassessing the Moral Economy of Post-War Reconstruction: The Terms of the West German Settlement in 1952', in Mark Mazower, Jessica Reinisch and David Feldman (eds.), *Post-War Reconstruction in Europe: International Perspectives, 1945–49* (Oxford: Oxford University Press, 2011), 47–69.

Trachtenberg, Marc, *A Constructed Peace: The Making of the European Settlement, 1945–63* (Princeton, NJ: Princeton University Press, 1999).

Tucker, Jonathan B., *War of Nerves: Chemical Warfare from World War I to Al-Qaeda* (New York: Anchor, 2006).

Turner, Ian, 'British Occupation Policy and its Effects on the Town of Wolfsburg and the *Volkswagenwerk*: 1945–1949', Ph.D. dissertation, University of Manchester, (1984).

Turner, Ian, (ed.), *Reconstruction in Post–War Germany: British Occupation Policy and the Western Zones, 1945–55* (Oxford: Berg, 1989).

Turner, Ian, 'British Policy towards German Industry, 1945–9: Reconstruction, Restriction or Exploitation?', in Ian Turner (ed.), *Reconstruction in Post-War Germany: British Occupation Policy and the Western Zones, 1945–55* (Oxford: Berg, 1989), 67–92.

Turner, Ian, 'Denazification in the British Zone', in Ian Turner (ed.), *Reconstruction in Post–War Germany: British Occupation Policy and the Western Zones, 1945–55* (Oxford: Berg, 1989), 239–268.

Tusa, Ann and John Tusa, *The Nuremberg Trial* (New York: Skyhorse, 2010).

Uttley, Matthew, 'Operation Surgeon and Britain's Post-War Exploitation of Nazi German Aeronautics', *Intelligence and National Security*, 17 (2002), 1–26.

Van Hook, James C., *Rebuilding Germany: The Creation of the Social Market Economy, 1945–1957* (Cambridge: Cambridge University Press, 2004).

Vig, Nicholas J., *Science and Technology in British Politics* (London: Pergamon, 1968).

Villain, Jacques, 'France and the Peenemünde Legacy', in Phillipe Jung (ed.), *History of Rocketry and Astronautics: Proceedings of Twenty-Sixth History Symposium of the International Academy of Astronautics, Washington, D.C., U.S.A., 1992* (San Diego, CA: Univelt, for the American Astronautical Society, 1997), 119–162.

Vogt, Timothy, *Denazification in Soviet-Occupied Germany: Brandenburg, 1945–8* (Cambridge, MA: Harvard University Press, 2000).

Von Homeyer, Uta, 'The Employment of Scientific and Technical Enemy Aliens (ESTEA) Scheme in Australia: A Reparation for World War II?', *Prometheus*, 12 (1994), 77–93.

Wachsmann, Nikolaus, *KL: A History of the Nazi Concentration Camps* (London: Abacus, 2016).

Wala, Michael, 'The Value of Knowledge: Western Intelligence Agencies and Former Members of the SS, Gestapo and Wehrmacht during the Early Cold War', in Camilo Erlichman and Christopher Knowles (eds.), *Transforming Occupation in the Western Zones of Germany: Power Politics, Everyday Life and Social Interactions, 1945–55* (London: Bloomsbury, 2018), 271–280.

Walker, J. Samuel, *Utter Destruction: Truman and the Use of Atomic Bombs against Japan* (Chapel Hill, NC: University of North Carolina Press, 2004).

Walker, Mark, *Nazi Science: Myth, Truth, and the German Atomic Bomb* (New York: Plenum, 1995).

Walker, Mark, 'The Nazification and Denazification of Physics', in Matthias Judt and Burghard Ciesla (eds.), *Technology Transfer out of Germany after 1945* (Amsterdam, Harwood, 1996), 49–60.

Watt, D. Cameron, 'British Military Perceptions of the Soviet Union as a Strategic Threat, 1945–50', in Josef Becker and Franz Knipping (eds.), *Power in Europe* (Berlin: de Gruyter, 1986), 325–339.

Weber-Newth, Inge and Johannes-Dieter Steinert, *German Migrants in Post-War Britain: An Enemy Embrace* (Abingdon: Routledge, 2006).

Weindling, Paul Julian, *Nazi Medicine and the Nuremberg Trials: From Medical War Crimes to Informed Consent* (Basingstoke: Palgrave Macmillan, 2004).

Weindling, Paul Julian, *Victims and Survivors of Nazi Human Experiments: Science and Suffering in the Holocaust* (London: Bloomsbury, 2014).

Welch, David, 'Priming the Pump of German Democracy: British Re-Education Policy in Germany after the Second World War', in Ian Turner (ed.), *Reconstruction in Post–War Germany: British Occupation Policy and the Western Zones, 1945–55* (Oxford: Berg, 1989), 215–238.

Welch, David and Jo Fox (eds.), *Justifying War: Propaganda, Politics and the Modern Age* (Basingstoke: Palgrave Macmillan, 2012).

Wend, Henry Burke, *Recovery and Restoration: U.S. Foreign Policy and the Politics of Reconstruction of West Germany's Shipbuilding Industry, 1945–1955* (Westport, CT: Greenwood, 2001).

Whalen, Robert Weldon, *Bitter Wounds: German Victims of the Great War, 1914–1939* (Ithaca, NY: Cornell University Press, 1984).

Willis, Roy F., *The French in Germany, 1945–1949* (Stanford, CA: Stanford University Press, 1962).

Winter, Jay, 'From War Talk to Rights Talk: War Aims and Human Rights in the Second World War', in David Welch and Jo Fox (eds.), *Justifying War: Propaganda, Politics and the Modern Age* (Basingstoke, UK: Palgrave Macmillan, 2012), 236–249.

Wittlinger, Ruth, 'Perceptions of Germany and the Germans in Post–war Britain', *Journal of Multilingual and Multicultural Development*, 25 (2004), 453–465.

Wolfe, Audra, *Competing with the Soviets: Science, Technology, and the State in Cold War America* (Baltimore, MD: Johns Hopkins University Press, 2013).

Wrigley, Chris (ed.), *British Trade Unions, 1945–1995* (Manchester: Manchester University Press, 1997).

Zallen, Doris T., 'Louis Rapkine and the Restoration of French Science after the Second World War', *French Historical Studies*, 17 (1991), 6–37.

Zimmerman, David, *Top Secret Exchange: The Tizard Mission and the Scientific War* (Stroud, UK: Alan Sutton, 1996).

Index